KITES

A Practical Handbook

SPECIAL INTEREST MODEL BOOKS

About the author

A life-long enthusiast for all forms of aviation, Ron Moulton was associated with the publication of model-making magazines for almost 40 years prior to his retirement as an Editorial Director. Kites were his earliest interest and, throughout his business life, he has been able to fly his kites on all of his overseas visits. He maintains contact with kite organizations worldwide and, since establishing the British Kite Flying Association in 1975, has organised three annual kite festivals.

About the artist

A devotee of all aeronautical subjects, Pat Lloyd is a skilled model-maker and describes himself as a "sporadic" kite-maker. His talent is evident in the 95 whole-page sketches he has specially produced for this book. Pat is internationally renowned for his standards of draughtsmanship through his many contributions to aeronautical and model-making magazines. These are part of his freelance activities, but it is similar to his regular job which is in the Patents, Designs and Trademark industry, where accuracy is of paramount importance.

On the cover

Steve Brockett of Cardiff, South Wales, specialises in creating unique kite shapes that symbolise 'unbound forms' with their imaginative shapes and decor. Always using traditional cotton sails, Steve's kites have gained worldwide acclaim for their shaded dyeing and painting. This pair are known as 'Dreamcatcher' and 'Ida the glider'. See the colour plates between pages 64 and 65 for his 'Vortigen' and the Welsh 'Dragon'.

KITES

A Practical Handbook

Ron Moulton & Pat Lloyd

Special Interest Model Books Ltd.
P.O.Box 327
Poole
Dorset
BH15 2RG

First published by Argus Books 1992
Second edition published by Nexus Special Interests 1997
This edition published by Special Interest Model Books 2004

ISBN 1-85486-143-3

Printed and bound in Great Britain by CPI Bath Press, Bath

www.specialinterestmodelbooks.co.uk

CONTENTS

INTRODUCTION

Over the past 15 years the art of kite flying has reached a watershed. For, while steadfast support sustains interest in the traditional shapes of the ceremonial and celebratory kites which have long been a hallmark of the East, kite activity today reflects innovation, new art forms and flight performance. In the same way, the pioneering working kites of the West, as devised by Baden-Powell, Samuel F.Cody, Alexander G.Bell and Charles J.Lamson, amongst others, still have their devotees. But, in the 1990s they are vastly outnumbered by a new wave of enthusiasts who have 'discovered' the hitherto unknown pleasures of modern kite flying.

Above: The kite has appeared many times on postage stamps but rarely so brightly coloured and distinctive as on the German 60 Pfennig issue in 1989.

Left: Samuel F Cody of Texas USA, who had this poster created to promote himself and the kites which still bear his name.

To the long-term dedicated kite flyer, the radical shapes, brilliant colours and extreme resilience of the structures are nothing short of phenomenal. Such has been the accelerated rate of progress, and the parallel growth of a supportive industry, that even the all-weather diehards have found difficulty in keeping pace. Hence, in our view, the need for an up-to-date book on kite flying. So, here we are *not* going to rake over historical detail. Instead we are dealing with the current situation and all the nuances which have attracted so many newcomers to this delightful hobby.

Kites are so much a part of everyday outdoor activity that they no longer reflect the eccentric, or become the object of ridicule. For that, we have to thank Schultz for his *Peanuts* cartoons, which are syndicated worldwide, or the Deutsches Bundespost CEPT Europa 89 stamps as typical examples of the subtle promotion that kites receive in everyday life. Of course, the publicity is not always favourable. For instance, sensational headlines announced that four flyers fell from the rooftops or were electrocuted during the frantic 1991 annual kite fighting festival at Rawalpindi. We are not advocating such zealous enthusiasm. Rather it's our aim to illustrate the delights of having your own world up there on a line, as typified by Charles Dickens in *David Copperfield*. His character, Mr.Dick, made a kite on which he had closely and laboriously written fanciful statements.

When flying *"he never looked so serene .... it lifted his mind out of its confusion and bore it to the skies. As he wound the string in .... he seemed to wake gradually out of a dream .... as if they had both come down together"*. Charles Dickens must himself have been a kite flyer to have penned that description of the euphoria, which is so well known to all who've flown with the wind and thrilled to the tension of a kite on a singing line.

The appeal of kite flying has always centred on that relaxing therapy which discharges the cares and worries into the elements. With the wind on one's back, and a taut line drawing the eye to a soaring, dancing frame, all else melts away.

New materials

So what exactly is *so* different about the present day kite? An immediate distinction is the universal adoption of ripstop nylon for the great majority of commercial and self-made kites. Less obvious to the newcomer is the change from plastics and for the fastest expanding sector of the hobby - the steerable aerobatic kites - the introduction of carbon-reinforced tubes with rigidity and strength greater even than any equivalent in steel.

Design evolution, from the basic two-line steerable of the mid-1970s to the incredibly fast and manoeuverable stunters of the 1990s, has been made possible through the use of carbon fibre tubes. Originally adapted from archery arrow shafts, demand for supply of longer lengths in greater quantities has, in itself, generated a new-found industry of considerable turnover. As availability increased, a galaxy of variations on the Delta stunter emerged from an increasing number of cottage industry manufacturers. The boom was enhanced by demonstrations of flying skills at the many fesivals and 'kite days' that began to fill the annual calendar.

The growth has been truly international. Major events promoted by seaside resorts and big cities around Europe accelerated demand for stunters and, in turn, these heavily-promoted meetings gave opportunity for other developments - the 'show' kites, the Teddy-bear para-droppers, the camera kites and the spectacular night flying which, with fireworks, laser beams and searchlights, gave new meaning to *Son et Lumiere*.

Coupled with this fast growing passion for variations in the modern kite and its applications was a new cordiality, brought about by the social ambience at kite meetings, especially where there was strong international participation. This has served to speed the generation of fresh concepts through the interchange of new ideas, many of which will be found in the pages that follow.

In this, the second edition of our book, we introduce eight new designs, bringing the total of dimensioned drawings for construction to 32, and through Pat's sketches we trace the developments in Sport kite design. Five years on from the earlier first edition, the pace of change has overtaken many aspects of kiting, and so there are many updates throughout and some deletions.

Most significant is the extension of Chapter 9 to delve into the early days of kiting with details of the original 'box kite; created over a century ago by that great inventor Lawrence Hargrave, and the similarly characteristic man-lifting kites by Samuel Franklin Cody. These, and the prize-winning, tri-wing kite by Charles Brogden, represent an era when bamboo and cotton were king. Now, there is an increasing number of kite-flying enthusiasts who are returning to those materials and, in many ways, unmatched achievements in the spirit of an unofficial 'Cotton Club'. We hope the new drawings will help to expand that interest.

As for variations in the modern kite, the remarkable 'Circoflex' created by Helmut Schiefer and Tom Oostveen of the Netherlands, and the no less stunning Viking Longship kite by the free spirits of the Sala Kite & Tango party in Sweden could hardly be bettered as examples of that well-worn axiom, 'simplest is best', hence their inclusion, thanks to the full co-operation of the designers.

Another addition is the extra eight-page section of colour illustrations (between pages 192 & 193). This is as much a tribute to the fantastic 9th International Festival of Kites at Dieppe as a reflection of how this bi-annual rendezvous of extremes from East and West has become a catalyst of creativity. Much is due to its amiable Director, Max Gaillard, and his team who have brought together the finest talents from all over the world, and established Dieppe beyond question as the capital of kites.

Organisation in 1996 involved provision for delegates from 30 nations in a kite village, and three vast demonstration areas taking over the entire seaside frontage. It was the greatest, the biggest, the most wonderful experience. If you ever have the opportunity, go and ride those Channel winds yourself. Enjoy the thrills and colour of a spectacle that have inspired us in producing this new edition.

Ron Moulton and Pat Lloyd, London, February 1997

Typical of the many kite festivals which are supported by cities throughout the world as a cultural outlet and an attraction for visitors to seaside resorts is the scene at Dieppe, France where representatives from up to thirty nations attend in September on even years.

CHAPTER ONE
o you want to fly a kite!

Let's start at the very beginning. You've seen some kites, they've impressed you and now you want one for yourself. For the time being, we have to assume that it's a single line kite you want (multi-line stunters come in Chapter Eight) and, without any prior experience, you haven't a clue where to begin.

Keep it simple

The easy option is to visit a kite shop, send for a mail order list or attend a kite meeting where there's a sales stand so that you can buy the whole outfit-handle, line and a ready-to-fly kite with, or perhaps without, a tail.

This first kite *has* to work if you are to be satisfied, so our first tip is for you to be conservative, and select one of the tried and tested shapes that give least trouble. In our experience, one's first kite eventually achieves a certain immortality. It becomes a favourite, always there in reserve whenever you tire of its successors, and forever regarded as 'Old Reliable'. Remember this, and select a Hexagon, a Malay ('Diamond') or a lightweight Delta. Make sure it has been made with ripstop nylon and has good quality spars and reinforced pockets to receive them. Deltas don't need tails, but the other two types should have one supplied. The line should be a full 60 metres regulation length, and the handle in robust plastic with extensions to retain the line when wound-in.

You might think that specification too simple. If so, you'll have to spend more and invest in a Flare or a Roller, which would be our second-stage recommendation. These are more involved in construction, slightly more complicated to assemble and will inevitably be much larger than the Hexagon or Malay. They'll cost more because they involve much more material and, in particular, more sewing; but, for all that, the Flare or the Roller are types that can be flown in a wide range of wind strengths, and are remarkably stable.

In any case, go for permanence. Plastic is bright and colourful but vulnerable, even when it has a glass fibre frame. If you must, then you can't do better than to buy a plastic 'Snake'; but we doubt if you'll have the same degree of reverence for it as for our other suggestions in nylon.

Making a 'quickie'

Perhaps the idea of buying a kite conflicts with your ecologically green view of the kite as being 'free as a bird' and therefore free on the

pocket. Or perhaps, like many of our correspondents who have sought assistance in working with young people at schools or in Community Art Centres, you're seeking a simple do-it-yourself project that brings quick results for minimum outlay. The answer here is a plastic Diamond. With variations, this very simple shape is widely used in kite workshops as a project that produces results, both in the flight performance and in the degree of satisfaction it gives to the constructor.

Pat has drawn the basic square, set on edge to make the diamond. We've also seen Butterfly, Delta and Indian Fighter profiles, each cut from shopping bag plastic with the aid of a hardboard template. Construction is the same for all - a straight spine, a curved cross-spar and a two-leg bridle. The tail is an essential, so too is flexibility in the cross-spar.

Start by cutting the square from lightweight plastic. Because the original dimensions were in imperial measure, either 12in or 15in on each side, we stick to these sizes since they've been tried and tested so many times. The metric equivalents are rather odd numbers at 305 or 380mm. Some workshops use white plastic off the roll so that it forms a plain background on which the constructor can sketch with a marker-pen. Then other marker-pen dyes can be used to fill in with a bright decoration to complete a pattern or cartoon figure.

If the ubiquitous black bin liner is to provide the covering, a decorative option is to cut a pattern out of its centre to suit a contrasting overlay. This can be taken from a plastic bag with a decorative design, or could be just a piece of plain colour cut into a shape. The patch is sticky-taped onto the kite blank, over the cutout, then the spine is prepared to a length equal to the corner-to-corner distance of the square. For imperial measure readers, this is 17in for a 12in sq. blank.

Both the spine and the cross-spar must be from resilient material. Split bamboo, shaped to a near round section as possible by scraping with the sharp edge of a piece of glass, or glass fibre rod of 2 or 3mm diameter, would be perfect (GRP rod is best cut using a 3-corner file around its perimeter). Now clear an area on a hard surface, such as a worktop. Lay down the kite blank and, using strong parcel tape or similar, stick the spine over the diagonal, and the whole to the work surface.

Above: For regular kite workshop use, templates made in plywood are more permanent. Note the different shapes, and use of a half profile for assured symmetry.
Left: The Diamond kite, using decoration from a plastic carrier bag.

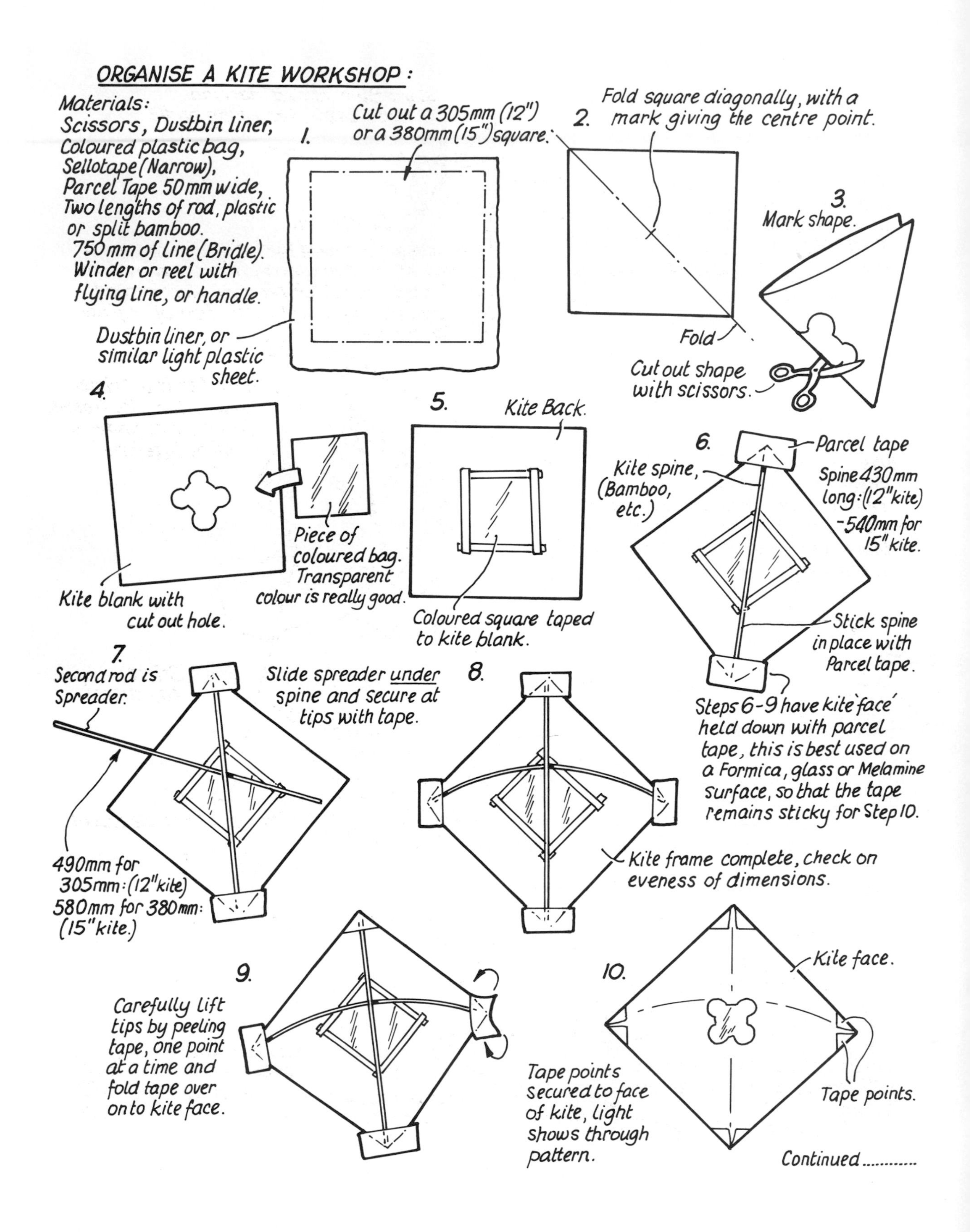
ORGANISE A KITE WORKSHOP :
Materials:
Scissors, Dustbin liner,
Coloured plastic bag,
Sellotape (Narrow),
Parcel Tape 50mm wide,
Two lengths of rod, plastic
or split bamboo.
750mm of line (Bridle).
Winder or reel with
flying line, or handle.
1.
Cut out a 305mm (12")
or a 380mm (15") square.
Dustbin liner, or
similar light plastic
sheet.
2.
Fold square diagonally, with a
mark giving the centre point.
Fold
3.
Mark shape.
Cut out shape
with scissors.
4.
Kite blank with
cut out hole.
Piece of
coloured bag.
Transparent
colour is really good.
5.
Kite Back.
Coloured square taped
to kite blank.
6.
Parcel tape
Kite spine,
(Bamboo,
etc.)
Spine 430mm
long: (12" kite)
-540mm for
15" kite.
Stick spine
in place with
Parcel tape.
Steps 6-9 have kite 'face'
held down with parcel
tape, this is best used on
a Formica, glass or Melamine
surface, so that the tape
remains sticky for Step 10.
7.
Second rod is
Spreader.
Slide spreader under
spine and secure at
tips with tape.
490mm for
305mm: (12" kite)
580mm for 380mm:
(15" kite.)
8.
Kite frame complete, check on
eveness of dimensions.
9.
Carefully lift
tips by peeling
tape, one point
at a time and
fold tape over
on to kite face.
10.
Kite face.
Tape points
secured to face
of kite, light
shows through
pattern.
Tape points.
Continued ....

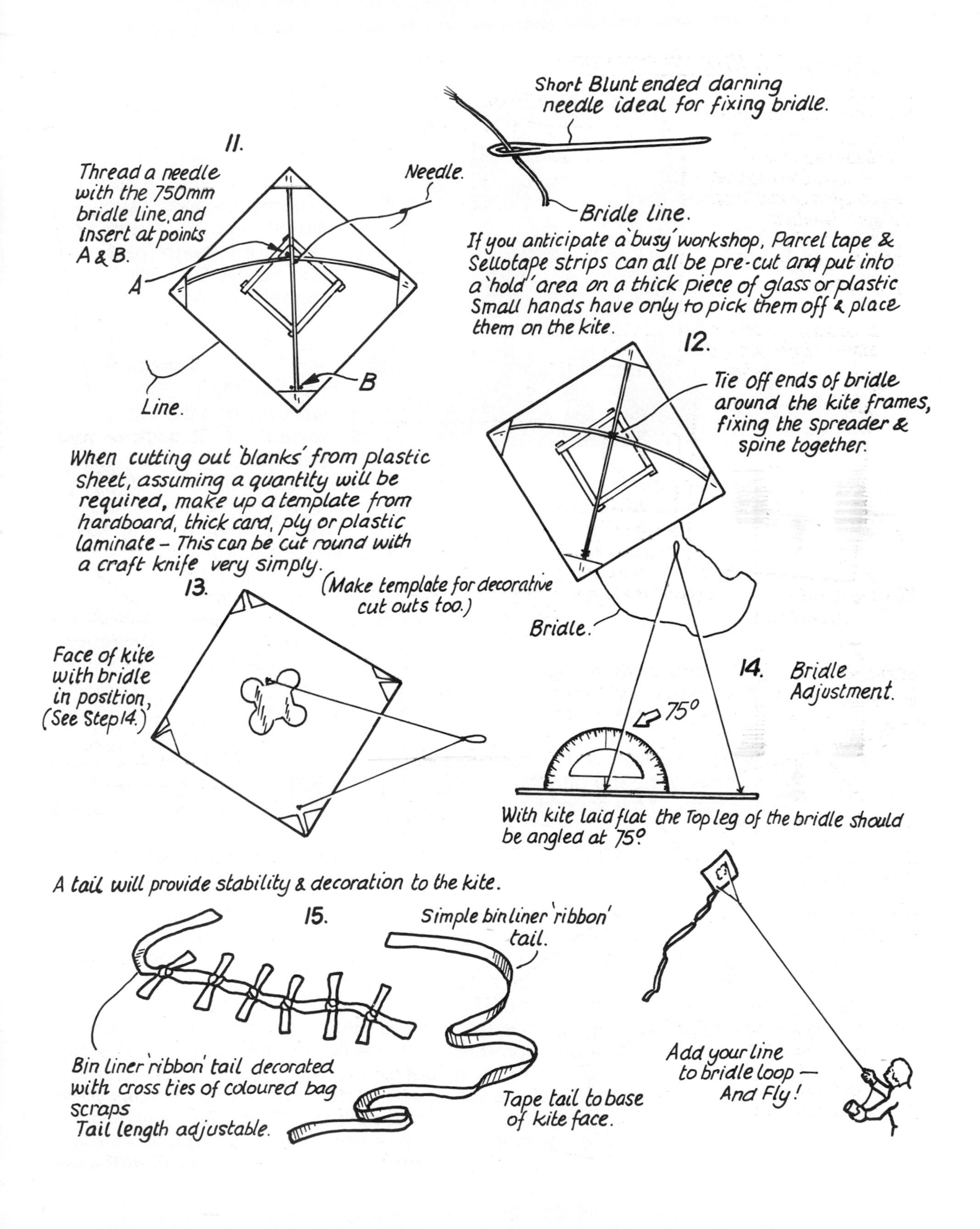
Short Blunt ended darning needle ideal for fixing bridle.
11.
Thread a needle with the 750mm bridle line, and insert at points A & B.
Needle.
A
B
Line.
Bridle line.
If you anticipate a 'busy' workshop, Parcel tape & Sellotape strips can all be pre-cut and put into a 'hold' area on a thick piece of glass or plastic Small hands have only to pick them off & place them on the kite.
12.
Tie off ends of bridle around the kite frames, fixing the spreader & spine together.
When cutting out 'blanks' from plastic sheet, assuming a quantity will be required, make up a template from hardboard, thick card, ply or plastic laminate – This can be cut round with a craft knife very simply.
13.
(Make template for decorative cut outs too.)
Face of kite with bridle in position, (See Step 14.)
Bridle.
14. Bridle Adjustment.
75°
With kite laid flat the Top leg of the bridle should be angled at 75°.
A tail will provide stability & decoration to the kite.
15.
Simple bin liner 'ribbon' tail.
Bin liner 'ribbon' tail decorated with cross ties of coloured bag scraps
Tail length adjustable.
Tape tail to base of kite face.
Add your line to bridle loop — And Fly!

Far-travelled Eiji Ohashi and his wife have raised their kite arches as a symbol of friendship all over the world. Their dedicated enthusiasm has crossed all language barriers and done much to help Western kite flyers understand the cultural signficance of the Japanese kite.

Prepare the cross-spar to the length indicated (about 60mm longer than the spine) and you will soon see why it is generally called a 'spreader', as you stick first one end down with more strong tape, then have to curve it, in order to fit it within the kite shape at the other end and onto the final corner. Our simple kite is almost complete. All that is required now is for each tip to be lifted in turn from the worktop, and the strong tape folded over to form a complete pocket at each corner.

Free from the worktop, we make a bridle with 750mm or 30in off the kite line and, with a darning needle, pierce the cover at the intersection of the spine and spreader, and tie with a secure knot (see Chapter Three). The other end of the bridle has to be tied in a similar way at the rear of the spine, but not at the very tip. To set the bridle provisionally, lay the kite face down and, with a protractor, arrange the division of the two halves of the bridle so that what will become the top leg is at approximately 75 degrees to the rear face of the kite. Secure the bridle adjustment with a half-hitch to form a small loop.

Now we need a tail. Again, coloured plastic shopping bags will offer a wide range of decorative strips which can be tied or glued to a central strip. Length should be more than 2 metres or 6ft, but is not critical. The tail aids stability by acting in the same way as a drogue which will keep the kite heading into wind, and its weight ensures that the centre of gravity is well behind the line attachment. This keeps the kite face at a good angle of attack to the oncoming airstream.

A tailless option

Larger, and involving a spot of heat treatment for the cross-spar, is the design by Eiji Ohashi, an internationally-renowned Kitemaster from Japan. It has an added attraction in that, if you want to make more than one, they can be joined with an interlinking line and flown in a 'train'. Eiji has made extremely long trains to form an arch, the biggest of which he was invited to fly in happier times at Kuwait.

This one can be made in almost the same way as the Diamond, except that the cross-spar or spreader has to be bent at the centre. This forms a dihedral angle of about 12.5 degrees either side of the kite. The upwards and outwards inclination of the kite either side of centre is what provides lateral stability. Making the bend in a bamboo stick is surprisingly easy over a candle flame or similar, but be careful not to burn it at this vital centre point of the kite!

Being larger (for imperialists, the height is 30in), the spars are much thicker,

THE EIJI OHASHI POLYTHENE KITE.

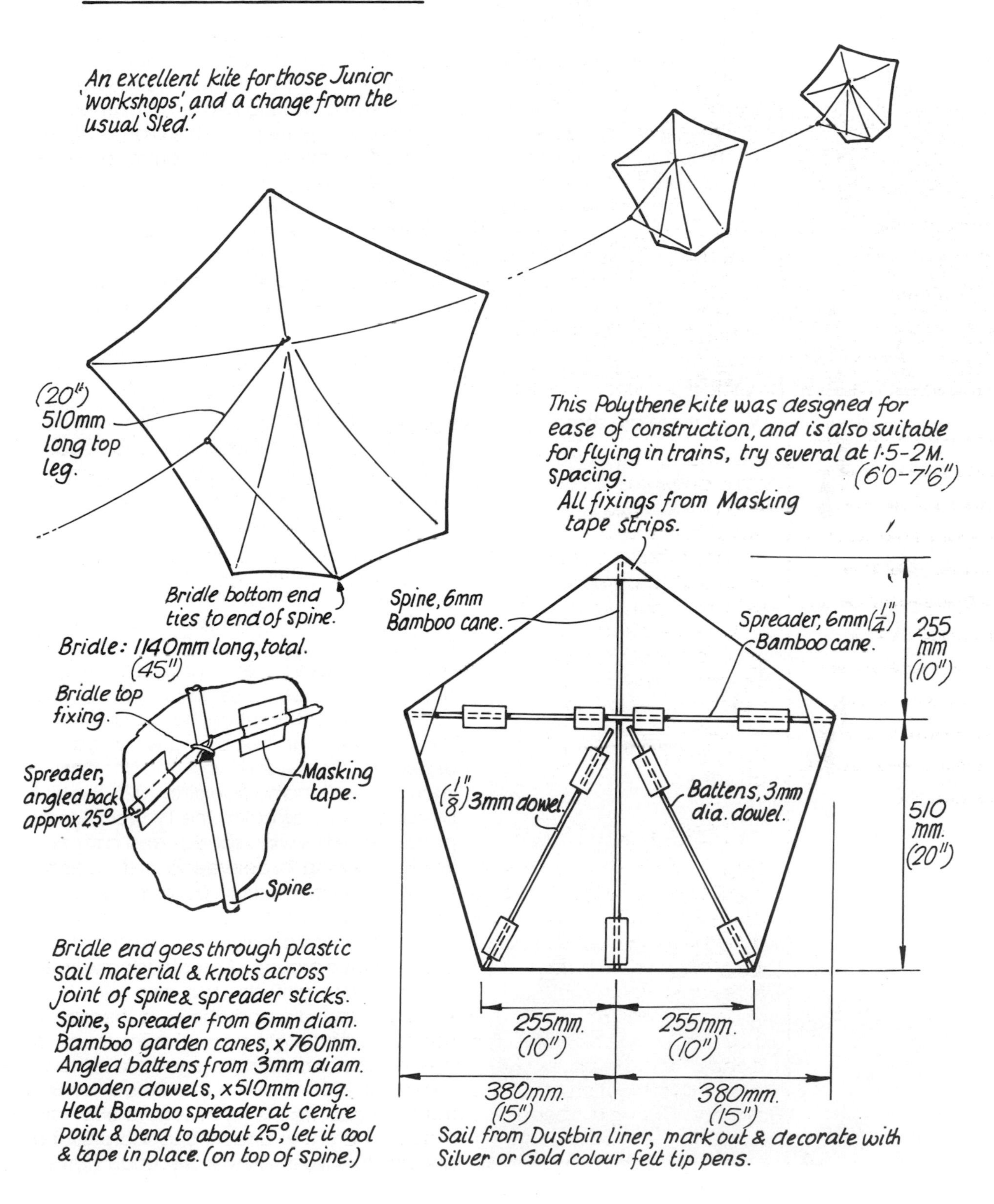

say 6mm bamboo or 3/16in dowel, and they do not have to flex. Instead, two much thinner battens are fitted to steady the lower apron and, being left free at the front or upper ends, they are able to maintain the sail shape while the rear end of the kite flexes in the manner of a soaring bird's tail feathers.

Sewing and tying the bridle is a similar exercise to that of the Diamond, and all the attachments of the four spars is by sticky tape. Of course, although we've said it is tailless, you can add one if you wish.

Assuming that you have acquired some braided nylon line and a suitable handle, we are now ready for flight *provided* there is enough wind and the direction is favourable for wherever we intend to fly. For a single line, stable kite, we need to catch the wind where it is constant and reasonably strong, free of turbulence and unaffected by obstructions. This is apart from considerations of safety.

Below: The joy of that first flight with a self-made kite is a treasured memory. Right: A workshop produces results! This group of French youngsters show their appreciation.

Wind

While there are some featherweight kites that will rise in calm air on thermal currents, or elevate when the flyer applies tension on the line by pulling as with the Eastern fighting kites, our concern is that there should be adequate wind strength. Wind provides all the energy we need for an anchored (llne held taut by the flyer) kite.

It comes to us by virtue of the 'Coriolis Effect' created by changes in atmospheric pressure. Regular forecasts by meteorologists on TV have taken most of the mystery out of isobars these days, so it should be easy to understand how, when the isobar lines are drawn close together, strong winds can be expected. Conversely, when the chart shows a large 'hole' and numbers like 1025, the high pressure promises calm. The TV weather experts also tell us wind speeds and direction. We'll be looking for between 5 and 15 mph (8 to 24 km/h) coming from a warm part of the globe.

The forecasts are given for a height of 10m or 33ft above ground. Actual speed will vary not only with altitudes but also with local terrain. As our diagram shows, a theoretical change with height can amount to an increase of six times from ground level to 60 metres or 200ft, or more realistically by 50% from the gust-free level of 10m to 60m. It is most important to understand the effect of gusts at low levels. Any obstruction upwind

will develop turbulence, be it a single tree or a block of apartments.

Trees can be useful indicators of wind variation with height, and the way in which leaves are turned, particularly on the lower branches, shows clearly how turbulent the air can be. Kite operation becomes educational, as experience gives better appreciation of air movement, and the way in which wind is a resultant from changes in barometric pressure.

So, in summary, we're looking for a clear site, exposed to wind, preferably on rising ground or on beach dunes where there are no buildings to break up the breeze.

The launch

Now we're ready for that first flight. No matter whether the kite is a simple Diamond in polythene, or a commercial Flare in ripstop nylon, it pays to check that the bridle is set at a reasonable angle before we commit the kite to the air.

Using a fishing snap-link on the end of the flying line, hook it onto the bridle (commercial kites should come with a ring on the bridle) and suspend the kite upside down. Take note of the angle at which the kite hangs. It must be with the tail hanging lower than the nose. The angle can be anywhere between 10 and 25 degrees at this stage, and only a flight test will show whatever change is needed. The tail need not be fitted for this test, but now's the time to attach it ready for the launch.

Have a helper hold the kite nose up, and walk out at least 15m (50ft) of line directly in-line with the wind - downwind, of course! When you can feel a steady breeze on your back, pull the line gently and call to the helper to release.

There are all kinds of releases. Kites need a smooth, upward and forward swing of the arms - not a throw, or a drop, simply a gentle cast-off. If all is well, the flyer is rewarded with an upward, arcing sweep of the kite through 60 degrees of line angle, and joy begins as it soars aloft, dancing with the elements.

However, some things simply do not work first time out. The kite might start to climb, then perform a curving sweep up to about 45 degrees, execute a beautiful turn and point head downwards until it crashes into the ground. In such a case, we would suspect that the bridle is setting the kite at too steep an angle. The cure is to untie the hitch, and rearrange with another loop that reduces the front leg and lengthens the rear. Check again. Better? If not, then we add to the tail. A quick remedy is to tie a light handkerchief to the end of the existing tail. This will soon prove whether more drag is needed to help stabilise the kite.

A sure sign that a tail is essential is a spinning kite which will not retain any heading and rotates around the line. Even a long crêpe streamer will straighten a spinning kite. A kite that climbs, then seems to slip sideways, twitching until it crashes on its side, needs to have a bowstring across the tips of its cross-spar or spreader. This induces a dihedral angle. Remember that we built this into the Ohashi by heating the bamboo and bending it?

Trimming a basic single line kite by bridle adjustment to adjust the angle of attack is dependent on the wind strength of the day. Lower windspeeds will need a longer front leg; stronger wind calls for a shorter front leg.

In the case of the Hexagon, you'll have three legs provided, but one of the delights of this shape is that only rarely does it need bridle adjustment. Delta kites come with a triangular keel into which an eyelet has been riveted for the line attachment. Because the Delta design incorporates leading edge spars which are not fixed to the spine, and can therefore flex, they are self-adjusting for a broad variance in wind speeds. Hence our recommendation for these easy-to-fly types for your first bought kite.

Self-launching

We suggested a 50ft or 15 metre line to start that first launch. This is to give the kite a chance to clear the rough air close to the ground. With practice, the self-launch is easy to achieve.

Pay out just enough line to have the kite clear of the ground when holding the handle high and, as the wind fills it out to its proper shape, then lifts, allow the line to unwind from the handle (or reel if you have one). Each time the line is paid out, the kite will begin to sink as its airspeed is decreased, so only release the line in stages and allow the kite to climb in these steps until you have the final height you want.

2-line Diamonds

There is one kind of kite that does not respond to paying-out the line(s) and that is the basic two-line stunter.

Not to be confused with the swept-wing Delta stunters, the 2-line Diamonds, known generically as Peter Powell types, although not always the size or the quality of Peter's long serving originals, are popular first kites. They are tolerant in their bridle adjustment (as distinct from the sophisticated 110 degree, swept-delta steerables to be found in Chapter Eight), and the newcomer is unlikely ever to require to alter the settings of the rings on the

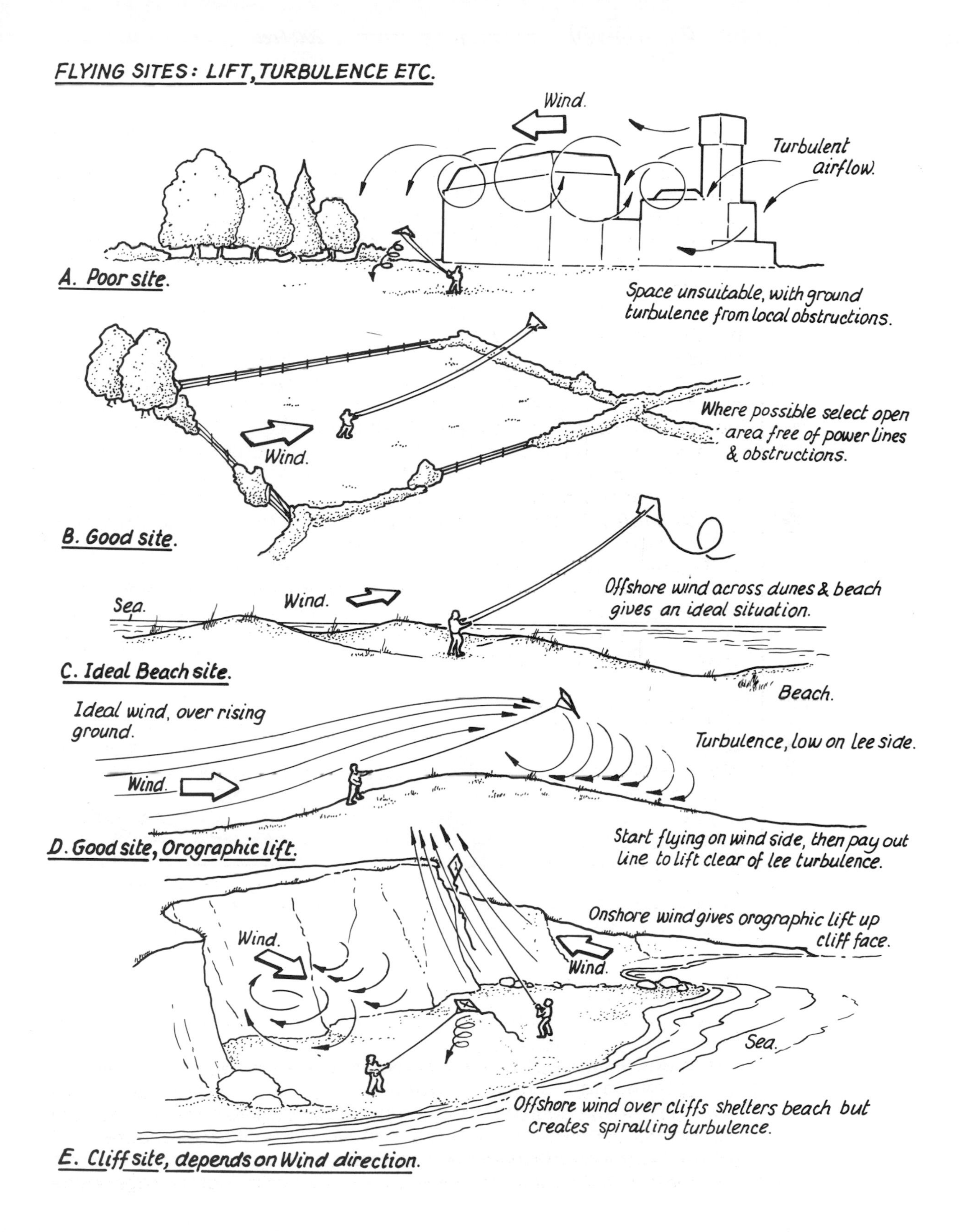
FLYING SITES: LIFT, TURBULENCE ETC.
Wind.
Turbulent airflow.
A. Poor site.
Space unsuitable, with ground turbulence from local obstructions.
Wind.
Where possible select open area free of power lines & obstructions.
B. Good site.
Sea.
Wind.
Offshore wind across dunes & beach gives an ideal situation.
C. Ideal Beach site.
Beach.
Ideal wind, over rising ground.
Turbulence, low on lee side.
Wind.
D. Good site, Orographic lift.
Start flying on wind side, then pay out line to lift clear of lee turbulence.
Onshore wind gives orographic lift up cliff face.
Wind.
Wind.
Sea.
Offshore wind over cliffs shelters beach but creates spiralling turbulence.
E. Cliff site, depends on Wind direction.

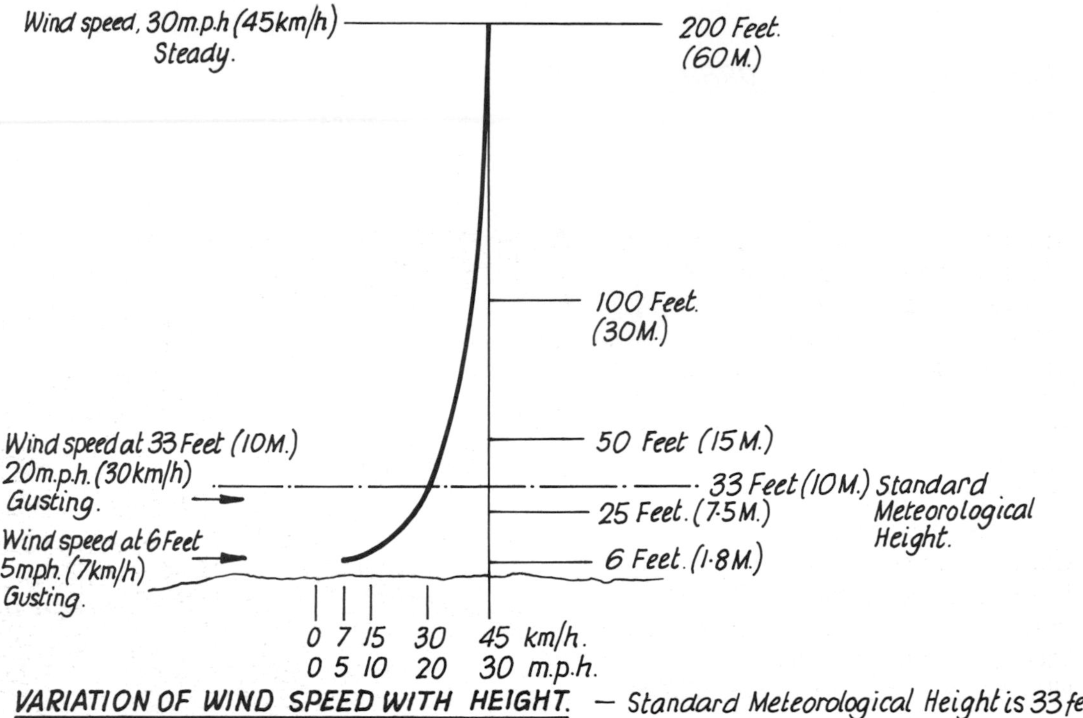

VARIATION OF WIND SPEED WITH HEIGHT. — Standard Meteorological Height is 33 feet: Below this conditions vary due to ground effect & terrain. Above this, Wind increases with altitude.

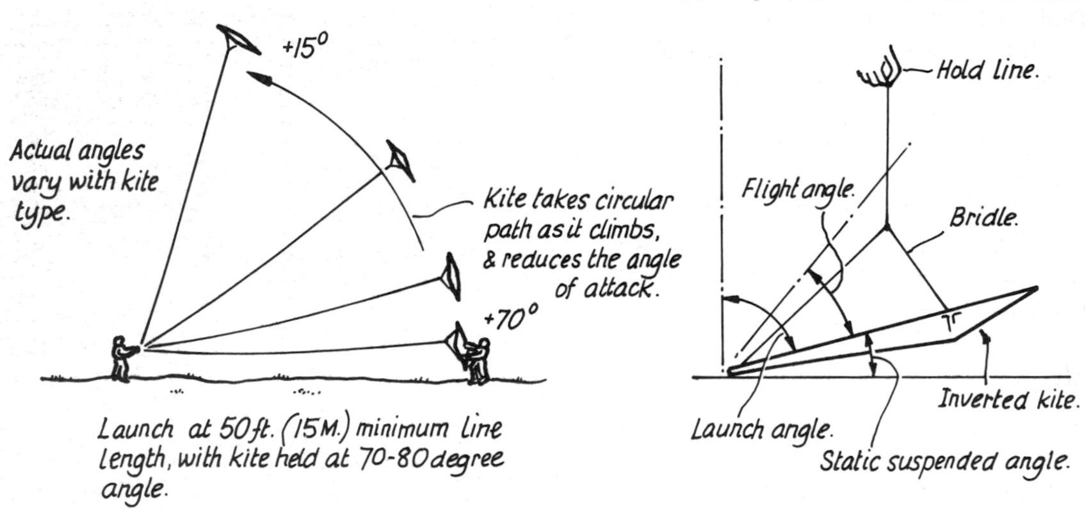

LAUNCHING.

BRIDLE CHECK. (Rough guide only.)

bridle legs. However, this doesn't mean that they are easy to operate!

A common fault is improper assembly. The front bridle legs have to pass over the cross-spar and be free to run in a straight line from the leading edge to the ring. The cross-spar has to be above the front face, and the spine below the rear face of the Powell kite when it is laying face up. The lines must be of equal length, and fully paid-out *before* launch. Most important is the 'hands together' technique for those vital first moments of flight.

These are but a few of the self-created problems we've seen around the festivals. Any of these difficulties can be avoided by a few patient moments of study with the clear and concise instructions that come with the Diamond stunters.

Once airborne, the technique of control comes just as easily as riding the proverbial bicycle... until that is, you fall off! Hands together, equal lines, and the Diamond is just another kite. Spread the arms, pull one hand then the other and you'll soon acquire the skill of mastering loops, low-level skimming the surface, or rapid pull-outs from vertical dives.

Peter Powell has flown trains of his kites from a fast launch up the Thames, from a rear-facing pillion seat on a speeding motorcycle, from the sunroof of a car, into the sea and out again, and has demonstrated these stunts all over the world; but do not expect to become a Peter Powell Mk.2 overnight!

What next?

So far we've only touched the basics and, although it's essential in our view to provide this introduction to kite flying and simple DIY designs, now it is time to get into more advanced material for the modern kite.

There is, however, one critical detail to observe, no matter whether we are raw novice, or long experienced and that is *safety*. Always remember that your lines can conduct electricity, cut skin, trip passers-by, or generally cause mayhem with grass mower blades. The kite itself can scratch cars, hurt people, and scare horses - so be careful out there; remember...

safe flying is no accident.

Who better to teach than Don Dunford, originator of so many designs for steerable kites.

aking your own

Now that we have a little experience in the joys of simple kite flying, it's time to look deeper into the whole subject and to understand more about those 'Modern Materials' which have emerged since the 1970s.

Plastic

Man-made fibres and plastic sheeting have long since replaced the porous cambric or fine cotton linen which had been the standard material for sails over seven decades.

In common use for so many domestic purposes, *polythene* has become part of everyday life. As well as the inescapable carrier bag, and the rubbish bin liner, either of which can provide large enough pieces for a kite sail, the material is available by the metre or on the roll. Sold by specialists for packaging or horticulture, the most convenient is the 250 gauge which comes 2 metres wide. Rolls are 150 metres long - enough for a lifetime - so it is easier to order by the metre. Being inexpensive, and lightweight, easily seamed and joined by using all-weather adhesive tape, polythene sheeting is fine for group projects where quick results are sought, and longevity is unimportant.

Snags? As Henry Ford said, you can only have it in black, or in this case, with one option - clear. So, as far as colour is concerned, sheet polythene is decidedly neutral! Moreover, polythene will not accept most paints, and, if you want to colour a large area, then you'll have to experiment with whatever plastic paints the local art store can offer. This is one subject on which we cannot give a firm recommendation. 250g is 63 microns thick, and this stock gauge can be found through mail order suppliers, or at specialist plastics supply houses in larger cities. The easy way out is to buy a pack of 'Bin Bags'. These refuse sacks come in various gauges, the most suitable being 225g. Standard size is 16 x 25 x 39in, which is 40 x 63 x 100cm, quite enough for a largish kite, and costing only a few pennies.

Although in itself durable, polythene sheet could hardly be regarded as pleasing in the long term, except where it has been colour printed on commercial kites. Most of the lower-priced Malay, Snake and Delta shapes, as well as the economic 2-line stunters, come with highly-decorated, printed sails that have proved the utility of this material; but, for the home builder, the limitations will now be obvious.

Paper

One material that will accept paints, especially water-based acrylic and polyvinyl acetates, is *Tyvek*. This extraordinarily strong paper-like sheeting is known as a Spunbonded Olefin. It comes in several types, as a paper or fabric substitute. The one in which we are interested is the Type 10 paper which is a white, rather stiff but extremely smooth surfaced bonding of fibres. Of many practical uses, the most common is that for indestructible envelopes. Its strength and abrasion resistance have made it ideal for shoe cleaning cloths, as provided by most hotel chains.

Although we have found that white wood glue (Poly Vinyl Acetate) or Aliphatic Resin will hold a seamed joint or hem in Tyvek, the real advantage of the material is its acceptance of sewing. With synthetic thread, polyester for example, and rather long stitches at 4 to the inch or 6mm each, Tyvek sheet can be treated as a fabric in the sewing machine.

Any increase in the pitch of the stitches will run the risk of setting up a perforated run if that seam is subjected to stress. Since Tyvek is so impervious as to be virtually waterproof and indestructible, the strength of seams becomes very important if the kite is large and to be flown in high winds.

A train of plastic sailed stunters, with tubular tails trailing to follow the flight path, as that great showman Peter Powell steers them from the sunroof of a moving car. The scene at Truro when winds fell to zero and Peter decided to create his own airspeed.

Snags? Availability is one. Tyvek is not easy to locate. Paper stores are the most likely source for small quantities. As it is supplied in bulk for re-processing in many industrial applications, you may be fortunate to have a local factory or printing establishment with offcuts to spare.

Another problem is its stiffness. Unlike a true fabric, Tyvek sheet tends to follow straight lines and does not billow, or curve under wind pressure. As long as this is respected, the rigidity can be used to good purpose. So Tyvek is better for kites that have outline frames and do not require flapping edges, as say, for a bird kite where unsupported edges are a feature.

Fabric

This leads us to the most widely adopted material for kite sails, *Ripstop Nylon.*

Developed for spinnaker sails, where tear resistance is critical and the ability to form into an aerodynamic shape is of greatest importance for efficiency, ripstop has become an ideal fabric for all types of kite. It gets its name from the strengthening threads which run across warp and weft to create small squares in the otherwise close-woven nylon. So a tear will be arrested, holes limited, and the fabric made infinitely constant in its run off the loom.

We see it in brilliant colours which flow through the whole spectrum, and extend to dayglo; but it starts life in a natural tone, initially floppy until the grippers or stenters pierce the outer edges to stretch it flat as it is heat treated, and dyed. A polyurethane coating then seals the pores, usually making the surfaces slick and shiny, one more than the other. At around 50 threads per centimetre (127 per in), it takes a powerful magnifying glass to distinguish the equal warp and weft; but look if you can, and it will give you a better appreciation of the quality of ripstop nylon.

Below: Ripstop nylon samples from Carrington Novare and Bainbridge as supplied to kite manufacturers for selection of colour and weight of spinnaker cloth.
Right: Checking the thickness of Ripstop nylon in microns.

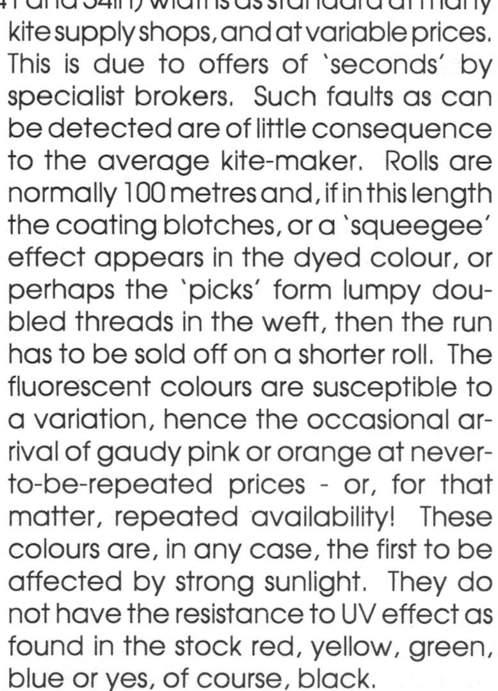

It comes in 92, 104, and 137cm (36,41 and 54in) widths as standard at many kite supply shops, and at variable prices. This is due to offers of 'seconds' by specialist brokers. Such faults as can be detected are of little consequence to the average kite-maker. Rolls are normally 100 metres and, if in this length the coating blotches, or a 'squeegee' effect appears in the dyed colour, or perhaps the 'picks' form lumpy doubled threads in the weft, then the run has to be sold off on a shorter roll. The fluorescent colours are susceptible to a variation, hence the occasional arrival of gaudy pink or orange at never-to-be-repeated prices - or, for that matter, repeated availability! These colours are, in any case, the first to be affected by strong sunlight. They do not have the resistance to UV effect as found in the stock red, yellow, green, blue or yes, of course, black.

Spinnaker ripstop must not be confused with its near relative, *Balloon ripstop*, which has a stretchy feeling to its touch, comes on 137cm (54in) rolls and does not crease. Ideal for the ram-air inflatable 'Soft' kites which depend on flexibility of the surface and absolute solidity of the non-porous sail material, it has another great advantage. This is its acceptance of a paint decoration without risk of 'show-through'. Spinnaker nylon is difficult to paint without having smudges where the colour runs through to the under surface. With the balloon ripstop, one can hold fine lines, and use the colours edge to edge without having to worry about spoiling the decoration with inter-mixing or any blurs.

However, its softness and crease resistance do not make it suitable for kites where the sail is stretched, as on a typical aerobatic type. Use balloon ripstop on a 110-degree narrow Delta steerable kite, and you will soon discover why the spinnaker nylon is best. In cases like this, one cannot run the risk of having one side of the kite sail stretch more than the other. It's bad enough trying to fly with unequal line lengths; having an asymmetric kite can be most frustrating! Remember that your balloon ripstop was probably made for the parascending sport, and you'll appreciate why it is so soft and at times stretchy.

Having stated that, the wide roll width, dense colour and crease resistance makes balloon ripstop superior for those artform kites which have become spectacular attractions at all major kite meetings. These highly specialised 'Sky Pictures' as we describe them in Chapter Four, are strictly 'one-off' types, reflecting the artistic skills of the maker.

In summary, the most likely material you will use for the kite sail is spinnaker ripstop of 43gm/sq.metre or 65gm/sq.metre grade (1.12 or 1.5oz/sq.yd), 65 or 100 microns thick. It will have been made for the sailcloth trade but, due to some minor fault in manufacture, was released to the kite trade instead, and at an economic advantage. These faults are rarely detectable so, in turn, the individual kite-maker benefits through being able to purchase quality material in small quantities from the kite shop. (Lighter fabric weights are quoted in the USA, but they are based on uncoated fabric, e.g. 1/2 or 3/4 oz/yd.)

Statistics on kite manufacturing volumes are not available anywhere. All we can do is to report observations at national and international meetings where the keenest, most dedicated enthusiasts gather. We would be quite safe in stating that over 90% of these kite flyers are using ripstop nylon of either type for their kite sails. With that level of recommendation, there really is no better option. Polyester ripstop has advantages of light weight and is used on many Sport kites. It has a 'harder' surface which crackles when flexed and is especially suited to indoor stunters.

Spars, spreaders and stand-offs

The framework of a kite is made of rigid or semi-rigid spars. Unfortunately for the newcomer, the parts are given a variety of names which frequently creates confusion, so we had better start by clarifying the variations.

Taking a standard Diamond or Malay shape which is formed by two spars at 90 degrees to each other, the 'spars' are known variously as 'sticks' or, the 'spine' and the 'spreader', or the 'keel' and the 'cross-spar'. Similarly, the spars in a Delta can be called the 'spine' and the 'leading edges'. In a Box kite, the four main spars are sometimes called 'longerons'. Where a spar has to be flexible, and is bent by a tensioning line, it is often known as a 'bow', or 'bender'. Aerobatic kites have vertical spars to tension the sail into its shape, and these are known as 'stand-offs', or 'whiskers', or sail stretchers'. Confusing isn't it?

Common usage has sifted these terms and left us with the most expressive. So for our purpose we will refer to all the frame components collectively as *spars*, and define their applications by calling them as follows: *spine, cross-spar, leading edge (l.e),* and *stand-offs*, assuming that such terms will be self explanatory.

Wooden dowel has been used for kite spars for as long as we can recall. It replaced square and rectangular section spar material sometime in the 1930s, and the only changes that have arisen are in the timber used and the sizes.

Pine, Ramin and Obeche are the most common dowels. Beech is comparatively rare nowadays, and in any case does not match the strength/weight ratio of Ramin or Obeche. Sizes are imprecise, as one might suppose any material would be if drawn through a hole with a cutting edge! In theory at least, the diameters range from 5 to 15mm (0.2 to 0.6in) and lengths are around 2 metres (78 in) which is enough for any kite spar.

When choosing dowel at any place other than a kite store which would have selected its stock, rotate the spar between thumb and forefinger of one hand while resting the dowel in the other hand and check that the spar is straight. When satisfied, look along the whole length and ensure that the wood grain does not run off the true length. Pine is weak this way, while the other timbers tend to have straighter grain. If undecided on the diameter you need, remember that tube joiners will be required somewhere or other and the fit will have to be a good one. This usually influences your choice.

Light alloy tubing also has its place in kite-making. Outmoded by the plastics for smaller kites, it comes into its own for the big stuff as, for example, replica Cody man lifters, or the large Baden-Powell 'Levitor' where, in turn, it replaces the bamboo spars in the original pioneer designs. But we are diverting from our purpose - on to the modern!

Glass Reinforced Plastic, otherwise known as GRP, or Glass Fibre, is available in solid rod, or as tubes over a broad range of sizes. Made from glass fibre rovings and polyester resin, it is normal for the rod to be white. However, by use of dye in the resin, manufacturers can produce pink, green or blue rods for quick identification in the assembly of commercial kites.

No less than 20 different diameter rods, from 1.5mm (1/16in) up to 31.8mm (1¼ in), are listed by the largest manufacturer in the UK, R.B.J. Reinforced Plastics Ltd. A similarly wide range of 23 sizes, from 0.86mm (1/32in) up to 38mm (1½ in), is made by Pultrex who were first on the kite scene, and who supplied many of the pioneer manufacturers of aerobatic kites with their white GRP rods. This company subsequently made and sold pultruding machines to many other plastics specialists, both in the UK and overseas.

So what, you may well ask, is *pultruding*? It's a word you are unlikely to find in the dictionary but, if you think of it as the *opposite* to 'extrude', meaning to thrust or push out, then the term becomes more understandable.

Left: Typical of the rigid frame using glass fibre tubing on the earlier steerable deltas is seen on this example.

Below: The range of glass fibre tubing, in this case moulded in various colours. The Vlieger Op range from Holland uses colour to identify sizes from 5.85mm in blue to 21.8mm in orange, many at constant wall thickness and at 2mm stages so that they are able to fit inside each other.

How the GRP rods are made

The raw material, glass fibre, is drawn from bobbins held at one end of the long pultrusion machine. There will be five or so such bobbins for each rod, and the pretensioned rovings are directed through guides down into a bath of polyester resin in much the same way as any glass fibre is prepared for setting into its shape by wetting the fibres with resin/hardener mix. In continuous motion, the now-wetted rovings have lost their natural glass colour and have been thoroughly saturated with the resin. This can be dyed, but is normally white.

As they move along the machine from the bath of resin, all the rovings are pulled together, and they disappear into a block of metal through a single hole which determines the final diameter. The block is a machined die, electrically heated to 150 degrees centigrade, so that, by the time the glass fibre emerges rather like minced meat from the kitchen grinder, it has already adopted the appearance of the final product.

Still hot, and completing its thermosetting process, the GRP rod continues on its way along the length of the machine in the company of another 24 or so rods which have passed through a parallel course. The actual number in the batch will vary, but this is a typical process for the popular 3mm size.

All the rods are being pulled by a tension of over 1000kg (about one ton) by means of a moving pressure pad. This makes contact with the batch of rods, then draws the cooling product along at predetermined speed for about 5m(16ft) before relaxing the pressure, lifting off, and allowing the machine operator to cut the batch free. Meanwhile, a second pressure pad has already taken up the tension for the next length of rods. The whole process is continuous, and the total run will amount to kilometres. In fact, for some industrial purposes, the rods are reeled on large diameter drums and can be made automatically without supervision. (A break in mid-run has been known to fill a factory with thrashing plastic like a demented clockspring which has disengaged from its retainer!)

All of this sounds easy. However, it is a far from simple manufacturing system and depends on the components, temperature, speed and the resin for ultimate quality. The GRP rod has its best application for kites in the range of diameters from 1.5mm to 10mm (1/16 to 3/8in nominal). Beyond that, the weight of solid GRP rod, up to 40% of which is resin, is too heavy for our use, and the tube becomes a better proposition.

Tubes

Our description of rod manufacture also stands for the tubes, except that there is a central bullet in the die to form the ultimate wall thickness. Tube sizes range from 5.5mm up to 22mm (7/32 to 7/8in) for practical use in kites. Many other sizes are made for an enormous range of industrial uses, but flexibility limits their application for kite spars, and in many cases the wooden dowel is superior.

Single-line kites can tolerate, even use to advantage, the flex of GRP rod and tube. Sold in lengths up to 2.5metres (8.2ft), it enables the kite-maker to construct large Delta or similar designs where spar length would otherwise call for joiners. Steerable or aerobatic kites require a rigid frame for top performance, and a first solution to the problem of obtaining a good stiffness factor against weight was to turn to another outdoor sport, archery.

Carbon fibre

Arrow shafts have for years been made to a very high standard, with spiral-wrapped glass filaments and with aligned fibre composites (AFC) using high modulus carbon. They are straight, consistent in diameter and have excellent rigidity to meet the demands of the sport. So they also met the specifications for aerobatic kites. There were, however, some limitations. Lengths varied according to source from 76 to 86cm (30 to 34in). In diameter, the standard is 5.5mm (7/32, or 0.22 in). Quality, quite naturally, was very good and matched by the high cost, but not really excessive for the job of making stunt kites perform to their best.

For many spines and cross-spars, the stock arrow lengths are adequate. Leading edges, the critical component for aerobatic kite stability, call for 125cm to 175cm (50 to 69in), preferably in one length to avoid joiners. So, while a thicker-walled metal joiner sleeve was used, kite manufacturers sought a supply of longer stock from the pultrusion companies. As a result, kite-makers have carbon tube in lengths of 5 metres from the pultruders, which is then cut as required for commercial designs, or halved to 2.5m (98in) for easier handling by the individual enthusiast.

Top to Bottom: A range of tubes by Pultrex, the 'Pullwound' type showing an inner layer; glass fibre tube and rod showing comparatively thicker wall in the tubing, accounting for greater weight; a mix of unidirectional carbon fibre tubes of the most popular type.

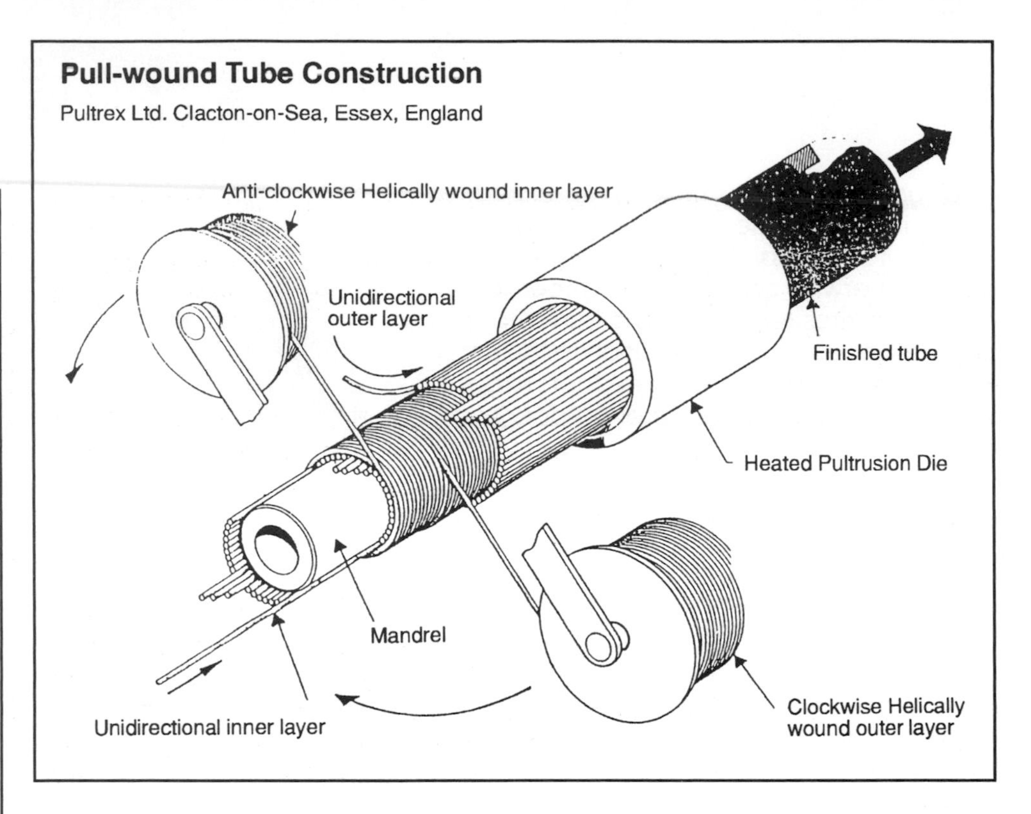

The same method of pultrusion as we have described for the GRP rod and tube is applied to carbon. Instead of glass rovings, the carbon tows are mounted on reels at the start of the process, and these are drawn through vinylester resin which has better wetting characteristics for the carbon. The tows then form the tube around a bullet or mandrel in the heated die, and the result is a tube of greater strength than the equivalent in steel! With the carbon fibres running *unidirectional*, end to end, this tube is crushable under pressure but not of any order likely to arise in kiting.

While the arrow diameter of 5.5mm remains a first choice, since so many long established and successful aerobatic kites have been designed around it, two larger sizes were introduced in the same unidirectional carbon tubing. These are 6.35mm (1/4in) and 8mm (5/16in) and many of the larger narrow Delta designs have taken advantage of the additional strength offered by such sizes in the 2.5metre lengths.

Pultruded tubes with unidirectional fibres have two recognition factors. A slight ridge can be felt along the length of the otherwise very smooth surface and, much more important, the wall of the tube tends to be thicker than with other production methods. It is often in ratio with the outside diameter, ranging from 1.2mm to 1.65mm wall thickness.

For most applications, this is of low significance, but there is a more involved method of making the tube which brings the wall to a constant 1mm over the same variation of outside diameters, from 6.35 through 7,7.5 and 8mm. This is *pullwinding*, as developed by Pultrex Ltd., and the result is a sandwich cross-section of unidirectional inner and outer fibres, with helical windings, clockwise and anti-clockwise between. Pullwound tubes can be made in glass or carbon fibre, have a high crush strength, and come in any of five colours in glass/polyester.

As with the pultruded tubes, the Pultrex Pullwound is made in adequate lengths, cut off as ordered by their trade customers. The 8mm has best strength/weight advantage and the 7.5mm is ideal for use as joining ferrules with its 5.5mm inside diameter to accept other tubes, including arrow shafts.

Shafts

Back to the beginning, and the archery shafts which led us into carbon tubes and the generation of so many new designs for steerable aerobatic kites.

While 5.5mm is the most widely adopted diameter for arrows, it is by no means the only thickness. Arrows come as slender as 4mm and as stout as 8.5mm in many makes. They range from the spiral wound glass fibre/epoxy type which, being black, are sometimes confused with the carbon shafts. More often, the carbon types have identifications, such as AFC 0.22in (USA) or Beman (France). Most easily identified are the Easton carbon/aluminium shafts with the unidirectional carbon fibres bonded to thin wall aluminium tube. These are claimed to be fatigue-resistant to cope with the high amplitude vibrations that are generated by fast stunters. Easton also produce ECS tube, an all-carbon shaft with the fibres spirally wound in opposite directions.

The choice will inevitably vary as time progresses and wherever demand exists. Long lengths are difficult to mail or send by carrier, so the kitemaker with a kite shop within reasonable travelling distance is better off. But, for the lone hand, those arrow shafts can be mailed by any supplier for a small supplement to cover packing.

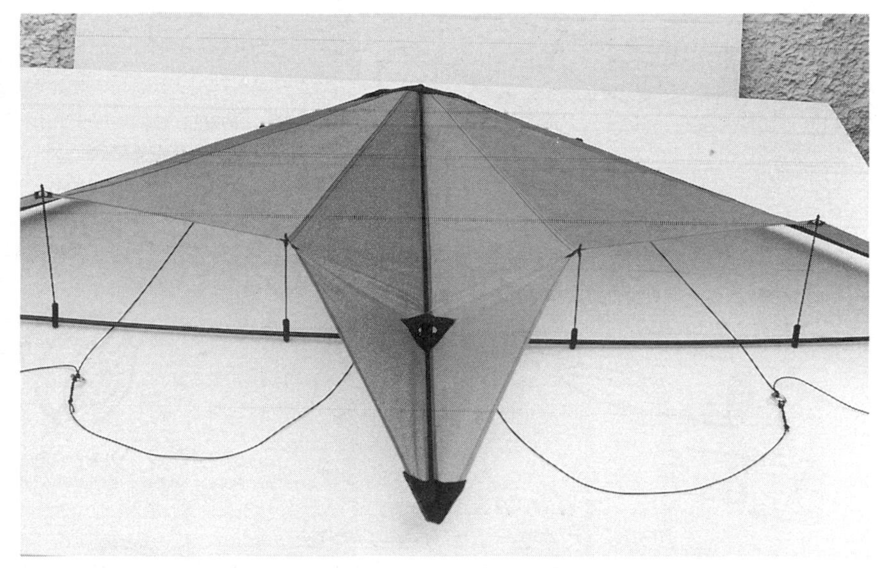

Rob Brasington uses four 'stand-off' struts on the Jester, this offers a flat topped cone in the sail.

Stand-offs

We mentioned rod largely in connection with kite frames. It has another important use as a support for the sail, either in the decorative sense to form shapes in profile, or more commonly to give depth to the sail for aerodynamic reasons. Hence the term to "stand-off" or lift the sail away from the plane of the framework on the steerable extended Delta designs.

These stand-offs apply tension, so they have to be strong. They fit into endcaps attached to the frame and the sail in order to force the sail into its 'Rogallo' cones. In other words, the sail rises from the spine and its trailing edge descends at the tips.

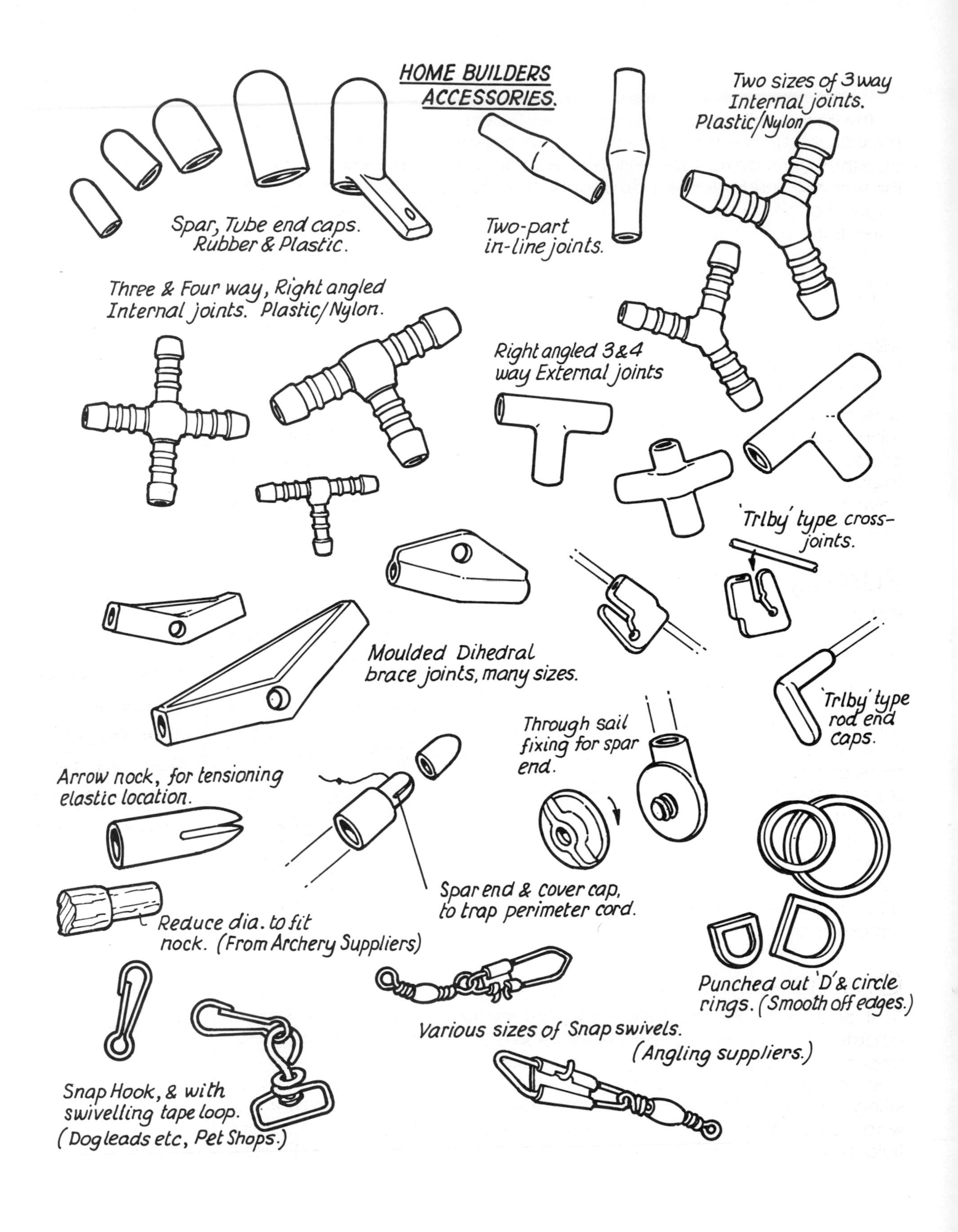
HOME BUILDERS ACCESSORIES.
Spar, Tube end caps. Rubber & Plastic.
Two-part in-line joints.
Two sizes of 3 way Internal joints. Plastic/Nylon
Three & Four way, Right angled Internal joints. Plastic/Nylon.
Right angled 3&4 way External joints
'Trlby' type cross-joints.
Moulded Dihedral brace joints, many sizes.
'Trlby' type rod end caps.
Through sail fixing for spar end.
Arrow nock, for tensioning elastic location.
Spar end & cover cap, to trap perimeter cord.
Reduce dia. to fit nock. (From Archery Suppliers)
Punched out 'D' & circle rings. (Smooth off edges.)
Various sizes of Snap swivels. (Angling suppliers.)
Snap Hook, & with swivelling tape loop. (Dog leads etc, Pet Shops.)

The frame connection is usually on the rear cross-spar, and the endcap on the sail will be in-line but, depending on the shape of the sail, the stand-off will be either rigid and straight or flexible and curved when in position. We explain this variation to emphasise the purpose of the stand-offs. They change the angle of attack from acute within the central delta, to near neutral over the outer triangular sail panel. Where additional stand-offs are used, either on the front cross-spar to vary the angle, or on the rear cross-spar to extend the Rogallo cone, the tension is just as important.

Where the sail extends behind the cross-spar, it may be necessary to use an over-length and curved stand off in GRP. When fitted, the glass fibre rod will compress into a curve, forcing the sail both backwards and upwards from the cross-spar. They can be 2 or 3mm (5/64 or 1/8in).

If the sail edge is overhead, or nearly overhead, the cross-spar at the point where the stand-off is fitted, then a carbon rod should be used. This can be thinner, down to 1.5mm (1/16in), so reducing drag. A further use of carbon rod is for sail battens to control flutter, or to improve sail shape. In each case, it has to go into a sail pocket and, to avoid wear through abrasion, the rod should have cappings or a Dacron polyester tape reinforcement at each end, or the fabric will soon become pierced.

Putting it together

Plastics come to the aid of the kite-maker when it comes to assembly of the frame. The marketplace offers a plethora of fittings that have been created for commercial kites and subsequently reach the accessory trays. They range from the basic spar endcaps we have already mentioned to dihedral joints, three and four way plug-in joiners, sail-clips and, for the carbon tubes, the arrow nocks and tube sleeves for joining.

As well as archery, two other sports - camping and fishing - offer useful bits and pieces to aid assembly.

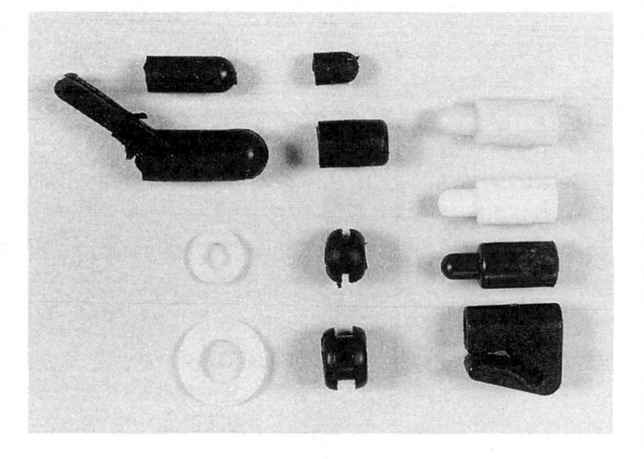

Fittings by Cochrane are sold as accessories and used by other kite manufacturers. This small sample shows end caps, stand-off caps, clips and grommets.

Grommets and sprays

Tent or ground-sheet screw-through fixings with a plug-on cap can be adapted for holding a spar, and similar screw-fixed grommets have their application on large ram-air kites for bridle attachments. Look around any good camping emporium and you'll find many worthwhile things to buy for kiting, not least the silicone protective spray-can to revive worn or weatherbeaten ripstop; or one of those hemispherical shelters to have on the field as a base in which to store your gear. The nylon material and GRP tube 'poles' will be familiar, so the shelter will not be out of place in the kite bag. In spite of having a range of injected plastic mouldings, there are always times when there just isn't anything for the size of a spar, the angle of fit, or sufficiently shock-absorbing to accept a crash, or to permit a kite to be folded.

Plastic tube joints

Transparent fuel pipe, as used on motorcycles and many cars, is easily obtained, economic and fulfils a variety of purposes as a spar joiner. It comes plain or braid-reinforced with an internal diameter that is a tight fit on 5.5mm (0.22in) carbon shafts.

Similar pipe in black neoprene, polyethylene or rubber can be adapted; in fact, on a carbon frame the neoprene looks all the more professional. All these pipes will cut with a sharp modelling knife, they can be drilled with care, or punched with a piece of tube that has been sharpened to act as a hole cutter. Pat has sketched the jig that ought to be used when punching holes. Note that you can 'work' the cutter to pierce only one side of the plastic tube by rotating the cutter. A wrap of plastic tape is advised to protect the hands while cutting with a screwing action.

It's a good tip to cut some thick 'washers' from the same plastic tube material. These are spaced on the spars to act as bridle stops, so that the bridle connection on, say, a leading edge is retained and the tendency to slip is controlled. If the washers are sliced on one side at an angle, they are even better for retention of a cross-spar joint.

Study the professionally-made aerobatic kites and you'll find innumerable inspirations for the use of plastic tube joiners.

Pockets and patches

We've only one area of kite assembly left, and that is the means of holding the frame of spars to the sail.

Dacron polyester tape is sold in 25, 50 and 75mm widths, usually in black only. It is the toughest material you can use for spar pockets, and the heavy sailmakers' webbing, or seatbelt Dacron makes perfect head or nose reinforcements for the stunters.

Always anticipate the abrasive capacity of metal tube, GRP or carbon spars within the pockets at their ends. Always round off, or preferably cap, the spars with a soft plastic fitting. Even so, a sudden 'arrival' on terra firma will inevitably enable that irritating spar to find a weak point and poke through a hole in the fabric. For such cases a field repair is soon effected with adhesive-backed sail repair tape, usually 50mm wide and in a heavyweight ripstop that might even be located in a matching colour!

Beware of fibres

One last word on cutting those GRP or carbon rods and tubes. A small 'Eclipse' hacksaw, or a fibre disc in a high-speed motorised modelling tool,

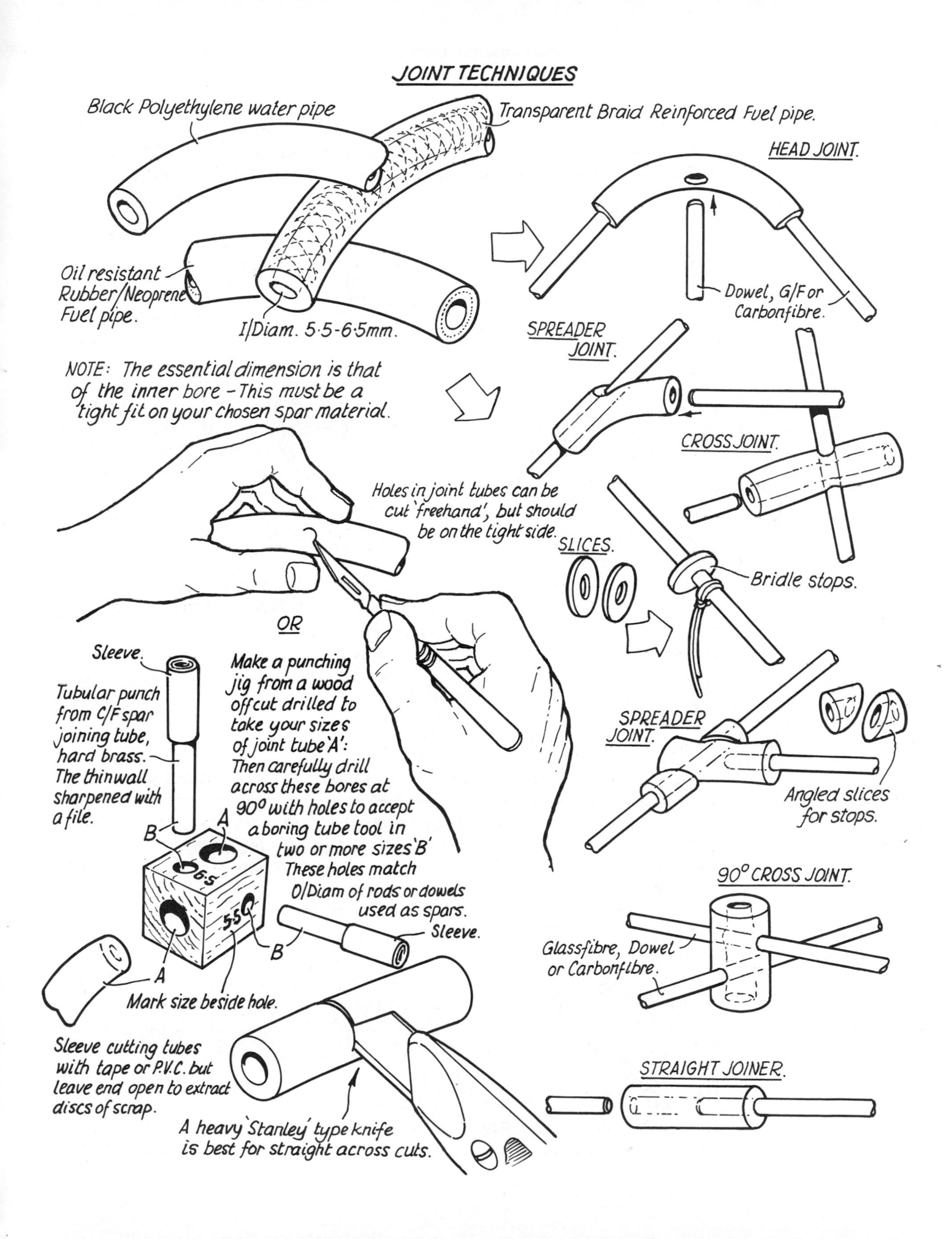
JOINT TECHNIQUES
Black Polyethylene water pipe
Transparent Braid Reinforced Fuel pipe.
HEAD JOINT.
Oil resistant Rubber/Neoprene Fuel pipe.
I/Diam. 5·5-6·5mm.
Dowel, G/F or Carbonfibre.
SPREADER JOINT.
NOTE: The essential dimension is that of the inner bore - This must be a tight fit on your chosen spar material.
CROSS JOINT.
Holes in joint tubes can be cut 'freehand', but should be on the tight side.
SLICES.
Bridle stops.
OR
Sleeve.
Tubular punch from C/F spar joining tube, hard brass. The thinwall sharpened with a file.
Make a punching jig from a wood offcut drilled to take your sizes of joint tube 'A': Then carefully drill across these bores at 90° with holes to accept a boring tube tool in two or more sizes 'B' These holes match O/Diam of rods or dowels used as spars.
SPREADER JOINT.
Angled slices for stops.
B
A
6·5
5·5
B
Sleeve.
90° CROSS JOINT.
A
Mark size beside hole.
Glassfibre, Dowel or Carbonfibre.
Sleeve cutting tubes with tape or P.V.C. but leave end open to extract discs of scrap.
STRAIGHT JOINER.
A heavy 'Stanley' type knife is best for straight across cuts.

will go through it like a knife in butter. A 3-cornered file is best for small diameters, *but* beware the fibre dust or shavings! There are few worse irritants for eyes and hand cuts. *Be careful when cutting fibre!*

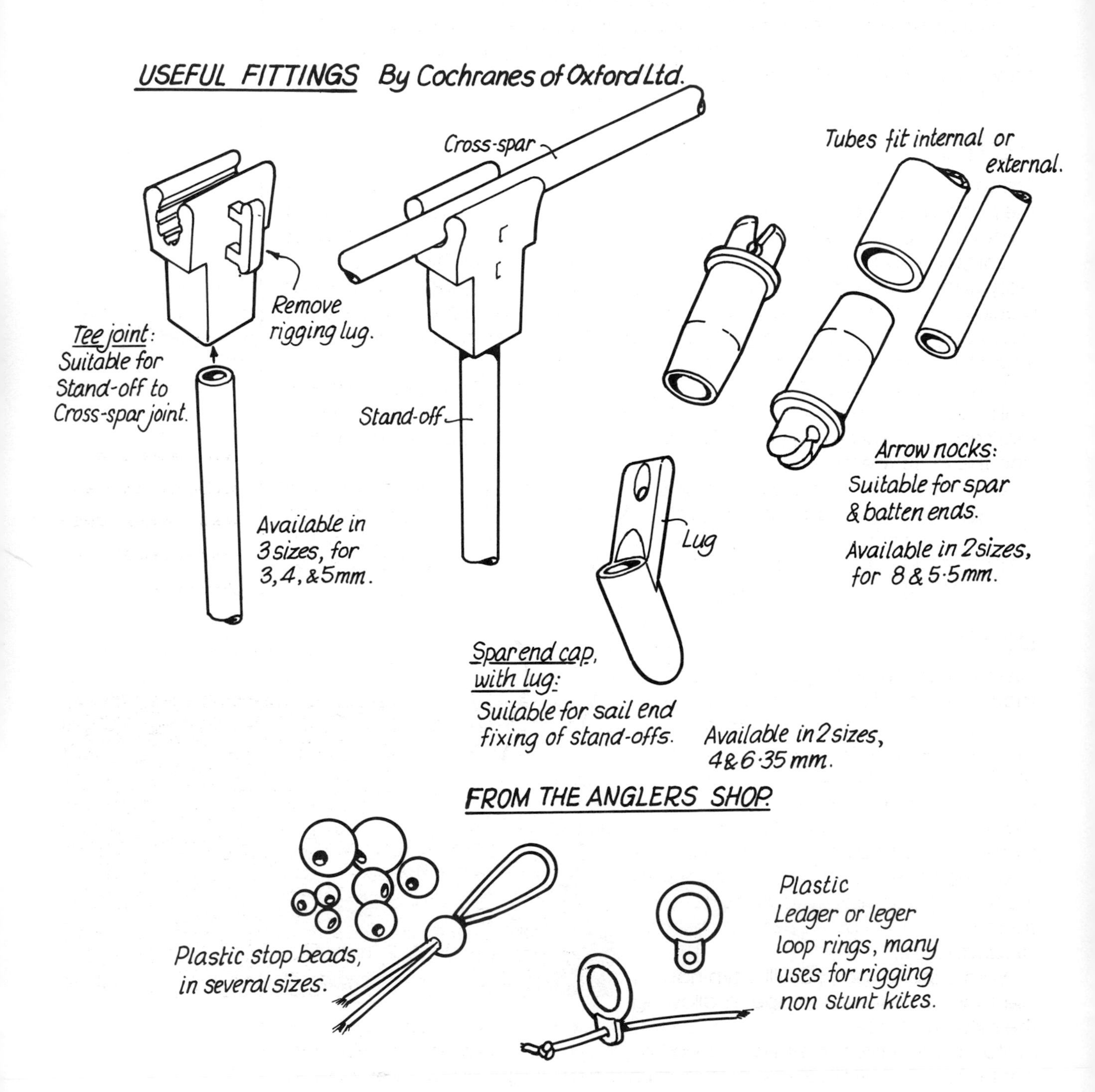

Sailmaking

We've reached the point of no return! The ripstop nylon *has* to be cut and *sewn* if we are going to reach our goal and have a kite to fly.

Dressmaking and sewing machines are totally foreign to many would-be kite flyers. Yet, once the bug has bitten, it is amazing how so many home-brewed designs appear, and most of them with excellent stitching in all its variations.

Patterns

First we have to know *what* we want to make. Then comes the fun of making the pattern - and do make sure that the dimensions you are going to work to are for the full pattern and not the kite frame shape. There's a real difference!

Large areas of paper will be required. Brown parcel paper, the back of wallpaper left over from that last bout of home decorating, or drafting paper would be ideal. At a pinch, use newsprint; but it will not help you to design a fancy project, nor will it help if the print rubs off onto your light coloured nylon when the 'plan' is used as a cutting pattern.

If the paper will not cover the area, then join sheets with masking tape or similar, so that it matches the size of the kite. Then draw out the kite full-size, marking on the spar positions and planning where the pockets will be needed and the reinforcements.

Depending on the type of kite, you'll now have to plan how the fabric is arranged to fit the shape. We're being rather glib here. You may be making a parafoil, or an inflatable animal shape, in which case the paper pattern will be more than just a flat plan. For basic shapes though, and it is better always to start off at that level, the paper pattern is the right approach.

Bias

Check for colour arrangement, select the fabric and then lay the nylon panels over the plan *using the bias,* and ensuring that the ripstop squares are symmetrical on each half of the kite. Bias in the fabric allows ripstop to stretch more along the diagonals of the square patterns made by the slightly heavier threads. So it pays to arrange the squares at right angles to the leading edges or at right angles to the spine. In other words, *do not lay the nylon at random.*

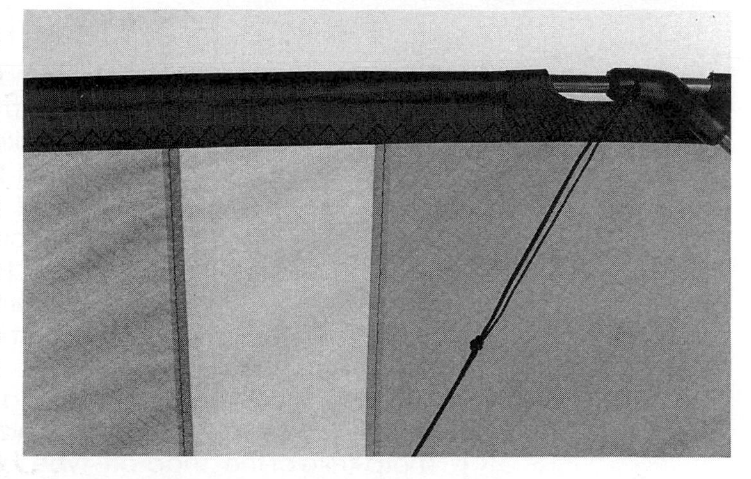

An example of good sewing, the single row of stitches becomes a double row on the topside in a Flat Fell seam.

With experience you will learn how to use bias to best advantage in allowing the nylon to 'billow' with a curvature that helps to create an aerofoil. Many professional stunters use strips of nylon or panels with the ripstop at different angles to achieve a curve upwards from the leading edge pocket. In such cases, all variations are symmetrical either side of the spine, and the whole plan has to be worked out very carefully before any nylon is cut.

Now to the cutting. If you want to make more than one kite then now is the time to transfer panel templates from the full-size plan. Card, plywood, sheet-metal or compressed hardboard are used in turn by all kite manufacturers but, for the independent enthusiast, another piece of stiffish paper will suffice. The panel pattern template is wanted primarily to enable you to mark the ripstop. If you feel able to do the same directly from the original plan, then by all means do so; but we're advising a *reversible* template so that each kite half is the same - and there's no guarantee that the first drawing will achieve that!

With a soft lead pencil, chalk, or dressmaker's marking wheel - but not marker-pen or wax crayon - transfer the panel shape to your ripstop. *Don't* cut yet! If you have the space, tape the ripstop to a flat surface first so that it is flat before the template is applied.

Below: A kite's trailing edge showing the double row of Flat Fell seam stitches plus reinforcements in Dacron for the spine end and the cross-spar support. Right: An industrial Singer machine, the end opened to reveal the run of thread.

Pockets and hems

Having marked the ripstop, stop to work out how much *extra* you will need to allow for the leading edge pocket, or any hems, either on unsupported edges or where joining to another panel. As a general guide, you will need 12 to 20mm (1/2 to 3/4in) for a hem or seam, much more for a pocket.

Think in advance about using Dacron tape for a spar pocket, and how it will fit into the plan of the kite, taking the ripstop into the pocket, not outside!

Cutting the patterns

As a man-made fibre, ripstop can benefit from a hot cutter that will seal the edges. Professionals in sailmaking use a tool that was specially designed for this task. A hobbyist's substitute would be a 25watt soldering iron with a shaped bit, or a woodburning tool. Some modelling shops can offer a Pyrograph, designed to reshape moulded polystyrene figures, that operates via a transformer. In all cases, the tip of the tool must be reshaped like a screwdriver blade, but with the edges rounded.

It isn't at all necessary to use a hot cutter. We know of one prolific kite-maker, whose kites are marketed everywhere, who uses no more than a sharp knife of the 'snap-off' type to cut dozens of panels in a laminated pack around a metal template.

If using scissors, a straight run is best cut by opening the blades a little then drawing the fabric through. Check with some scraps for practice first! Around curves, use small cuts with the tips of the scissors. Because it is ripstop, and fraying is therefore limited, pinking shears would only be decorative and not at all necessary.

The most satisfactory cutting method for one-off home-built kites is to use a sharp modelling knife or 'roller-cutter' against a straight edge on ripstop that is secured to a firm base by tapes. A half sheet of hardboard from your DIY store can be used over and over as a base, but it does blunt the blade. A graphic artists' cutting board will cost a lot for the large size required so, unless it is part of your daily function, forget it. The simplest substitute is corrugated cardboard. You can only use this once! Whatever you use , try to retain the template, and don't cut it along with the ripstop. It is bound to be needed again.

So sew

There are so many machines, and so many good books, to deal with the subject that it would be foolhardy to attempt any detailed sewing instructions here. But there are sound basic tips to observe where the slippery ripstop is concerned. It's no good to produce neatly cut panels then spoil them with distorted sewing, so the very first concern must be to understand the adjustments for thread tension.

As the material is synthetic, use a matching thread - 50 denier is ideal and it comes in many colours, although strangely very few kite-makers seem to want other than black. Cotton thread can produce a pucker in the sewing with shrinkage. Use a polyester thread. If in doubt see if it melts with heat, then you are sure you have the right stuff.

The sewing needle is next. Singer size 14 (90 in continental sizes) is a good general purpose needle but, if you want one finer, then use a 10 (70) or a 12 (80) and, for heavier material an 18 (110) needle size. You can also use a twin needle on most domestic machines. This clever device produces a parallel double row on one side and a single row of zigzag stitches on the other. It makes a stronger seam of two panels and is a trademark of some of the best sewn, professionally-made kites.

A general rule is to use a higher number thread with a lower number needle so, for a 70 needle use 80 denier thread, and for a 110 needle use a 40 thread. The standard thread used in the UK industry is Polyfil 120.

The needle plate should be matched to the needle. If the hole in the plate is much larger than the needle shaft, then the needle will tend to push some of the fabric down the hole with it. The result is a variation in stitch formation.

SEWING HINTS & IDEAS

Whichever type of machine you use to sew your kites with, you will by experience have learnt the basics required by your particular machine. The hand type machine is very basic and slow, occupying both hands fully - The treadle machine frees both hands and is quicker, but the ultimate is the electric machine having the facility for zig-zag stitching. All these use the basic technique of supplying two threads (from above & below) which join fabrics with a looped stitch in a continuous line. Each thread has its own tension adjustment which is vital to the correct stitch.

A C D F B E Masking tape guides.

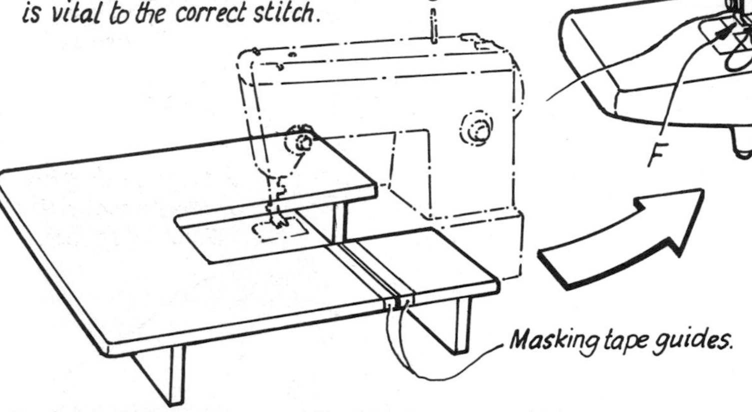

Useful D-I-Y extended work table for any machine can be made up from scraps of chipboard, or best from melamine faced scraps. Use wide packing tape to fix temporarily (if its not solely a kite making M/c).

Typical Zig-zag machine:
- A: Presser foot pressure adjuster. (Feeding of fabric through needle).
- B: Internal adjustment of bobbin tension. (Some older machines.)
- C: Needle thread tension adjuster.
- D: Stitch length adjustment.
- E: Electric machines have foot operated Speed controller.
- F: Presser foot, interchangeable with special types for hemming etc.

Stitch Length: (Stitches per Inch).

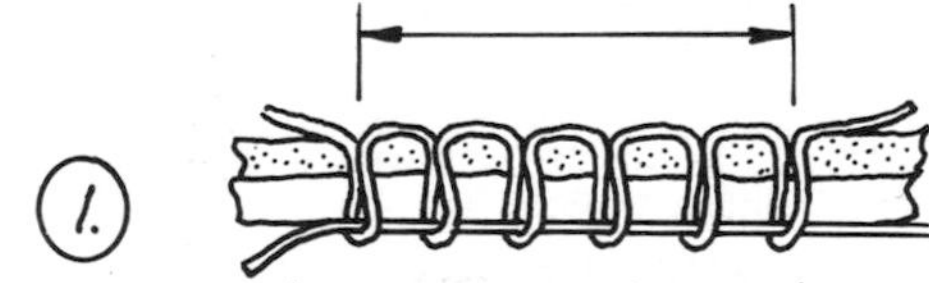

1. Wrong: Top thread looping, Lower thread flat on surface.

2. Wrong: Top thread flat on surface, Lower thread looping.

3. Correct: Both threads 'drawn-in' equally and linked between material layers.

Basic Guidelines for Ripstop.

Needle sizes: No 14; No 18 For thick fabrics.

Stitch length: 8 Stitches per Inch (25mm).

Thread: Use only Polyester thread, Code: 50 - 60.

To correct 1. The lower (bobbin) thread tension is too great, this can be adjusted by a screw on the bobbin case. Old type hand/treadle machines with a 'boat' shaped shuttle have this same screw. NB. This adjustment will only be necessary if a very different fabric is used.

To correct 2. The upper (needle) thread tension is too great, this is adjusted by the knob 'C' (top) Generally this is the most common adjustment.

See F above, the Presser foot helps as a needle guide, and as a fabric transport across the bed of the machine. It has a pressure adjustment above its shaft (vertical) either a wheel, or threaded collar.

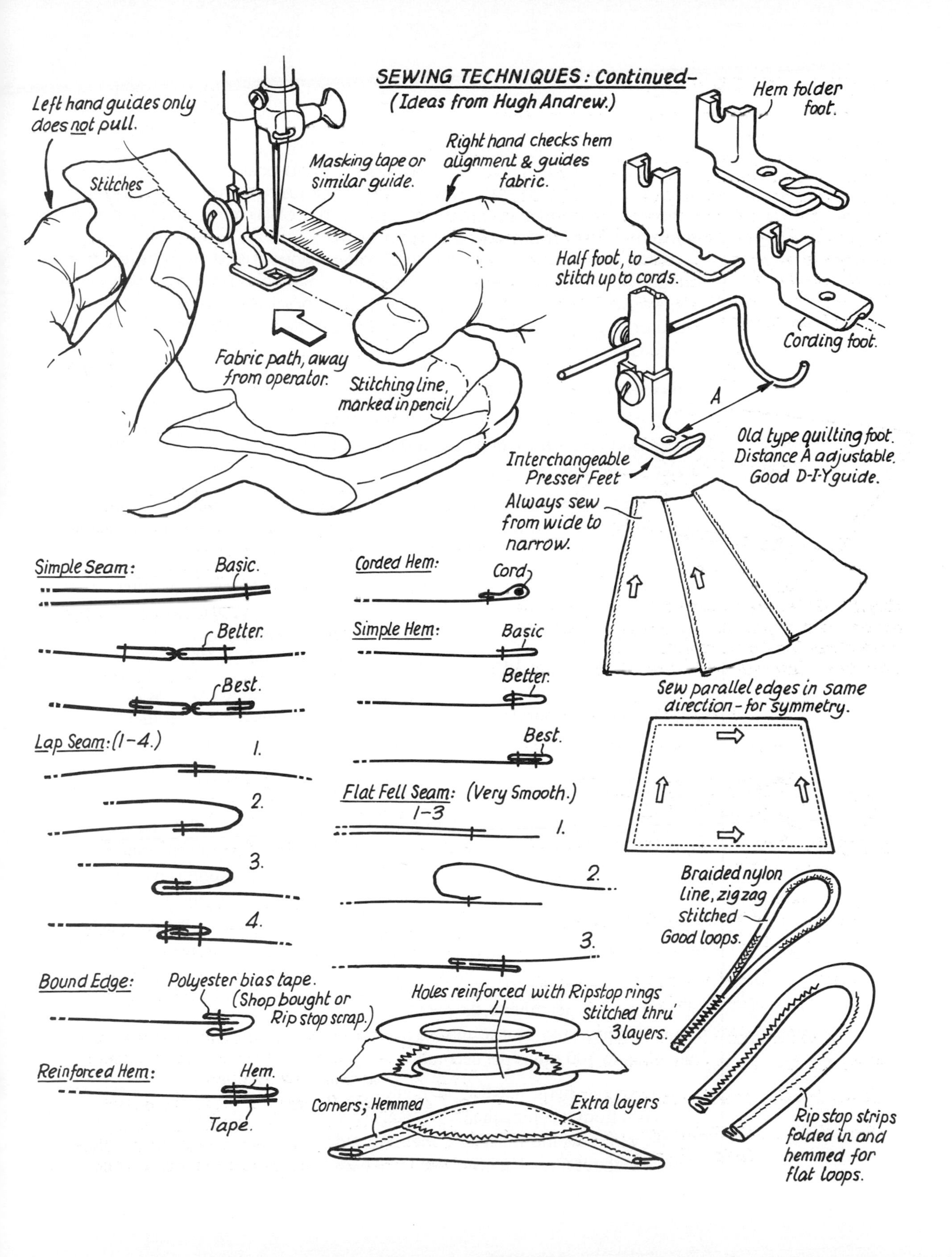
SEWING TECHNIQUES: Continued-
(Ideas from Hugh Andrew.)
Left hand guides only does not pull.
Stitches
Masking tape or similar guide.
Right hand checks hem alignment & guides fabric.
Hem folder foot.
Half foot, to stitch up to cords.
Cording foot.
Fabric path, away from operator.
Stitching line, marked in pencil
A
Interchangeable Presser Feet
Old type quilting foot. Distance A adjustable. Good D-I-Y guide.
Always sew from wide to narrow.
Simple Seam:
Basic.
Better.
Best.
Lap Seam: (1-4.)
1.
2.
3.
4.
Corded Hem:
Cord
Simple Hem:
Basic
Better.
Best.
Flat Fell Seam: (Very Smooth.) 1-3
1.
2.
3.
Sew parallel edges in same direction - for symmetry.
Braided nylon line, zigzag stitched Good loops.
Bound Edge:
Polyester bias tape. (Shop bought or Rip stop scrap.)
Holes reinforced with Ripstop rings stitched thru' 3layers.
Reinforced Hem:
Hem.
Tape.
Corners; Hemmed
Extra layers
Rip stop strips folded in and hemmed for flat loops.

Thread tension is the key to lockstitching. The machine manual should go into detail on this but Pat's sketch shows the two possibilities for going wrong, and explains the correct way with notes on adjustment for a typical domestic machine. Top (needle) thread and bottom (bobbin) threads have to meet in the middle of the materials being sewn.

Some understanding of the internal workings might help here, and we bring in John Clarke whose working life has involved all forms of sewing. As a well-known London character with a penchant for miniature kites, and a lecturer on sewing standards, his advice supplements what one finds in the manufacturers' manuals.

Advice from an expert

The underthread for the lockstitch is provided by the bobbin, and the needle thread has to perform a loop to go around the bobbin which is held in a bobbin case under a slide plate.

The speed at which the loop is formed is such that the bobbin is suspended, so eliminating any need for lubrication despite the continuous rotation of the bobbin carrier.

To make a stitch 1/8in (3mm) long, the needle thread has to be drawn from the top spool, through the guides and top tension discs to the eye of the needle. It is caught by a hook on the bobbin case carrier at a point just above the eye of the needle when on its upward stroke. Take a moment now to look at the needle.

This remarkable tool, which costs so little, is fashioned with several features other than its point and the flat which locates it in the needle bar. Hold it to the light, with the eye pointing up and down, and you should observe that the thickness of the shaft is scalloped away by 50% just above the eye on the rear face. This is called the 'scarf'.

The purpose of this relief is to permit the bobbin case hook to pass the needle without touching it. The loop is now formed around the bobbin case, and the thread drawn back through the needle by the take-up lever usually found on the front of the machine. This forwards and backwards movement of 4in (102mm) of thread occurs no less than 32 times if the stitches are 1/8in (3mm)! Additionally, the needle has a long groove in its front face. This is to allow for dispersal of heat - yes, heat!

Sewing imposes very severe conditions for the threads. In a machine operating at 6000 stitches per minute, the needle thread has to accelerate to a speed close to 100 mph (160kph), stop, and accelerate backwards to the same speed, and all at a cycle rate of up to 100 times per second!

This groove in the needle both guides the thread and, provided it is not too thick, protects it from touching the fabric. It also has the virtue of dispersing heat from the needle which would otherwise melt the nylon or polyester. At 400 stitches per minute without thread, a needle can reach 260 degrees centigrade! This is another reason why many dressmakers sew through a paper strip on the top surface. Just feel the heat in the thread when you next do some sewing. You will be surprised to discover just how hard those needles work .

(Of course, not all of us can, or need to, sew at such a high speed, John!)

Some kite-makers stick edges together for location when sewing. In such cases use a Blukold needle with phosphated surfaces to avoid gumming up. A better plan is to use softened soap or 'balance notches' every 4in (10cm). This is a term for small vee cuts through both pieces of fabric on the outer edge. As the fabric is fed through the machine foot, use hands to hold these location notches together. This method is used throughout industry. Another method is to pin the seam, especially on a straight run. The pins must be at right angles to the presser foot, and should be removed before they come into contact with the feed. Remove the pins as they approach the needle.

When turning a corner always do so with the needle down.

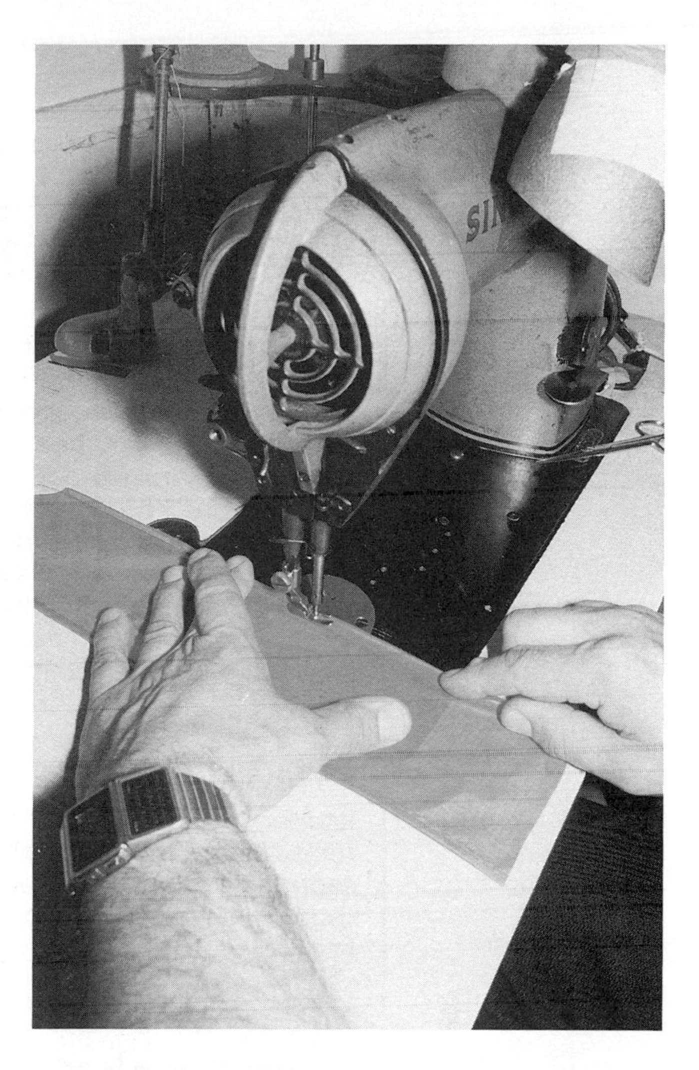

Recommended hand positions for accurate sewing as the fabric is fed through the machine.

Holding the ripstop at the rear as well as the front of the presser foot helps to keep any location keys together. If edge stitching, always use a guide made up of layers of coloured tape stuck to the machine base. Avoid looking at the needle - watch the fabric running along the guide. It is restful to sit at an angle so that the right arm is in a straight line with the needle and at right angles to the machine so that the wrist is also straight.

Use of the ball-ended or 'yellow band' needles is not generally advised. They can split on impact with the high twist yarns of ripstop nylon, and are really intended for sewing knitting fabrics where the ball forces the yarn apart, stretching and borrowing yarn from adjacent loops in the fabric without cutting through as with the normal pointed needle.

These wise tips come from John Clarke, and Pat's sketches of ideas from Hugh Andrew should help you appreciate the wonders of sewing. The subject is limitless - we have not touched on appliqué, edge bindings, attachments for folding hems or the complexity of parafoil sewing. These are separate topics in their own right, just as sewing itself is a hobby that has its own literature which we commend if you wish to learn more.

Chapter Three

Flying Lines, bridles,

Whatever the design of kite, the one common feature is the line by which it is flown and controlled. Yet so very often, it is the least considered item when kite flyers shop for new materials. Many a kite has been lost out to sea from a beach site, or into treetops from the local park, as the result of an over-stressed and broken line.

Cheap line is a false economy. After all, if this hobby is to become a regular habit, maybe bordering on obsession, then the flyer will have invested in good quality reels. What's the point of loading the reel with an unbranded twist that 'looked right', when you can be sure that at some time or other the varying tensions on this material are going to produce lively snarls and inevitably weaken it, so that a line break happens just when you least want it.

Types of line

Kite line can be in any of three forms - *Monofilament, Twist or Braid.* These terms are best understood by visualising the monofilament as a solid, circular section flexible rod, while twist is a combination of fine threads, wound together as found in a Hawser. Braid is similar, except that the threads are interwoven or plaited so that they are unable to unwind or separate.

Line materials

All three of these types of line are obtainable in the variety of man-made synthetic materials available to kite flyers. These are nylon, polyester, (including Dacron by DuPont), polyethylene (including Spectra by Allied Signal Inc. and Dyneema by DSM), and Aramid (Kevlar, DuPont trade name) and polypropylene.

Other materials, made from natural fibres such as cotton, hemp, flax, and even silk, have played their part in the past, and that is the operative term, for, apart from the traditional and ceremonial kites where authenticity is upheld to the highest degree, such materials are no longer good enough for the modern kite.

Choosing the line

Faced with a rack of up to twenty different spools in the kite shop, you must first consider the three main parameters which govern the suitability of the line. These values are: tenacity, elasticity and cross-section. In other, perhaps better understood, terms - strength, stretch and drag.

The second consideration is the purpose to which the line will be applied.

knots and reels

This could range from a small lightweight kite to fly on a single line, through to a four-line aerobatic kite. Size of the kite is most important. In general terms, the greater the area of the sail, the higher the tension on the line. So we have to respect the quoted 'Breaking Strain'. This will be stated on the product. Trust it to 50% of the figure and, as a rough guide, work to 15kg per square metre of sail, or 28lb/square yard, and you should be safe ...except for accidental line cuts and gust loads, of which more later.

Breaking strains are established by the Quality Control departments of the line manufacturers. Rigid standards are set by National Standards Institutions and expensive equipment such as the Instrom pneumatic tensionmeter is used to check random samples during manufacture.

However, this is done in the ideal conditions of the laboratory. Outdoors, on all kinds of surfaces where the lines can be abraded as they are dragged, in a range of temperatures and exposed to ultra-violet rays, and often made wet from dew in the grass, all the lab tests become suspect. Each of these hazards can reduce the tensile strength by several percent. Hence the advisability of 'playing safe' and allowing a 50% margin on the quoted strain - and we haven't yet mentioned the weakening factor of the knots!

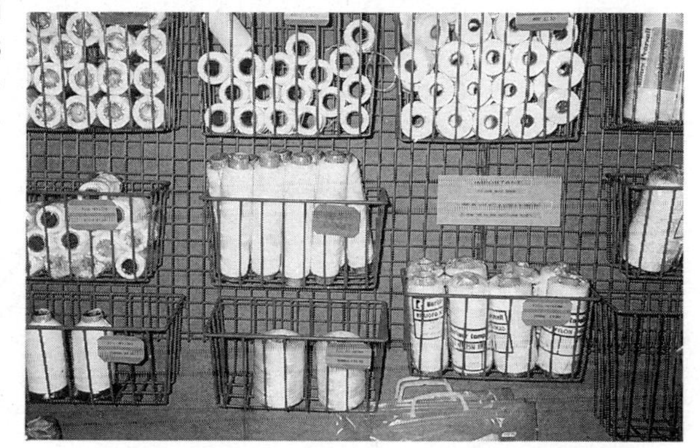

The range of kite line is vast. These 'kops' of nylon and polyester braid average 122m (400ft) for single line sport kites.

Elasticity

Monofilament is used for fishing. It reels easily, rarely tangles, has acceptable cross-section for its breaking strain, and it has a lot of 'give' so it can handle variations in tension. It is also used for model glider towlines for these reasons, and in particular, nylon monofilament has enough elasticity to catapult a model glider to greater height than the line length when launched with a strong tension. Nylon makes a 100% recovery to its original length but does so rather slowly. When under full tension and stretched, it has a reduced cross-section and consequently a reduced breaking strain.

So, if you don't want rigid control over the kite, and the cushioning effect of monofilament is to advantage as for example with a lightweight thermal-seeking Delta, then fishing line nylon is ideal. But, for almost any other type of kite, it is unsuitable, even in the other monofilament materials.

Control over the kite is critical and, while a degree of elasticity is desirable for single line types, particularly those with a large surface area, it is not

USEFUL KNOTS.

ROUND TURN & TWO HALF HITCHES.

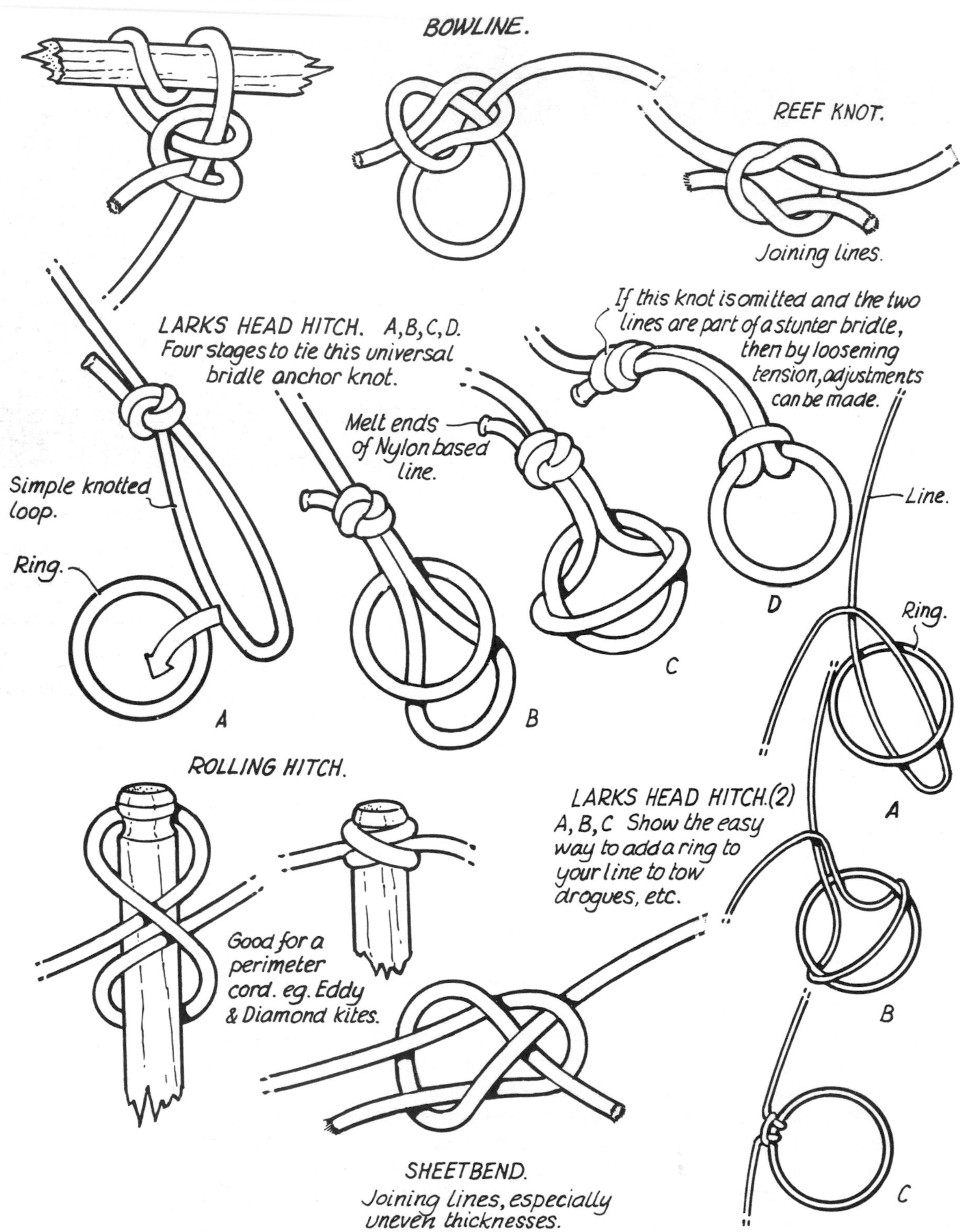

wanted at all with the multiple line aerobatic kites. So the choice comes down to twist or braid, and the synthetic material from which it is made.

In a way, the selection is made for you if using nylon, as the highest breaking strain in nylon twist is usually 40kg or 861b. In polyester or Dacron (which have less stretch and a partial recovery of length), you can find higher breaking strain twist but then another factor emerges with these lines. The term 'twist' implies that a degree of torque has been applied to the line during manufacture. When wound onto a reel from a straight run, the torque is balanced and, when unreeled again, the line will lay straight.

Derived from the Indian reel, with extensions to allow the line to run free as it spools out with the ends spinning in the hands, the plastic reel is to be recommended.

Snarls

But, as every kite flyer discovers early in the hobby, if you wind onto a static reel over its edge, or do a similar action by winding around a handle, then the line becomes 'lively' and wraps itself into snarls.

This is due to upsetting the balance of the torque in the twisted line. The only easy cure is make the winding-on in alternate directions, say ten turns each way. Of course, the ideal is to invest in a spinning reel, where a swivel in the line connection to the kite is all that is needed to remove any unwanted twist or torque.

These lively snarls occur mostly with low breaking strain twist lines, as supplied with small kites. Only rarely do they arise with braided line, or in the sophisticated Kevlar or Spectra twist. Braided line is generally of less cross-sectional area than its equivalent in twist for the same breaking strain. It does not stretch as much as nylon or polyester twist and it knots easily. Sometimes a 'Kop' or tapered reel of braided line will be labelled '100%'. This means that it is wholly braided as distinct from the type which has a central core of twist.

The Kop is an economic way of purchasing polyester braid, as a half kilo or 18oz reel of 25kg (551b) breaking strain can provide as much as 920metres or 3000ft to play with! Although more resilient, and easier to handle as well as coming in a far higher range of breaking strains (up to 200kg or over 4001b), braid is still vulnerable to kinks and twists.

Take care with all lines. If possible, have the line pass through a bare hand *slowly* - these synthetics can burn or cut - and feel for any irregularities. In most instances, a kink will produce a slight burr on the surface which can be smoothed out. A worn braid feels rough and is best cut away and the line joined by a blood knot.

The choice

But we are jumping ahead. Right now we are choosing the line and we've defined the limited use of monofilament, the disadvantages of twist and the advantages of braid, so the next stage is to gauge the material against the type of kite. In summary, we come to the following conclusion:

(i)Nylon or polyester twist is fine for single line kites up to an area of 3 sq.metres or 30 sq.ft which is big enough for any sport kite.

(ii)Braided line is superior for kites up to a whopping 12 sq.metres or 120 sq.ft. Above that, you're into boat chandlers' supplies, and if you are building that big, then you'll know more about the subject than we do!

Newtons and all that

Assuming that by now we've chosen between twist and braid, we now come to the matter of cross-section and the drag it creates. Mostly, the drag and attendant weight of the line actually improves stability of a single line kite. As the line is paid out, and the kite ascends on the wind gradient with improving lift as it gains altitude, so the additional weight and drag of the line compensates, rather like the ballast on the keel of a yacht. So the cross-section of the line is not so critical for the average sport kite.

It is when we get into the high performance area that the ratio of cross-section to strength comes into play. The manufacturers refer to 'Newtons/ Decatex'. We don't really need to know these finer points. Suffice to have the benefits of their research, which has thus far produced the ultimate in extreme strength for smallest diameter. We refer of course to Kevlar, Spectra, Dyneema, and derivatives.

Multiple lines

These practically inextensible lines have best application on multiple line aerobatic kites. They are not cheap, and have their own share of problems; but the advantages greatly outweigh disadvantages and, as a result, they have become indispensable for the competitive stunt kite flyer.

Kevlar (DuPont) has fallen out of favour for reasons of safety. It can cut badly, either flesh or other lines, even itself! For the critical ability to slide line-over-line in aerobatic manoeuvres, it should be waxed.

Spectra (Allied Signal Inc) or Dyneema (D.S.M.) comes with a smooth surface, braided or twist, slides with multiple line wraps and is even stronger than Kevlar by up to 50% on equivalent thickness. No wonder that the thin 40kg (851b) Spectra has become universal for the plethora of medium-size stunters. Its low drag offers lines that are taut and straight throughout a broad wind window.

There is one perverse problem. Spectra has a low melt point. When it rubs against a 'foreign' line, such as a polyester, the result is an instantaneous break. So steer clear of other kites being flown on single lines. Conversely, and unlike Kevlar, it has no suicidal self-cutting tendencies and will maintain good control even in Quad-line team mix-ups where 12 or even 16 lines have been known to wrap together through complicated manoeuvres.

Derivatives with various brand names are Spectraline, Hot-Line, Speed Line, Spiderline and the separately produced Dyneema by the Dutch State Mines, which is marketed in the UK by the Britannia Braiding Company. They all share the requirement for 'sleeving' at line ends, or for that matter over any joint in the line.

We mentioned that most braids are marked '100%', meaning that they do not have a core. There is one line *with* a Kevlar core and a Dacron braid around it, known as Skybond. Rather thick by comparison, it is not in the top competition class but, by combining the virtue of Kevlar strength with the protection of the polyester Dacron, it is popular for sport flying in the USA.

Sleeving

Spectra or Dyneema lines are slippery and very difficult to knot permanently, so the fishing technique of 'sleeving' has been adapted for this and each of the similar materials. Sleeving kits are sold by the kite shops and consist of a length of Dacron coreless (hollow) braid, a piece of thin wire and basic instructions.

You can cope otherwise with a fine shoe lace, or picture cord and thin piano wire (which must be at least double the length of the lace). The wire has to be bent back tightly upon itself so that it can be a pull-through to transport the Spectra along the centre of the sleeve.

We have found that a lot of newcomers to Spectra line have been confused by the very term 'sleeving'. A simple way to appreciate what it means is to visualise the Spectra line being passed down a hollow tube (the 'sleeve') until it appears at the other end. It is then fused by heat so that the Spectra and the tube or 'sleeve' are irretrievably joined as one synthetic material *at the extreme end.* The Spectra is otherwise free within the sleeve.

The tricky part is getting that thin, slippery Spectra line through a 300mm (12in) length of unwilling hollow braid. The wire 'pull-through' is best made from a straight length of .030in or 0.8mm piano wire. It must be at least double the length of the sleeve, say 26in or 66cm, so that the shorter end doesn't enter the sleeve and get stuck. Bend the wire back on itself as close as possible.

Sealing

Now turn to the sleeve material which has to be sealed from fraying at each end. Use a steel bodkin to probe the central core and hold it open, then swiftly pass a lighted match under the sleeve extremity and you'll find that the synthetic has melted into a tiny trumpet-sectioned outlet when the bodkin is withdrawn.

We now have to pass the wire all the way through the sleeve so that its bend or fold emerges at the other end. A trick here is to bunch the braid. This opens the weave, shortens the length and opens up the diameter of the

SLEEVING FLYING LINE.

A braided sleeve is used to 'buffer' the effect of the 'Spectra' line cutting through itself.

'Spectra' Flying line is perfect for Stunt kites as its strength and smooth surface allows multiple twists not to affect control.

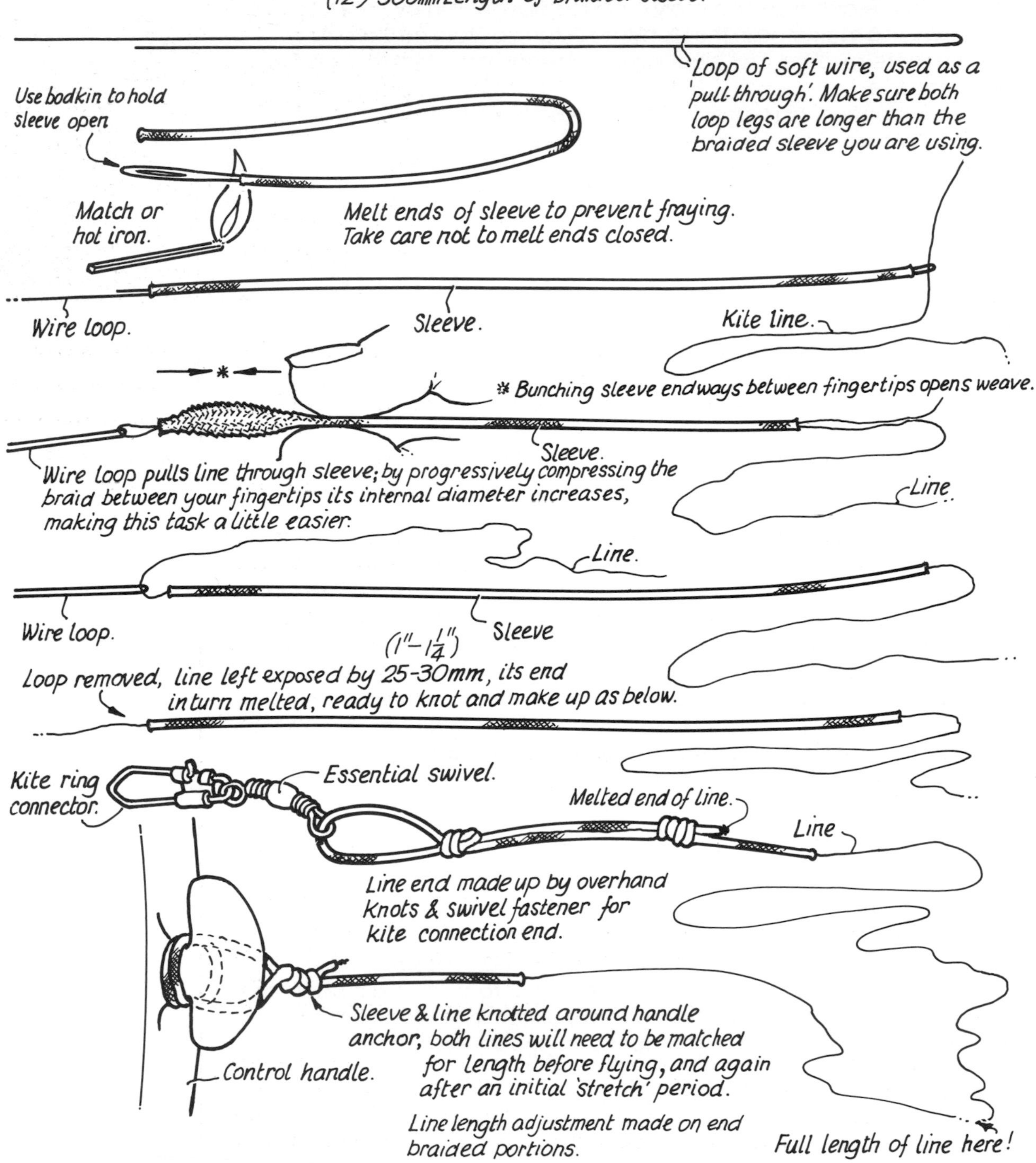

braid. Work that wire all the way, look upon it as none other than a super-length needle with the eye in the bend of the wire. Use that 'eye' to hold the end of the Spectra line as it is pulled back through the length of the braid until an inch or 25mm emerges.

The matchbox comes into action again, this time to melt the Spectra to the sleeve so that we have the two components fused as one, just as we described earlier. Why have so much sleeve? You'll soon find out when you try to prepare two lines of equal length. Fit the swivel plus connector, and lock the sleeving onto itself with two separated overhand knots. Another method of securing the connector to the sleeve is by a Lark's Head Hitch; but the loop on the connector has to be large enough to cope. Whatever you do, *be sure to use a swivel* for the kite end of the line. If there isn't a kite shop nearby, try a fishing tackle shop for suitable snaps and swivels. Use link or lockable connectors, not the 'squeeze' type.

Winding side of the Maurice Sawyer Del-Saw 'Bright Sky' reel, made from tough multiple layer combi-ply.

Make up each of the two kite ends first, then use a screwdriver, tent peg or skewer or similar stake to hold the extreme ends positively to the park ground. Few kite flyers have a 50metre or 150ft clear run in their home yards, so this operation has to be completed in public, hence the need for speed *and* precision.

You will, by now, have had the experience of making at least two sleeves, so the pair for the handle ends of the lines can be made up 'in the field' quite easily with just that wire needle and a box of matches to hand. The bodkin trick should have been completed beforehand to make the sleeve ready.

With the lines stretched out, have a friend apply tension on the ends to be prepared and scissor them to equal length. Then go through the sleeving procedure and, according to type of handle, apply knots to secure, but no 'Granny' knots please as you will possibly have to make adjustments later. Remember that knots *reduce* line strength by 20 to 40%.

Bridles

Most kites are dependent on their bridles for performance but, there are exceptions. Some commercial Diamond shapes have a single connection direct to the spine. Box kites can be flown without a bridle as can the Malay or Cutter shapes with a ventral keel. These have direct line connections.

Sometimes the fabric keel is replaced in outline by a pair of bridle lines, shorter at the front, longer at the rear. Such a twin bridle has the advantage of easy trimming by means of shifting a Lark's Head knot in the attachment ring, fore or aft.

The bridle is a vital contributor to the stability of the kite. It regulates the relationship of the attachment point to the centre of pressure on the kite sail. A basic check on bridle settings is to assess the 'angle of dangle'. Lift the kite by the attachment ring, let it hang upside down, and the rear, or trailing part

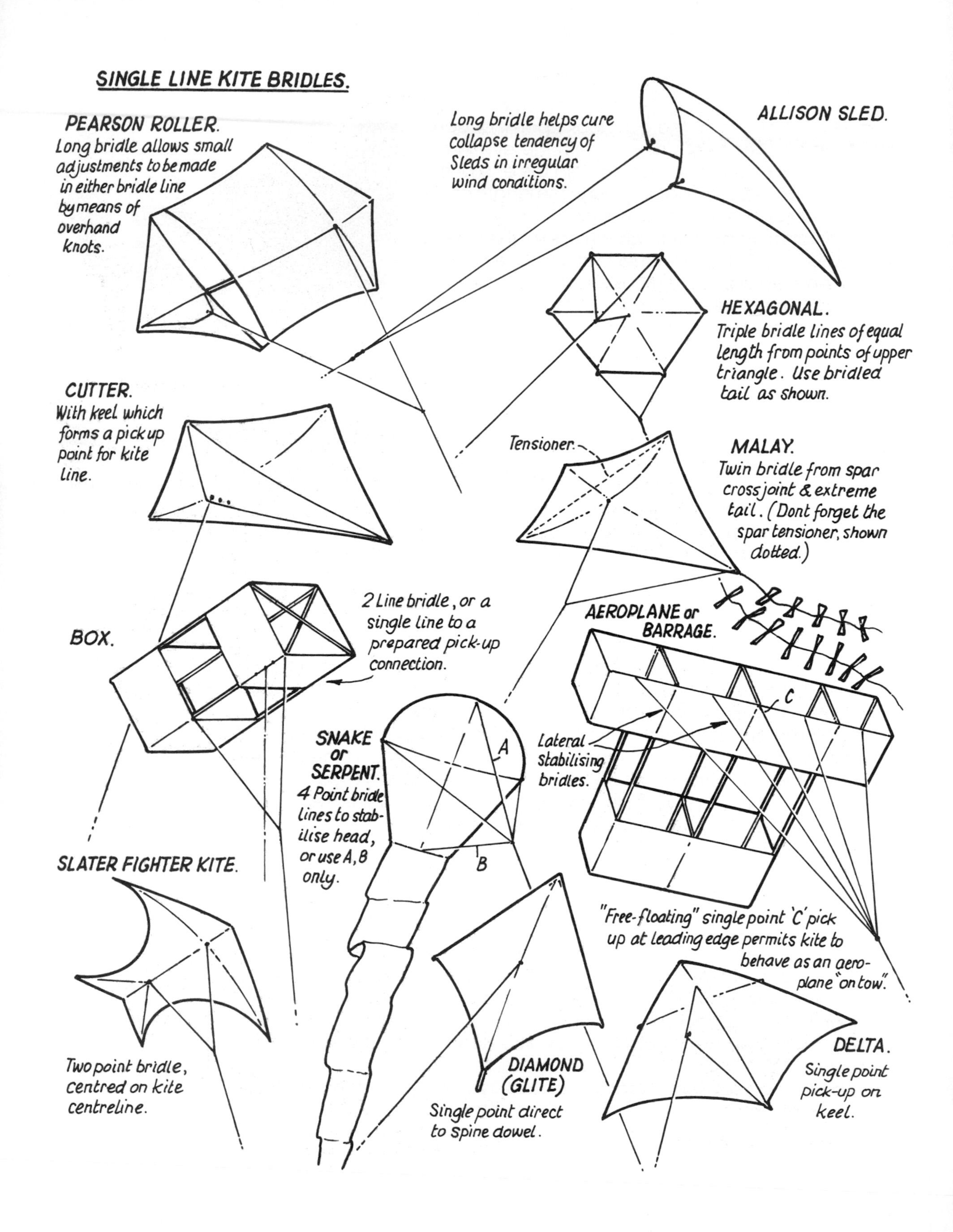
SINGLE LINE KITE BRIDLES.
PEARSON ROLLER.
Long bridle allows small adjustments to be made in either bridle line by means of overhand knots.
Long bridle helps cure collapse tendency of Sleds in irregular wind conditions.
ALLISON SLED.
HEXAGONAL.
Triple bridle lines of equal length from points of upper triangle. Use bridled tail as shown.
CUTTER.
With keel which forms a pick up point for kite line.
Tensioner.
MALAY.
Twin bridle from spar cross joint & extreme tail. (Dont forget the spar tensioner, shown dotted.)
2 Line bridle, or a single line to a prepared pick-up connection.
BOX.
AEROPLANE or BARRAGE.
C
Lateral stabilising bridles.
SNAKE or SERPENT.
4 Point bridle lines to stabilise head, or use A, B only.
A
B
SLATER FIGHTER KITE.
"Free-floating" single point 'C' pick up at leading edge permits kite to behave as an aeroplane "on tow".
Two point bridle, centred on kite centreline.
DIAMOND (GLITE)
Single point direct to spine dowel.
DELTA.
Single point pick-up on keel.

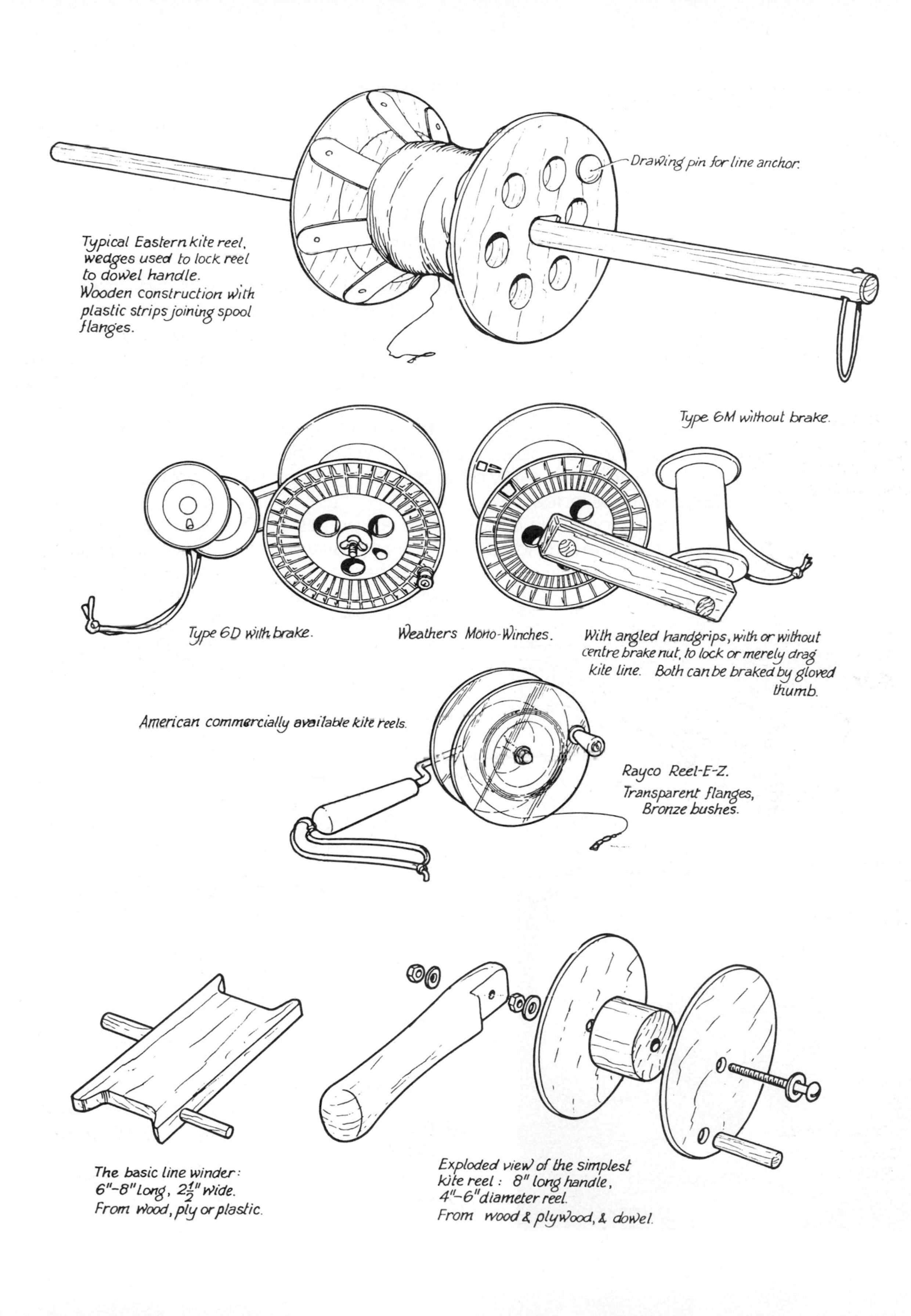

Drawing pin for line anchor.
Typical Eastern kite reel, wedges used to lock reel to dowel handle. Wooden construction with plastic strips joining spool flanges.
Type 6M without brake.
Type 6D with brake.
Weathers Mono-Winches.
With angled handgrips, with or without centre brake nut, to lock or merely drag kite line. Both can be braked by gloved thumb.
American commercially available kite reels.
Rayco Reel-E-Z.
Transparent flanges, Bronze bushes.
The basic line winder: 6"–8" long, 2½" wide. From wood, ply or plastic.
Exploded view of the simplest kite reel: 8" long handle, 4"–6" diameter reel. From wood & plywood, & dowel.

of the kite should 'dangle' at an angle of 15 degrees or more along the spine.

This rudimentary check is for a static condition; it has no consideration for the disposition of the surface area of the sail, or the lifting component of any separated sails, as in a Box kite or Roller. So, while the 'angle of dangle' can give one a start in setting up a bridle, it should not be regarded as a universal solution.

On the contrary, there are kites, such as the Barrage with its aeroplane-like wing and tail configuration, that will fly best in trail. The bridle in this case is a lateral one with lines to either side of the central pick-up point, and all at the very leading edge of the kite which then flies much like a glider on its towline. This same lateral braced bridling is found on the flexible Parafoils or Stratoscoop types.

Most kite designs specify the bridle line lengths and the attachment points. What is not always obvious is that the bridle line itself, and its connections to the kite frame, are every bit as important as the kiteline ends over which far greater fuss is made.

Our personal rule is to use braided polyester (Dacron) which offers only moderate stretch and can be knotted positively without risk of slipping. It can be of higher breaking strain than the kiteline itself, as weight and drag will have only minimal effect on performance. Additionally, because the bridle will be wrapped in with the kite, and handled frequently, it has to withstand more abuse than the kitelines. This is specially important for aerobatic types, of which more advice will be found in Chapter Eight.

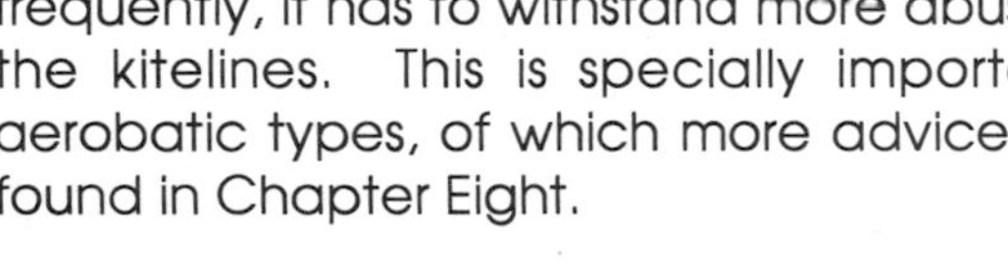

The John Clarke reel, turned from the solid in selected mahogany and remarkably compact.

Reeling it in

Why spend a week's wages on a kite, a lot more on the line(s) then insult the lot by wrapping those lines around an old cocoa tin - or worse, around the handle in a bulky bundle? We're referring mainly to the 500ft or 150 metre brigade who seem to find the cost of a reel prohibitive.

Reels are an investment. Get one with a cycle bearing shaft and you'll never regret that purchase. We regard the reel as an essential tool for single line flying. It will hold all the excess line you are ever likely to need, and its very concept of *reeling* will preserve the line from those snarls we mentioned earlier, because the line has little option but to reel back in the same form as it was reeled out.

If the reel is still just beyond the range of the pocket, then make a basic winder from a flat panel, as sketched on page 53. All you will need

HANDLES FOR STUNT FLYING.

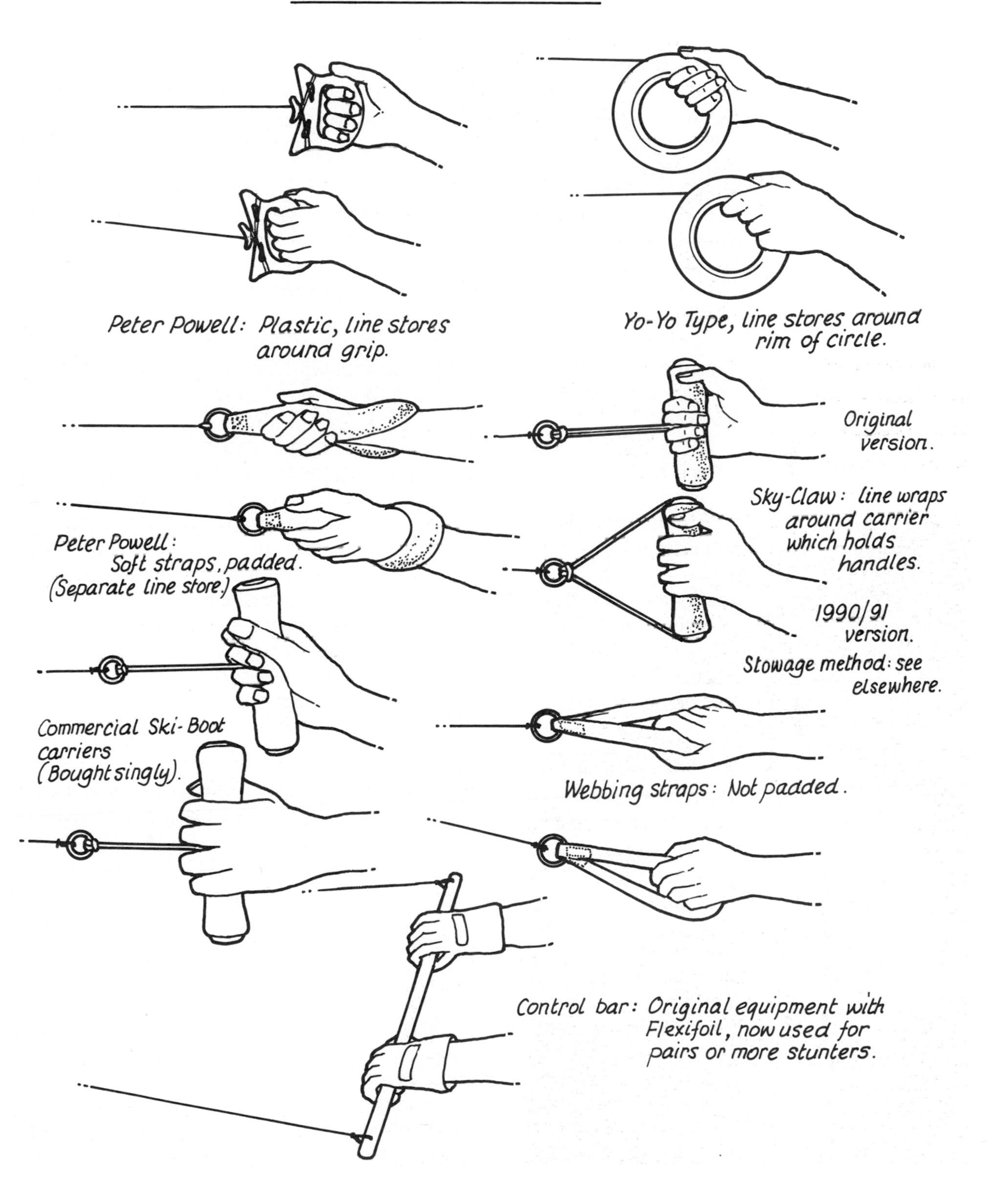

is a piece of plywood, plastic or plain wood about 65mm x 200mm (2 ½ x 8in) and a couple of short lengths of dowel. The latter form handles for winding in the line and the handles can be at opposite ends, or have one central to act as a pivot, rotating in the free hand.

Stunt handles

As the control lines for aerobatic or steerable kites are rarely longer than 50 metres or 150ft, there is no special requirement for a 'winding' reel. There are several options.

First, and presuming that the lines are going to be completely detached from the handle, you can use a plastic line store of the 'Yo-Yo' type. Spectra or Dyneema is best wound on the spool individually, and not as a pair (or foursome) of lines. Always treat these lines with the respect that their cost demands! Wrap them together and wind them in and you'll run the risk of abrasion.

Secondly, if the handle has provision for containing the line, such as the Peter Powell or Shanti Sky-Claw, then keep the line permanently attached to the handle and use the handle as a core for winding-in. The Sky-Claw has a patented means of retaining a pair of handles on a dowel with locating cross-pegs. The dowel then becomes a spinning pivot, rotated in the hands, and the lines are wound in straight. This obviates the risk of inducing twist or torque in the lines with attendant 'liveliness' or snarls.

The Peter Powell type of stunt handle has to be used individually. Although the line can be wrapped overarm onto the handle easily, there is always the risk of generating that 'twist'. So we recommend the method of winding in straight by turning the handle over and over.

Wrappers or socks

Line protection is important. When we first saw Mark Cottrell's fabric wraps around his handles/lines we wondered why no-one else had ever thought of the same idea. After all, what does the average stunt kite flyer do with his or her lines when packing up for home? They fling them in with all the other paraphernalia, only to have the line exposed to the hard bits, hooks, swivels and carbon tubes in one glorious fight for survival! So we invested in one knitted, stretch acrylic waistband as supplied for tracksuits, sweaters or windcheaters. They come in colours, open out from the doubled width and can be cut into stretchy lengths sufficient to wrap around the combined PP handle and line. A Velcro strip completes the job and the result is excellent protection of the line.

Later, having made enough sets for all our lines, we discovered a 'quickie' alternative. Discarded socks can be cut off at the ankle bend to make a similar protective sleeve. All you have to do in this case is to stop the fraying at the cut by zig-zag stitch.

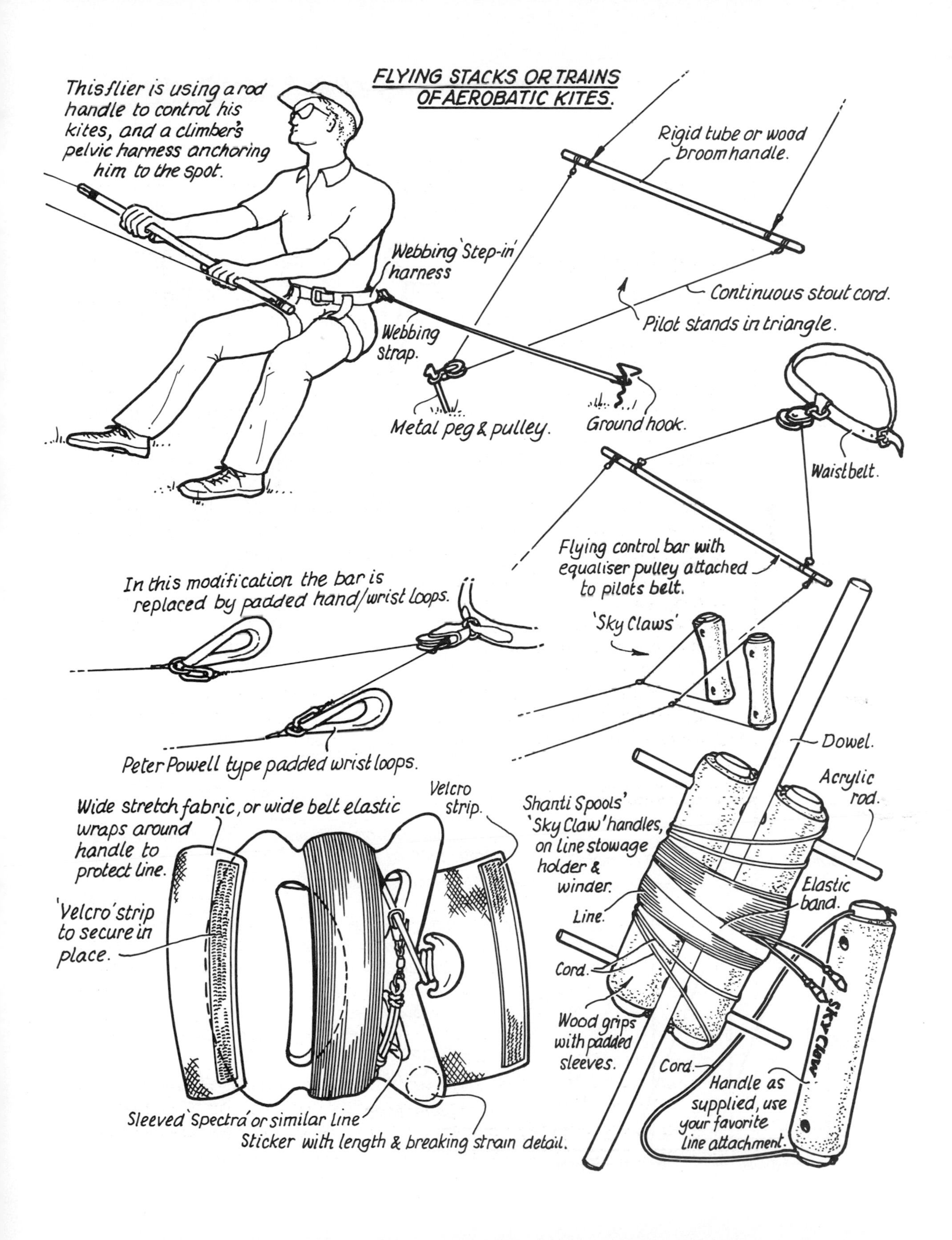
FLYING STACKS OR TRAINS OF AEROBATIC KITES.
This flier is using a rod handle to control his kites, and a climber's pelvic harness anchoring him to the spot.
Rigid tube or wood broom handle.
Webbing 'Step-in' harness
Continuous stout cord.
Pilot stands in triangle.
Webbing strap.
Metal peg & pulley.
Ground hook.
Waistbelt.
Flying control bar with equaliser pulley attached to pilots belt.
In this modification the bar is replaced by padded hand/wrist loops.
'Sky Claws'
Peter Powell type padded wrist loops.
Dowel.
Acrylic rod.
Velcro strip.
Wide stretch fabric, or wide belt elastic wraps around handle to protect line.
'Velcro' strip to secure in place.
Shanti Spools' 'Sky Claw' handles, on line stowage holder & winder.
Line.
Elastic band.
Cord.
Wood grips with padded sleeves.
Cord.
Sky Claw
Handle as supplied, use your favorite line attachment.
Sleeved 'Spectra' or similar line
Sticker with length & breaking strain detail.

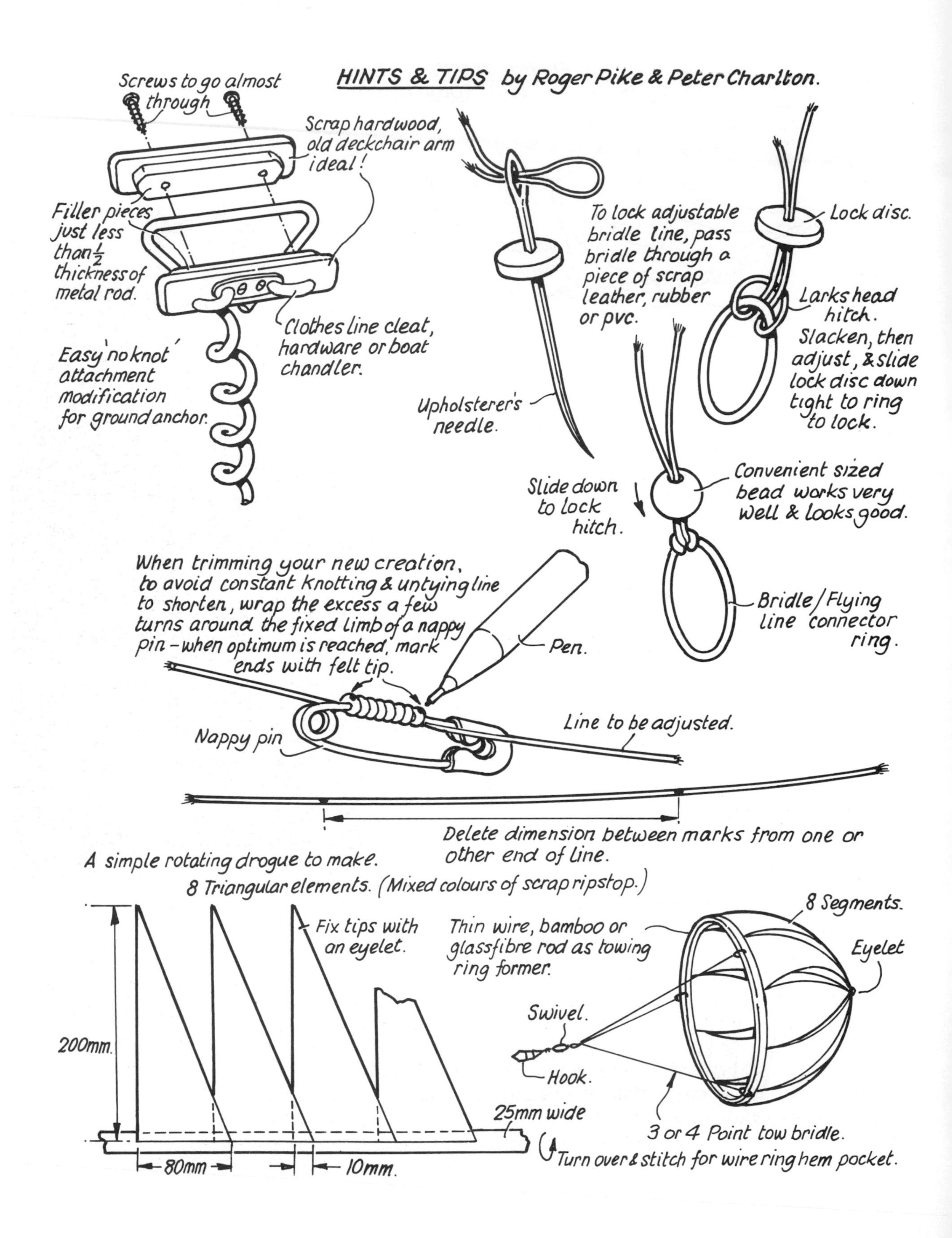
HINTS & TIPS by Roger Pike & Peter Charlton.
Screws to go almost through
Scrap hardwood, old deckchair arm ideal!
Filler pieces just less than 1/2 thickness of metal rod.
Clothes line cleat, hardware or boat chandler.
Easy 'no knot' attachment modification for ground anchor.
Upholsterer's needle.
To lock adjustable bridle line, pass bridle through a piece of scrap leather, rubber or pvc.
Lock disc.
Larks head hitch.
Slacken, then adjust, & slide lock disc down tight to ring to lock.
Slide down to lock hitch.
Convenient sized bead works very well & looks good.
Bridle/Flying line connector ring.
When trimming your new creation, to avoid constant knotting & untying line to shorten, wrap the excess a few turns around the fixed limb of a nappy pin – when optimum is reached, mark ends with felt tip.
Pen.
Nappy pin
Line to be adjusted.
Delete dimension between marks from one or other end of line.
A simple rotating drogue to make.
8 Triangular elements. (Mixed colours of scrap ripstop.)
Fix tips with an eyelet.
200mm.
25mm wide
80mm
10mm.
Thin wire, bamboo or glassfibre rod as towing ring former.
8 Segments.
Eyelet
Swivel.
Hook.
3 or 4 Point tow bridle.
Turn over & stitch for wire ring hem pocket.

Breaking-in

Even when a set of stunt lines have been very carefully matched to each other, it is inevitable that line tension will stretch one more than the other. Our preference is to have the lines so matched that, when the two handles are held together, the stunt kite soars overhead and is as stable as a single line kite. This also becomes a test for unequal length.

Using a polyester line without sleeving allows easy adjustment by re-tying the knot after shortening a line or, in the case of minor differences, adding an overhand knot or two to take up a couple of centimetres. With Spectra or Dyneema, we have to adjust in the sleeving, and here a little extra care pays off dividends.

Apart from the satisfaction of being able to combine the handles and hold an otherwise very lively stunt kite up there in reasonably static position, the technique is the best way of finding out exactly how much adjustment is needed. Take the kite downwind, to the centre of the 'wind window', and make a vertical climb. Then put your hands together. You'll find that after a few accumulated hours of stunting, one handle will have to be held back a little from the other to hold the overhead soaring position of the kite.

Note roughly how much you will have to take up on the longer line. Then play around some more, pulling some sweeping horizontal passes with lots of back tension from your arms, throw the kite through tight loops, pulling on each line in turn, and resume the overhead check again to reassess the difference in length.

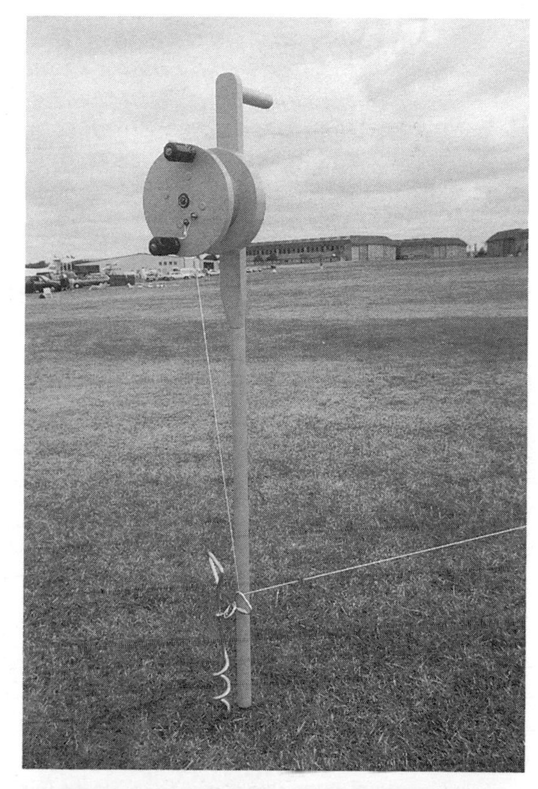

Hugh Andrew's reels, inspired by John Wilding, the Deep Sky type on a supporting pylon and the Strato-spool with a belly band to hold the bar to the waist and permit a free hand to apply leverage on the end of the bar.

Make 'em equal!

You might say "Why bother, my arms aren't of equal length anyway".

Well, the short answer to this comes in one word, *experience*. If the leading exponents of International Stunt competition follow this practice, there has to be a good reason. It's because we are not flying in just one direction. Our unequal arms might well take up the difference in line lengths when flying one way; but the perverse nature of our natural reactions doesn't allow the same correction when flying the other way.

Maybe you will have found that, no matter how well the bridle line lengths have been equated, all your left-hand loops are larger than your right-hand loops. Or perhaps you've found that you cannot make a square turn to the right like those you find so easy to the left. Making sure that the flying lines are equal is a good start to perfecting the symmetry of all your manoeuvres.

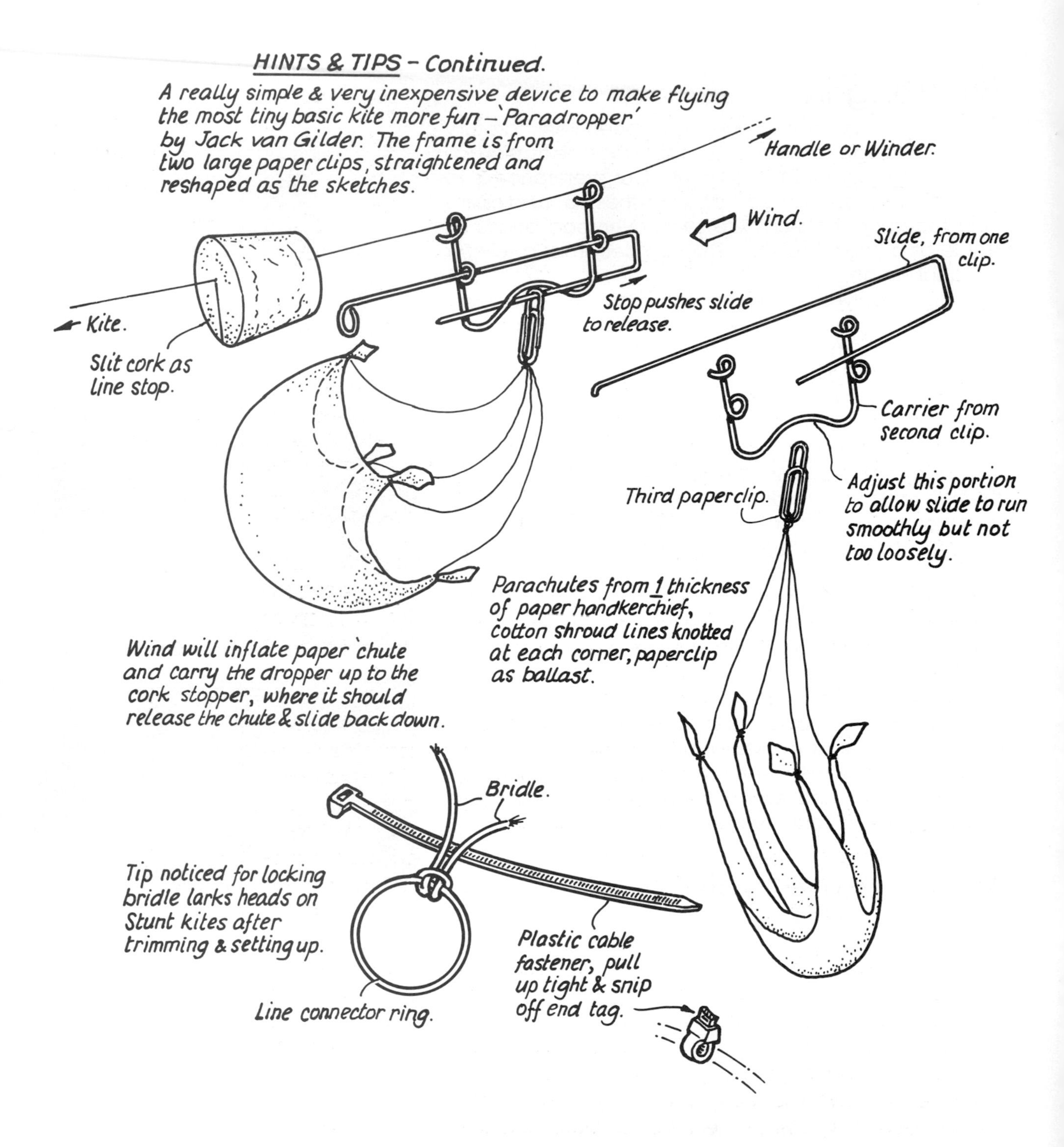
HINTS & TIPS - Continued.
A really simple & very inexpensive device to make flying the most tiny basic kite more fun - 'Paradropper' by Jack van Gilder. The frame is from two large paper clips, straightened and reshaped as the sketches.
Handle or Winder.
Wind.
Slide, from one clip.
Kite.
Stop pushes slide to release.
Slit cork as line stop.
Carrier from second clip.
Third paperclip.
Adjust this portion to allow slide to run smoothly but not too loosely.
Parachutes from 1 thickness of paper handkerchief, cotton shroud lines knotted at each corner, paperclip as ballast.
Wind will inflate paper 'chute and carry the dropper up to the cork stopper, where it should release the chute & slide back down.
Bridle.
Tip noticed for locking bridle larks heads on Stunt kites after trimming & setting up.
Plastic cable fastener, pull up tight & snip off end tag.
Line connector ring.

Handy tips

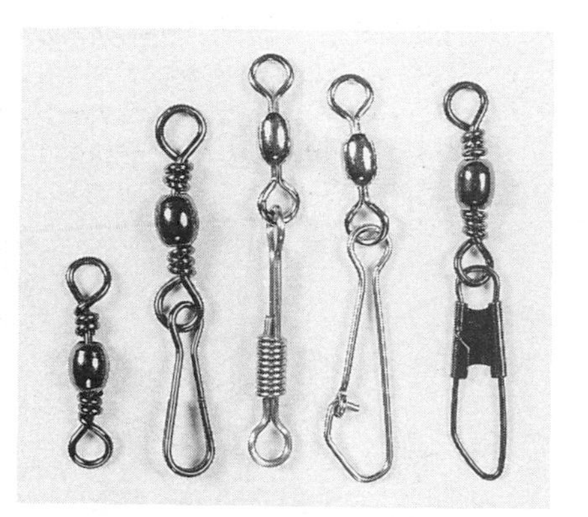

A range of swivel links. That on the right is NOT recommended for stunt kites.

Already we are encroaching on the specialist areas that deserve separate attention in following Chapters. Suffice to say at this stage that the relative importance of the flying line should never be underestimated. The line can offer as much fun as the kite to which it is linked. It is, in its way, the flyer's own symbolic wand of happiness, and we should never dismiss it as just another necessary component.

Two of the Essex Kite Group flyers have a selection of hints and tips to offer. Better known for his inflatable Dragon kites, Roger Pike, and his clubmate Peter Charlton, show how to make a very simple paradropper, or a twirling hemispherical drogue. For those who like to fly several kites at a time on tethers, they suggest a simple addition to the standard Goat or Dog corkscrew which provides a cleat to lock the line and winch or reel when sufficient line is paid out. Pet shops are a source of supply, alternatively try camping or climbing stores.

There is one type of kite that makes best use of a corkscrew anchor, or maybe two anchors for extra safety, and that is the traditional and super stable hexagon or roller, which will maintain the same position for hours, given a steady wind. We are referring to the larger sizes, of course, most likely to be made with heavier dowel spars and joints which are much the same as were employed fifty years or more ago.

With braided polyester line available in up to 440lb or 200kg breaking strain for the heavy kites with large sail areas, the remaining question for the kite-maker is how to construct the frame that will be both dismountable and very strong.

On the next page we offer a practical system which reduces the spar lengths for a hexagonal kite to a little less than the radius of the kite. Assuming the hexagon is to be 5ft to 7ft across (1.5m to 2m), the largest size of Ramin hardwood dowel generally available in kite shops – 12mm or ½in. – will be the automatic choice. The central core of the kite will then be cut from the plywood of the same thickness as the dowel, i.e. 12mm as sketched. Slots accept the spars, and stout discs form the sandwich outside layers. As shown, these are best cut from aluminium, but not necessarily so. By cutting a flat on each spar end to accept a D ring on each point of the sail, the assembly is simply a matter of ensuring that the sail is sufficiently tensioned to stretch just enough for opposing D rings to fit over the spar ends.

Before we leave this section, there are new thoughts on joints that are made possible by use of the cyanoacrylate adhesives, better known as superglue. Many kite-makers have used cyano to seal knots in Spectra or Dyneema lines. Available from all model shops in various grades for porous, non-porous or standard applications, cyano is an ideal locking, or quick-fix solution to assembly of joints either with use of wire inserts or splints with tube and rod in the smaller diameters. See sketches on the following pages.

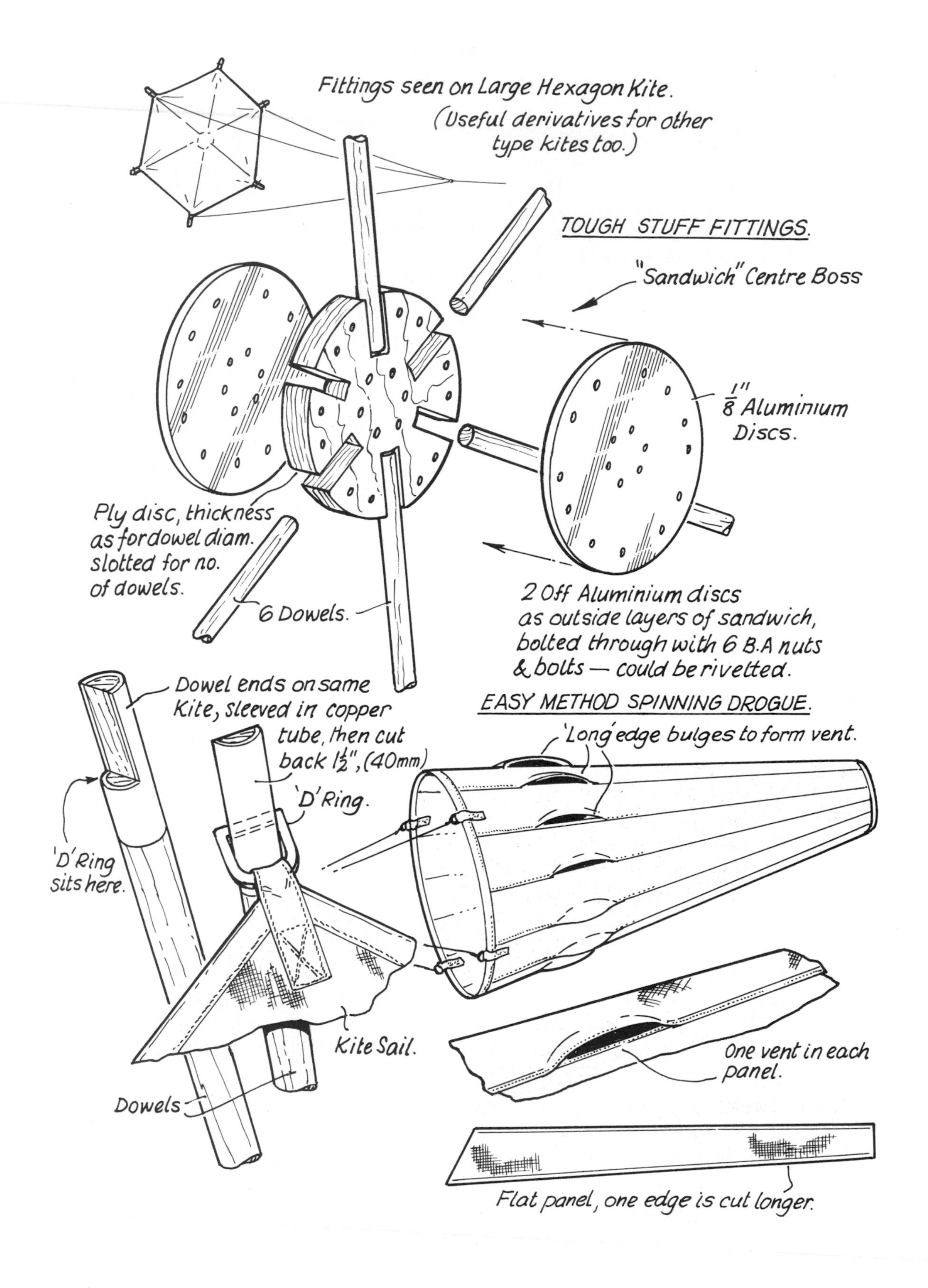
Fittings seen on Large Hexagon Kite.
(Useful derivatives for other type kites too.)
TOUGH STUFF FITTINGS.
"Sandwich" Centre Boss
1/8" Aluminium Discs.
Ply disc, thickness as for dowel diam. slotted for no. of dowels.
6 Dowels.
2 Off Aluminium discs as outside layers of sandwich, bolted through with 6 B.A nuts & bolts — could be rivetted.
Dowel ends on same Kite, sleeved in copper tube, then cut back 1½", (40mm)
'D' Ring.
'D' Ring sits here.
Kite Sail.
Dowels
EASY METHOD SPINNING DROGUE.
'Long' edge bulges to form vent.
One vent in each panel.
Flat panel, one edge is cut longer.

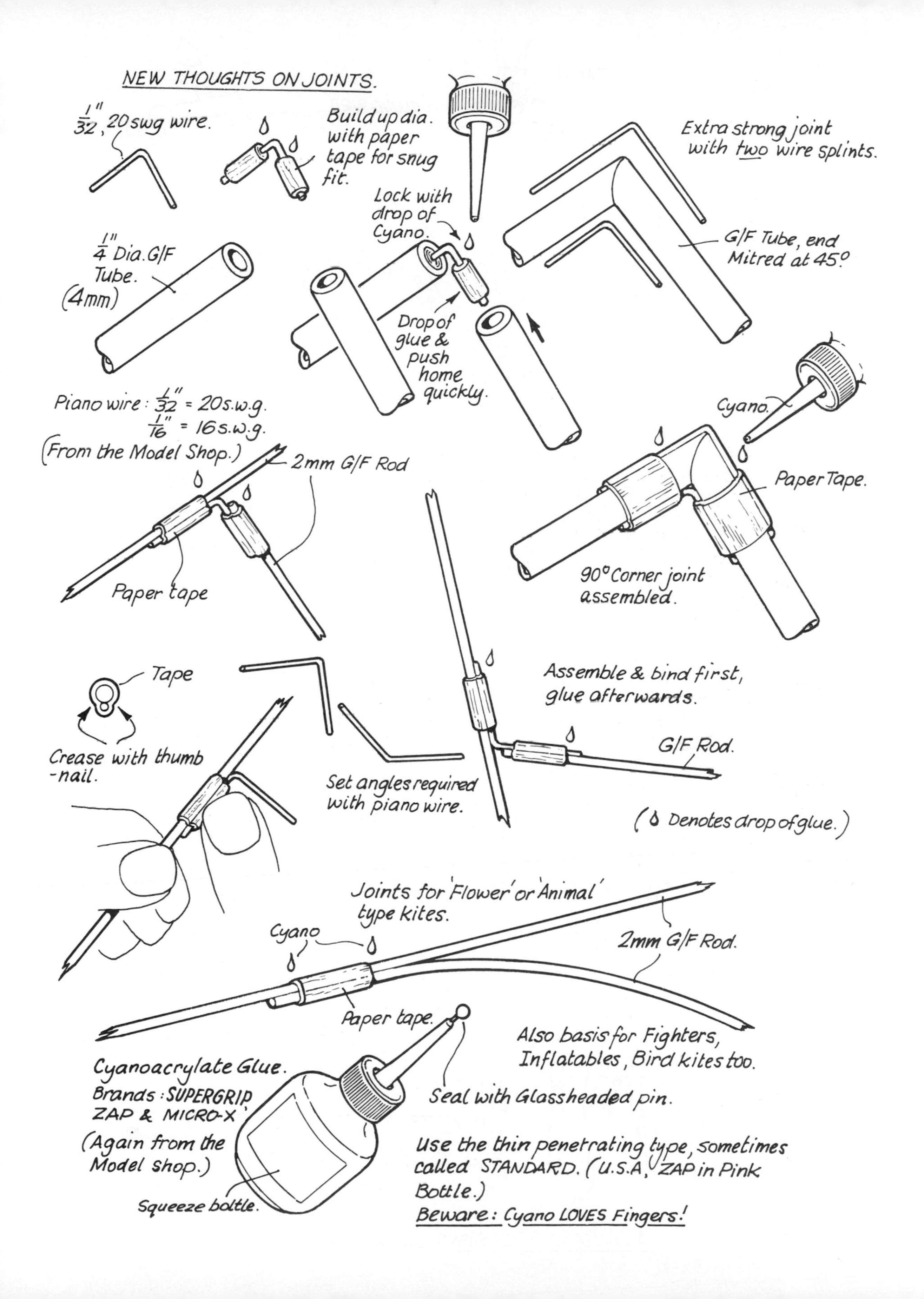
NEW THOUGHTS ON JOINTS.
1/32", 20 swg wire.
Build up dia. with paper tape for snug fit.
Extra strong joint with two wire splints.
Lock with drop of Cyano.
1/4" Dia. G/F Tube. (4mm)
G/F Tube, end Mitred at 45°
Drop of glue & push home quickly.
Cyano.
Piano wire : 1/32" = 20 s.w.g.
1/16" = 16 s.w.g.
(From the Model Shop.)
2mm G/F Rod
Paper Tape.
Paper tape
90° Corner joint assembled.
Tape
Crease with thumb -nail.
Assemble & bind first, glue afterwards.
G/F Rod.
Set angles required with piano wire.
(Denotes drop of glue.)
Joints for 'Flower' or 'Animal' type kites.
Cyano
2mm G/F Rod.
Paper tape.
Also basis for Fighters, Inflatables, Bird kites too.
Cyanoacrylate Glue.
Brands : SUPERGRIP ZAP & MICRO-X'
(Again from the Model shop.)
Squeeze bottle.
Seal with Glassheaded pin.
Use the thin penetrating type, sometimes called STANDARD. (U.S.A, ZAP in Pink Bottle.)
Beware : Cyano LOVES Fingers!

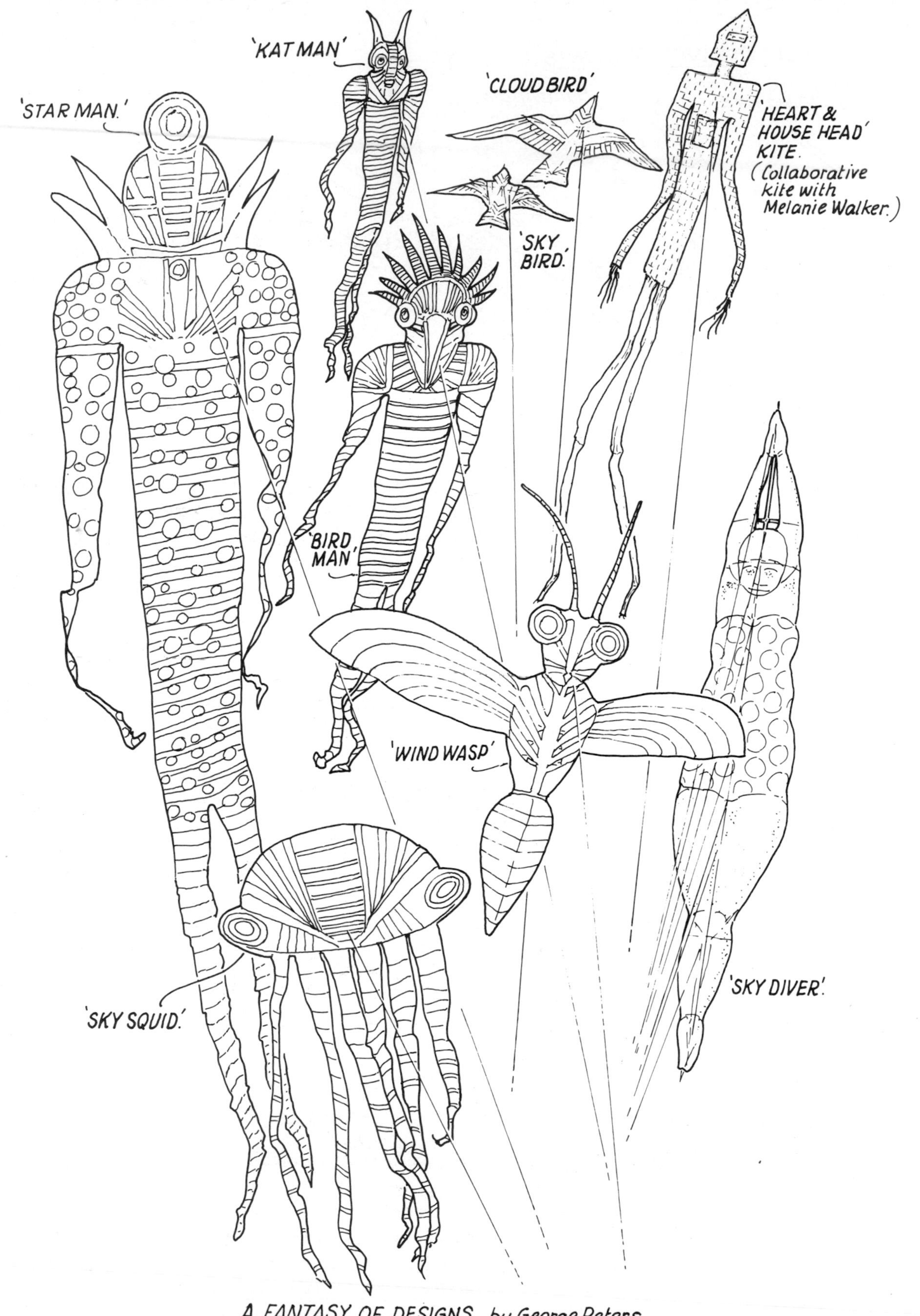

A FANTASY OF DESIGNS by George Peters

Main picture: one of Peter Lynn's huge Manta Ray inflated shapes from New Zealand dominates the seafront at Weymouth during a Kite Society of Great Britain convention. Upper left is George Peters' Mantis Man from Colorado, floating high at Bristol, and below left, to show that almost any shape can be made to fly as a kite, is a locomotive, its tender and carriage made by Jan Pieter Kuils of Holland, in the air at Dieppe.

Painting the Sky

Main picture is of Joel Scholz's multiple train of colourful Parrot kites from Texas, seen flying here at Dieppe. Below, doyen of French kite flyers and internationally renowned for their complex assemblies of multi-coloured panels are the André Cassagnes flying crowns.

Below right is a Korean fighting kite design seen at Blackheath, London, remarkable for the many strips of ripstop carefully blended in rainbow tones. Right, Steve Brockett's creations from Wales with their abstract painted art and form, each aimed to represent the Welsh legend of Vortigen and the Dragons.

The Detail

Top right: the clever use of black zig-zag stitching, coupled with appliqué colour panels introduced a new form of line art as seen by Randy Tom's Patrick Nagel 7-Sisters kite which won the Grand Championship at the AKA convention, Oregon in 1990 (G Maurer).
Above we see how the experts arrange an 8-way polyblock junction, as on Don Mock's Sun Eagle. A similar structural detail from the Korean Fighting Kite on the previous spread shows another approach.
Opposite we see details from creations by André Cassagnes which help to explain his ingenuity.

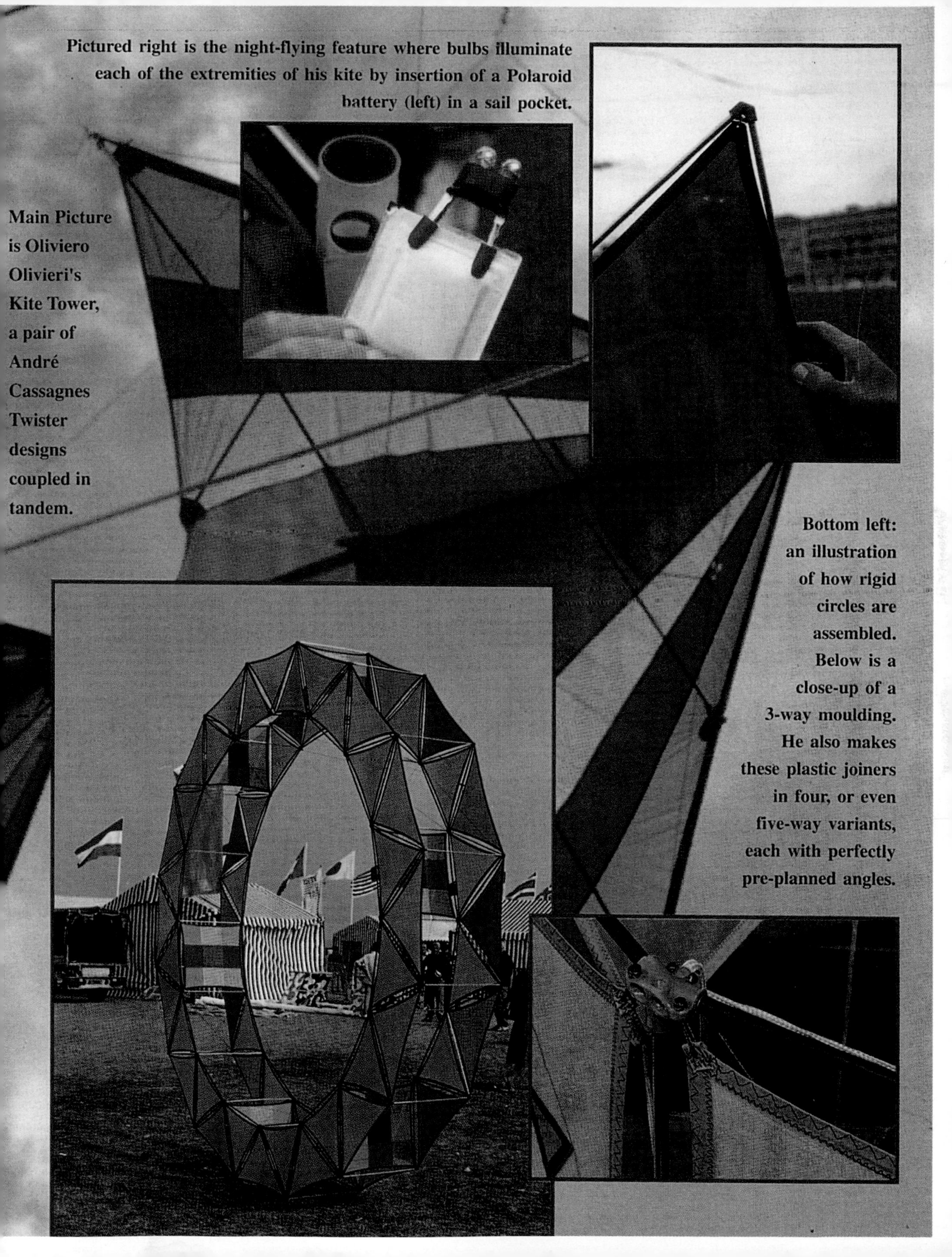

Pictured right is the night-flying feature where bulbs illuminate each of the extremities of his kite by insertion of a Polaroid battery (left) in a sail pocket.

Main Picture is Oliviero Olivieri's Kite Tower, a pair of André Cassagnes Twister designs coupled in tandem.

Bottom left: an illustration of how rigid circles are assembled. Below is a close-up of a 3-way moulding. He also makes these plastic joiners in four, or even five-way variants, each with perfectly pre-planned angles.

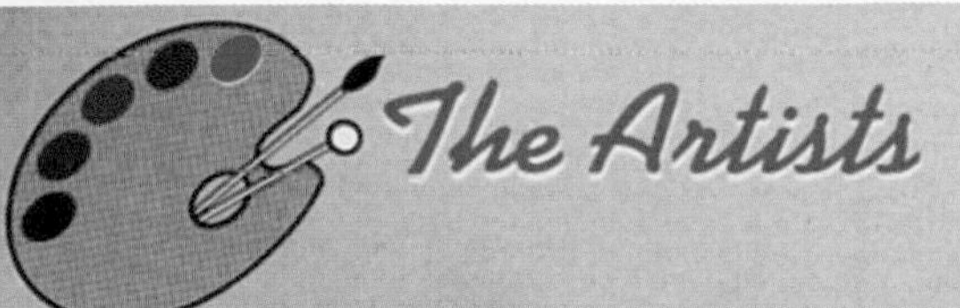
The Artists

Main Picture shows artistic Steve Brockett and his fantastic Dragon kite which incorporates the head in its bridle system. Upper left is Adrian Conn, a Grand Prize winner at the annual AKA conventions, with his long-tailed Flexwings (G Woodcock). Bottom left is Randy Tom showing all of those Seven Sisters as shown in close up on the previous spread (G Maurer). Below right is André Cassagnes in person, centred in his flying wheel that utilises taut cords as spokes to a central hub, self-made as ever.

The Artists

Main picture: world travelling Martin Lester of Bristol flying his enormous 20m (65ft) pair of legs. Maestro of the inflatable shapes, Martin makes legs in smaller sizes - 2.4m and 9.6m long - as well as sharks and birds.
Below, from New Zealand, Yvonne de Mille uses appliqué to create her ecological message for peace and goodwill. The pattern incorporates a New Zealand Cabbage tree.

Above is George Peters, another world travelling kite flyer, with his Birdman of Paradise, another very long kite well suited to International events.

Chapter Four

Painting the sky

A major contribution to public awareness of the modern kite has come from exceptional kite flyers with artistic ability to create multi-coloured artforms. These bear little relationship to any shape that might be termed traditional, they are generally very large and they have been in demand at seaside beach displays, city-organised festivals and internationally at sponsored functions.

So impressive are these abstract figures that their creators have travelled the world to be part of what almost amounts to an International Circus of skilled performers, all by invitation and with expenses paid! This really is a well-deserved modern development in kite flying!

An alert artistic temperament is the first requirement for the creation of these so impressive two dimensional figures. Moreover, to be 'accepted' they have to fly ,and to fly well, on an anchored line without further attention for hours on end. Although some of the structures owe a little to the age-old methods of the Chinese, with their curved bamboo spars to form bird frames, the fascinating aspect of these artforms is the introduction of techniques that are absolutely new. One cannot help but reflect on how much all of this emerging artistry stems back to the exhibitions of Tom Van Sant's work in the USA and UK during 1976.

GF and ripstop

Until the Van Sant collection arrived in London for display in the ICA on Pall Mall, and we were able to see his creations fly by the Round Pond in Kensington Palace Gardens and at the Festival Hall on the River Thames, most of our kite-thinking came in straight lines. Tom had caught the kite-bug when working in the Far East. On return to his Californian studio, he determined to make a Centipede just like those he had seen in his travels. An accomplished sculptor, he sought materials that might be easier to work with than scrape-tapered bamboo and hand-made papers. He chose to use sailmakers' ripstop nylon and tapered fishing rod units in glass fibre. So the curvaceous shapes were born. Tom developed multi-wings, bird shapes, ellipses, all taking the flexibility of the GF rods to extremes.

He laced the sails rather than use pockets, as now preferred, and introduced colour blends that stimulated entirely fresh approaches to kite design. We owe a lot to Tom who still has a fondness for his kites, although he is better known in the 1990s as the creator of the GeoSphere Project, a spectacular image of the Earth produced from over 2000 satellite images.

Polyblocks and plastic rods

As use of GF rod and tubing expanded rapidly, and ripstop nylon became available off-the-roll through the 1980s, new concepts were fertilized by exponents of the multi-cellular type of kite, such as the Tetrahedral. However, the simple method of adapting plastic tubing for joints was woefully inadequate where 3 or 4 rods converged.

Enter the HDPE block. This is High Density PolyEthylene which can be cut, drilled and machined very easily, using average home workshop facilities. Polyblocks are the solution for any shape that requires multiple spar connections. Taking this a long step further, André Cassagnes (France) has his own special polyester fittings moulded by a friend M.Minot. These enable him to create the most spectacular shapes with biplane 'crowns' and tetrahedrals.

Inflatables

The Parafoils, created by Canadian Domina Jalbert and subsequently used for parachuting and paragliding, established that a properly-rigged, fully flexible shape would fill with ram air and fly as a kite. The famous 'Flexifoil' is a prime example.

This principle was not lost on the creative. We were soon to see dragons, fish, giant insects and even trouser legs suspended in mid-air. Martin Lester's largest set of legs to date are no less than 20 metres (65ft) long! He has also made the 'Chorus Line' of paired sets of dancing legs and even a soccer tackle with contrasting team shorts on the legs plus a suspended ball! New Zealand's Peter Lynn's fertile mind created a giant octopus with a body 7 metres (22ft) across and tentacles stretching another 20 metres in continuous motion. It was inevitable that his Ninja mutant turtle followed!

The inflatable kite demands an extra skill. Marking out and matching the fabric is critical, so too is the internal bracing, as might be required for a whale or shark shape. Martin Lester of Bristol has produced a Boeing 747 Jumbo jet, a Space Shuttle, countless sharks and even a flying Daedalus. Venting is important to sustain the shape with an internal airflow. The vent is sometimes of the kangaroo pouch ventral type, if not realistic at the nose. Internal baffles or risers restrain the fabric gores to hold shape where practical, otherwise a central spine and spar structure is employed as we show for the Shark.

Martin uses balloon fabric, as it does not crease as easily as spinnaker ripstop. All patterns are made in paper to try-cut the parts, and each seam is pre-stretched in turn as the shape is built up. Then comes the rigging.

All of these special shapes require very special bridling to ensure stable flight. There are no standard procedures, in fact, by now you will have realised that the 'flying fantasies' are unique to their creators and, although some sell theirs on a commercial basis, it is a case that, if you want to make one, then you have to devise your own creation.

The 'Show' kite can be as diverse in its form, as the inspiration of its creator.

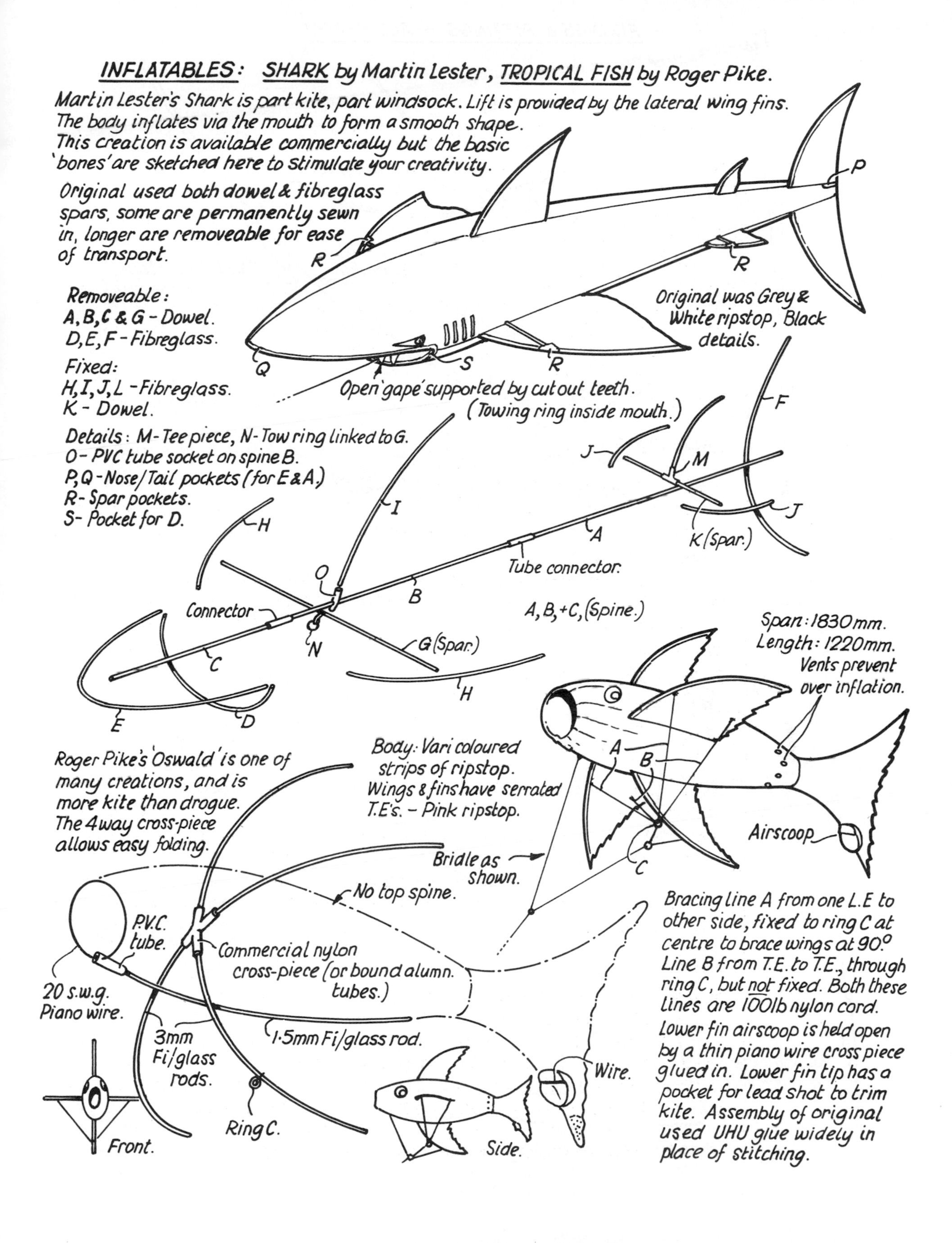

INFLATABLES: SHARK by Martin Lester, TROPICAL FISH by Roger Pike.
Martin Lester's Shark is part kite, part windsock. Lift is provided by the lateral wing fins. The body inflates via the mouth to form a smooth shape. This creation is available commercially but the basic 'bones' are sketched here to stimulate your creativity.
Original used both dowel & fibreglass spars, some are permanently sewn in, longer are removeable for ease of transport.
P
R
R
Original was Grey & White ripstop, Black details.
Removeable:
A, B, C & G - Dowel.
D, E, F - Fibreglass.
Fixed:
H, I, J, L - Fibreglass.
K - Dowel.
Q
S
R
Open 'gape' supported by cut out teeth.
(Towing ring inside mouth.)
F
Details: M - Tee piece, N - Tow ring linked to G.
O - PVC tube socket on spine B.
P, Q - Nose/Tail pockets (for E & A.)
R - Spar pockets.
S - Pocket for D.
J
M
I
J
H
A
K (Spar.)
Tube connector.
O
B
Connector
A, B, + C, (Spine.)
N
G (Spar.)
C
H
E
D
Span: 1830mm.
Length: 1220mm.
Vents prevent over inflation.
Roger Pike's 'Oswald' is one of many creations, and is more kite than drogue. The 4 way cross-piece allows easy folding.
Body: Vari coloured strips of ripstop.
Wings & fins have serrated T.E's. - Pink ripstop.
A
B
Airscoop
Bridle as shown.
C
No top spine.
P.V.C. tube.
Commercial nylon cross-piece (or bound alumn. tubes.)
20 s.w.g. Piano wire.
3mm Fi/glass rods.
1·5mm Fi/glass rod.
Ring C.
Front.
Side.
Wire.
Bracing line A from one L.E to other side, fixed to ring C at centre to brace wings at 90°. Line B from T.E. to T.E., through ring C, but not fixed. Both these lines are 100lb nylon cord.
Lower fin airscoop is held open by a thin piano wire cross piece glued in. Lower fin tip has a pocket for lead shot to trim kite. Assembly of original used UHU glue widely in place of stitching.

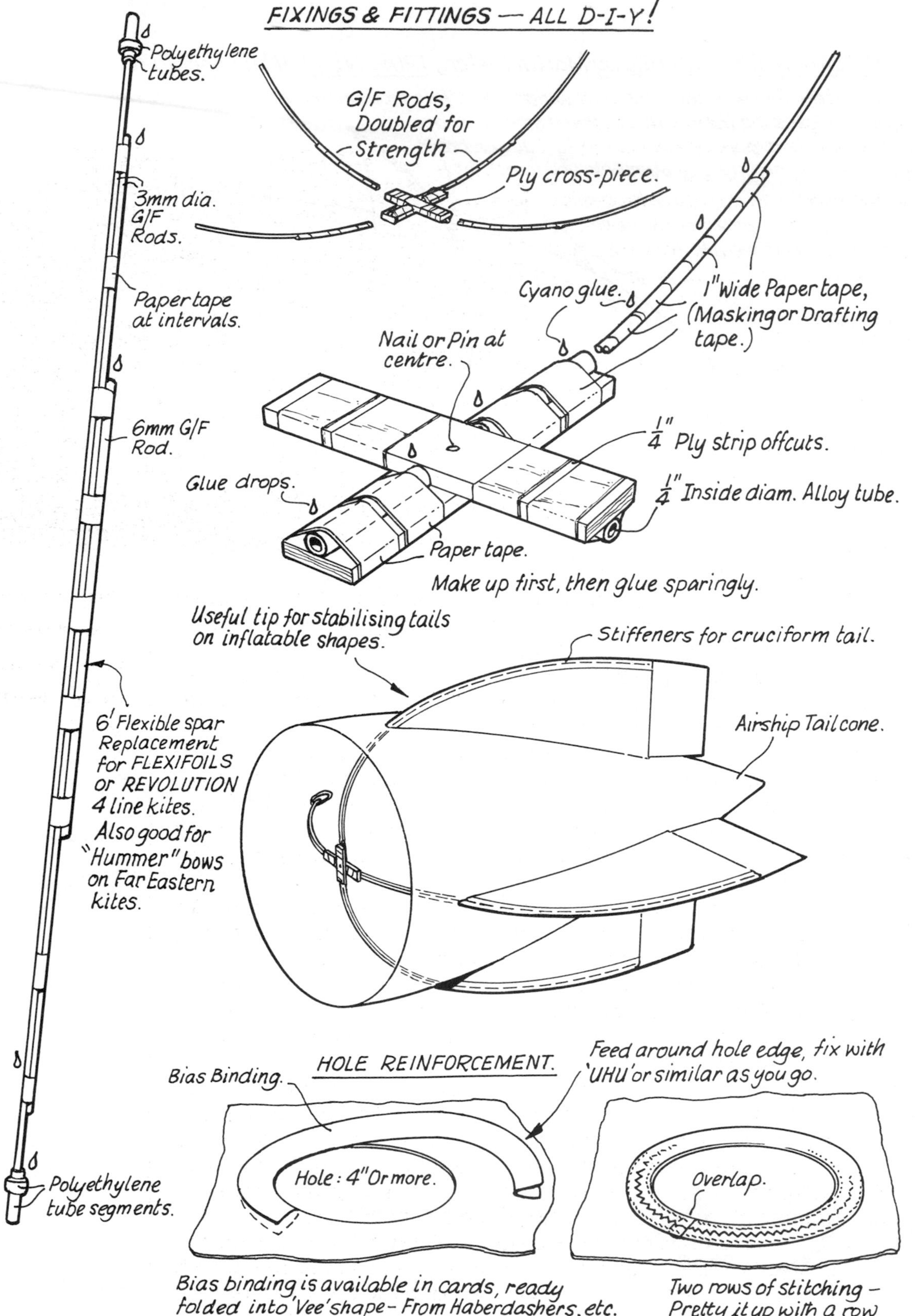
FIXINGS & FITTINGS — ALL D-I-Y!
Polyethylene tubes.
G/F Rods, Doubled for Strength
Ply cross-piece.
3mm dia. G/F Rods.
Paper tape at intervals.
Cyano glue.
1" Wide Paper tape, (Masking or Drafting tape.)
Nail or Pin at centre.
6mm G/F Rod.
1/4" Ply strip offcuts.
Glue drops.
1/4" Inside diam. Alloy tube.
Paper tape.
Make up first, then glue sparingly.
Useful tip for stabilising tails on inflatable shapes.
Stiffeners for cruciform tail.
Airship Tail cone.
6' Flexible spar Replacement for FLEXIFOILS or REVOLUTION 4 line kites.
Also good for "Hummer" bows on Far Eastern kites.
HOLE REINFORCEMENT.
Feed around hole edge, fix with 'UHU' or similar as you go.
Bias Binding.
Hole: 4" Or more.
Overlap.
Polyethylene tube segments.
Bias binding is available in cards, ready folded into 'Vee' shape - From Haberdashers, etc. In matching or contrasting colours.
Two rows of stitching - Pretty it up with a row of Zigzag.

There are no positive guidelines for structure or design, and in that respect there is little that this book can contribute, other than to illustrate those impressive examples we have seen. Maybe they will trigger some latent talent to encourage even more extensive artforms.

Three striking heads on George Peters' Flying Colours birdman designs are each 84in (213cm) wide.

The artists

We've mentioned some of the leading lights in this fascinating development and you will have noted how widespread are their origins - New Zealand, France, Canada, USA and the UK. It would be unjust if we did not mention other specialists and their particular techniques.

Steve Brockett from Cardiff, South Wales, uses the kite as a flying canvas on which he paints startling picture stories, often based on mythology or the art of the North American Indian. Each a study in itself, with titles like 'No two feathers', 'Butterflyman', and 'Angel and the Dreamer', they have been flown as an aerial ballet to classical music, wind harps and synthesisers. Steve's work has done much to erase the stigma of kite flying as a childish diversion.

In much the same way, George Peters of Colorado, USA, has gained acceptance of the kite as contemporary art. George wanted to obtain shapes that were visually stimulating in the sky and has been so successful that, since his start in Hawaii, he has travelled the world, influencing many other like-minded free spirits. Two of his very long and colourful kites grace the cover of this book.

Among fellow Americans, Joel Scholz is another outstanding artist/designer best known for his long train of delicately coloured birds, and his flying 'Geisha Girl'. It's said that he first traced the outline for his ladylike creation around a tall woman friend. He then sewed the sail using invisible thread for the appliquéd dress and body colours before he added a frame. After many trials with cross-spars and bridles, the Geisha flew well enough to win Joel a trip to Japan! It has since performed around Europe.

Bill Lockhart and Betty Street have also achieved international fame for the quilted effects on their Texan creations, where the sewing skills are admired as much as the medley of geometric colour patterns. Others, like Scott Skinner, have extended the artform of appliqué to convey bird pictures, while Tom Casselman has used commercial sign painting techniques to produce superbly intricate pictorial decoration on his Rokkaku kites.

The sky picture kite is just as uniquely individual in Europe. Silvio Maccherozzi and Franco Guibilini have taken the multi-faceted design to extremes with their Italian shapes. In Germany, Peter Malinski has done the same, except

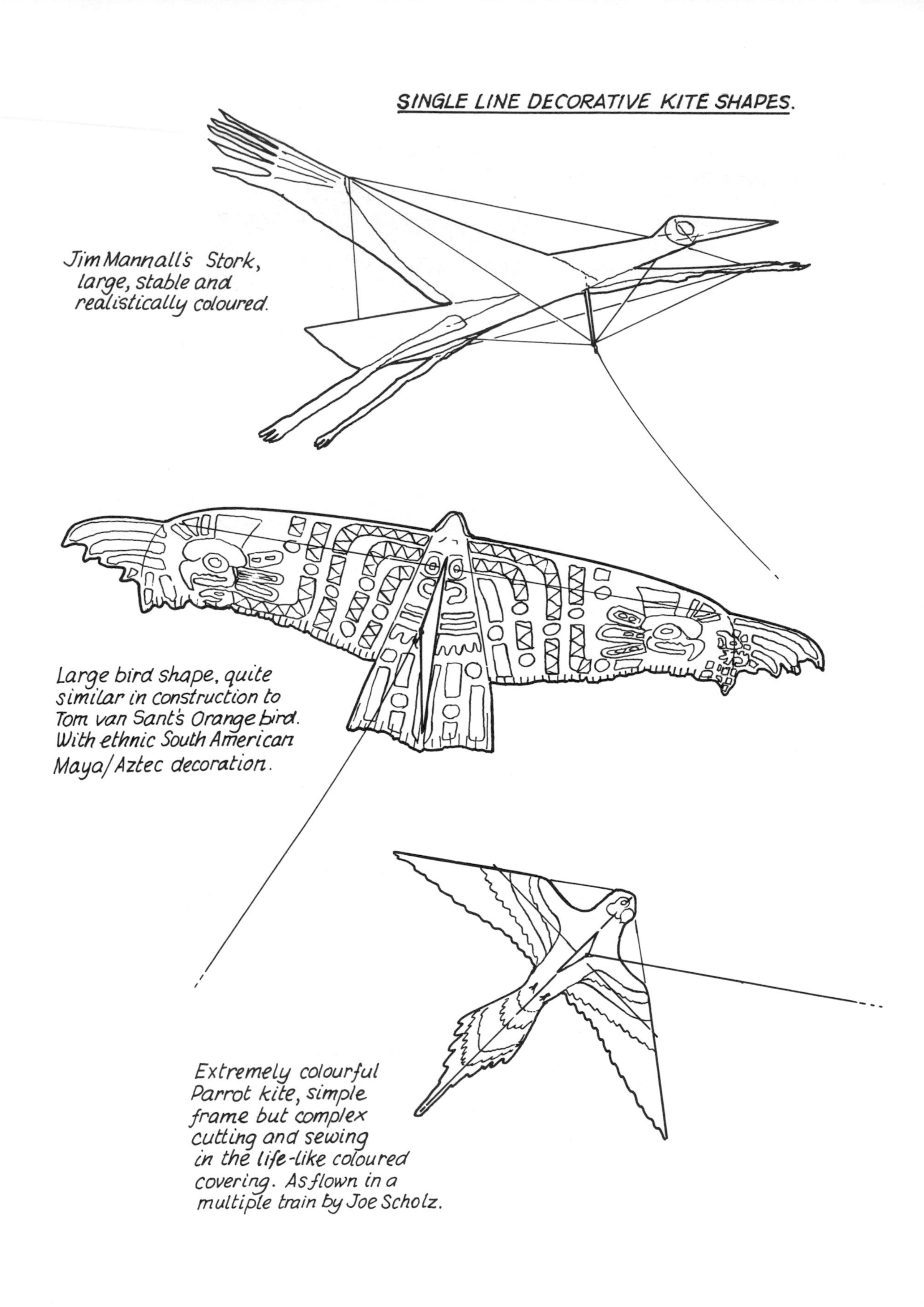

SINGLE LINE DECORATIVE KITE SHAPES.

Jim Mannall's Stork, large, stable and realistically coloured.

Large bird shape, quite similar in construction to Tom van Sant's Orange bird. With ethnic South American Maya/Aztec decoration.

Extremely colourful Parrot kite, simple frame but complex cutting and sewing in the life-like coloured covering. As flown in a multiple train by Joe Scholz.

that, whereas the Italian kites fly in the vertical plane, Peter's are very wide span horizontal designs. French artistry abounds, as one might well expect. It ranges from the enormous inflated centipedes and winged insects created by Pierre Fabre to Michel Gressier's aerial sculptures which include gigantic rotating tubes.

By virtue of their size, and with a view to portability, these show kites are mostly 'soft' and do not use spars. The exceptions are the Cassagnes/Malinski types where polyblocks or moulded fittings permit easy assembly with relatively short spar units. The field of opportunity is wide open for anyone with the gift of originality and an artistic outlook.

On the other hand, if the flying fantasy is out of reach, there is always the wind sail, or windsock, you can make from scrap leftovers to erect on a mast as a bright locator for your base on the field.

Go on, create your own splash of colour!

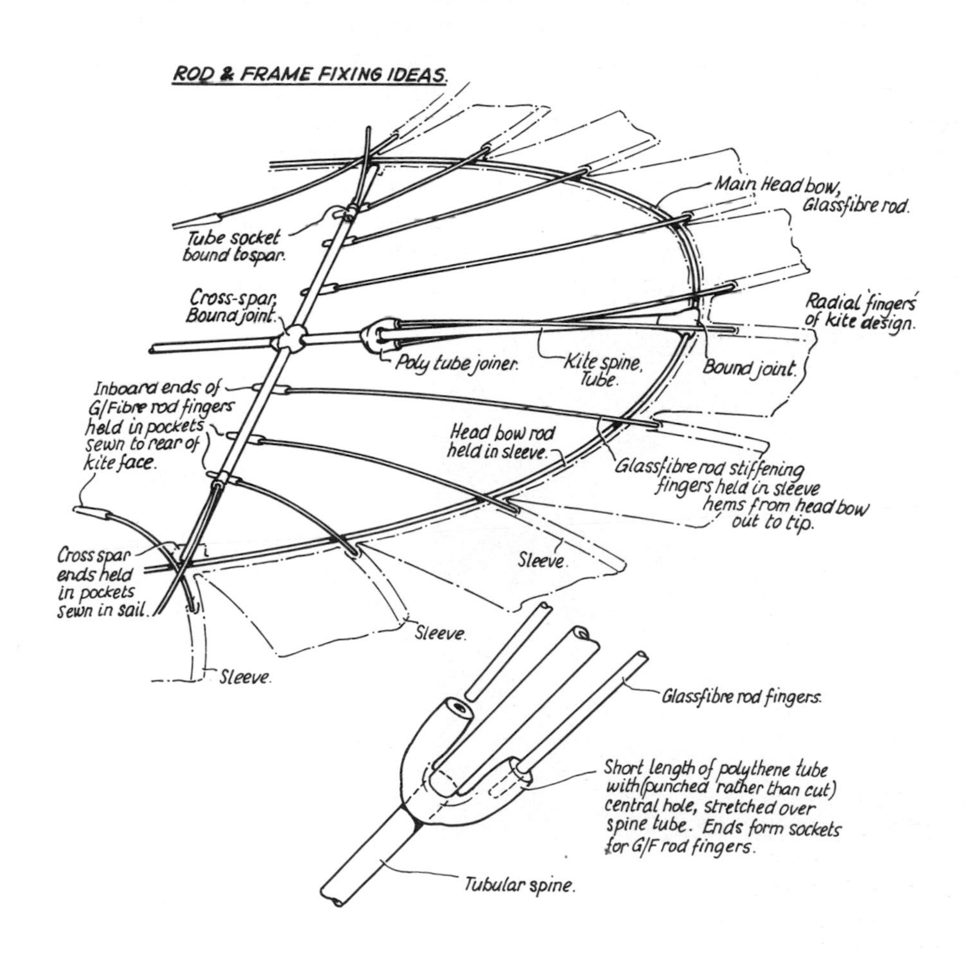

Chapter 5 Aerial photography from

Just over one hundred years ago, a Frenchman, Arthur Batut, established the possibility of aerial photography when his successful experiments were described in *La Nature* of August 1888. The concept was quickly taken up by others, notably Emile Wenz, then by William Eddy and Gilbert Totton Woglom who took photographs of the American cities of Boston and New York in 1895/6.

Initially a technique for scientific research and surveys, the equipment was heavy due to the use of plate cameras, and the method of mounting, the type of kite and means of triggering the shutter have long since been surpassed. However, the basic principles remain little changed. A kite or train of kites is used as an aerial platform to suspend the apparatus. The camera can be preset or remotely controlled to cover the subject or area to be photographed, and provision made for a sequence of pictures or a single exposure, according to the camera type and its shutter release.

The modern kite, coupled with the vast range of photographic equipment which is available, opens up the prospect of aerial snapshotting for anyone. Amateur archaeology, agricultural crop survey to check disease, low-level views of landscaped gardens, housing or industrial buildings and souvenir photographs of your local flying site are all possibilities which make the camera-kite combination so attractive.

Getting started in aerial photography is sometimes a formidable hurdle. At the outset, you must first decide just how serious you want to be. Is it to be only for fun and general amusement, or are you a perfectionist, seeking impeccable exposures and absolute control over the photography?

Camera

Compact, SLR or larger format? These cameras come in many variations and consideration of the risk involved may well prove to be a deciding factor. As we shall describe, provision can be made for camera survival; but for most

kite flyers the camera value, not to mention associated control systems, will be a major consideration.

Weight is important as well as value. The unit is likely to form a heavy and self-destructive lump in the (unlikely) event of a line break. There is also the question of camera vulnerability through general usage, swinging around on a line and likely to be the first arrival on the ground when the kite line is reeled down.

So let's appraise the pros and cons of the camera types we have mentioned.

Compact

This is the most popular. It's a package that can incorporate desirable features like auto-focus, auto-wind, flash, wide-angle lens (28 or 35mm), DX coding for auto film speed setting, and usually has a standard screw thread hole for mounting in the base.

Some of these features are not essential. A single focal length, fixed-focus is often sufficient. The flash has to have a manual override to be useful, as it will serve only to tell the operator when a photograph has been taken. The DX coding may affect aperture, whereas we will need a fast shutter speed of 1/250th or faster for shake-free results.

The disadvantages of the compact are a tendency to overexpose the negative due to reflections from the ground, and the ergonomical designs of the camera bodies, which often place the firing button where it is awkward to arrange a plunger. They do not have fittings for a flexible shutter release cable. So beware the curved top compacts.

In summary, a compact with auto-wind, flash manual override, fixed-focus, DX coding and a tripod mount socket would be fine for a basic unit.

SLR

The abbreviation stands for 'Single Lens Reflex' which is of no concern to us! It means that the photographer can actually look through the lens with the aid of mirrors and so view exactly what the photograph will frame. As we will not be ascending with the kite, such a feature is superfluous. However, the

Above: Doyen of the aerial photographers in Britain is Tom Pratt, seen here using his twin reel system.

Left: Skyview photograph of a social centre, interestingly with a Parafoil in flight by the lowest point of the building.

Tom Pratt's carrier to be pre-adjusted for angles. The flash unit signals when the shutter has been operated

average SLR incorporates other advantages when compared in general with the Compacts.

These fall into the category of superior quality in the lens and the shutter operation. They can be manually set or use automatic sensing for exposure. Detachable lenses enable you to experiment with various fields of view. Filters can be used for polarisation, or to cut haze. They have sockets for flash and remote shutter operation. Some have auto-exposure compensation. Lens aperture can be preset on an aperture-priority camera, and auto-wind is a universal feature, even if a supplementary 'add-on'.

Although not 'SLR', there are many other 35mm cameras which have *some* of these advantages and are lighter; in particular, the clockwork, selfwinding Ricoh or Lomo types. Unfortunately they are no longer produced but are to be found in specialist used-camera stores, and their lens quality is often superior.

Large Format

This camera is for the professional operator. Everything about it spells cost. Film, camera body, lenses and ultimate processing will add up to several times that of the more commonplace 35mm unit. This is the tops in quality; but not really within the capacity of our average kiteflyer. However, not to discourage those who seek absolute perfection, we suggest starting with a secondhand camera of 6 x 6cm (2 ¼ x 2 ¼ in) format and roll film.

The carrier

While it is possible to loft the camera within the kite itself, and the earliest experiments were all conducted by use of this system, we thoroughly advise that the deadweight of the equipment be hung, pendulum fashion, at a reasonable distance from the lifting kite.

Only where rigid frame kites are used, for example the Box, Cody or more recent Dunford Flying Machine, have any really practical results been obtained when the camera has been an integral part. The advantage is that the kite itself is free of the high frequency vibrations which occur naturally in all kite flying lines; but a fast shutter speed takes care of this problem in most instances.

So we recommend that the carrier be secured to the flying line by means of a rigid strip with capstans at each end, and from which a pendulous slat or similar will hold the camera. Pat has shown the very basic applications. The steadying windvane is a help, and the alternative triangular frame idea shows a way of using weight itself as a stabilising force. In each case, the mode of shutter release suggested is by radio control, of which more later.

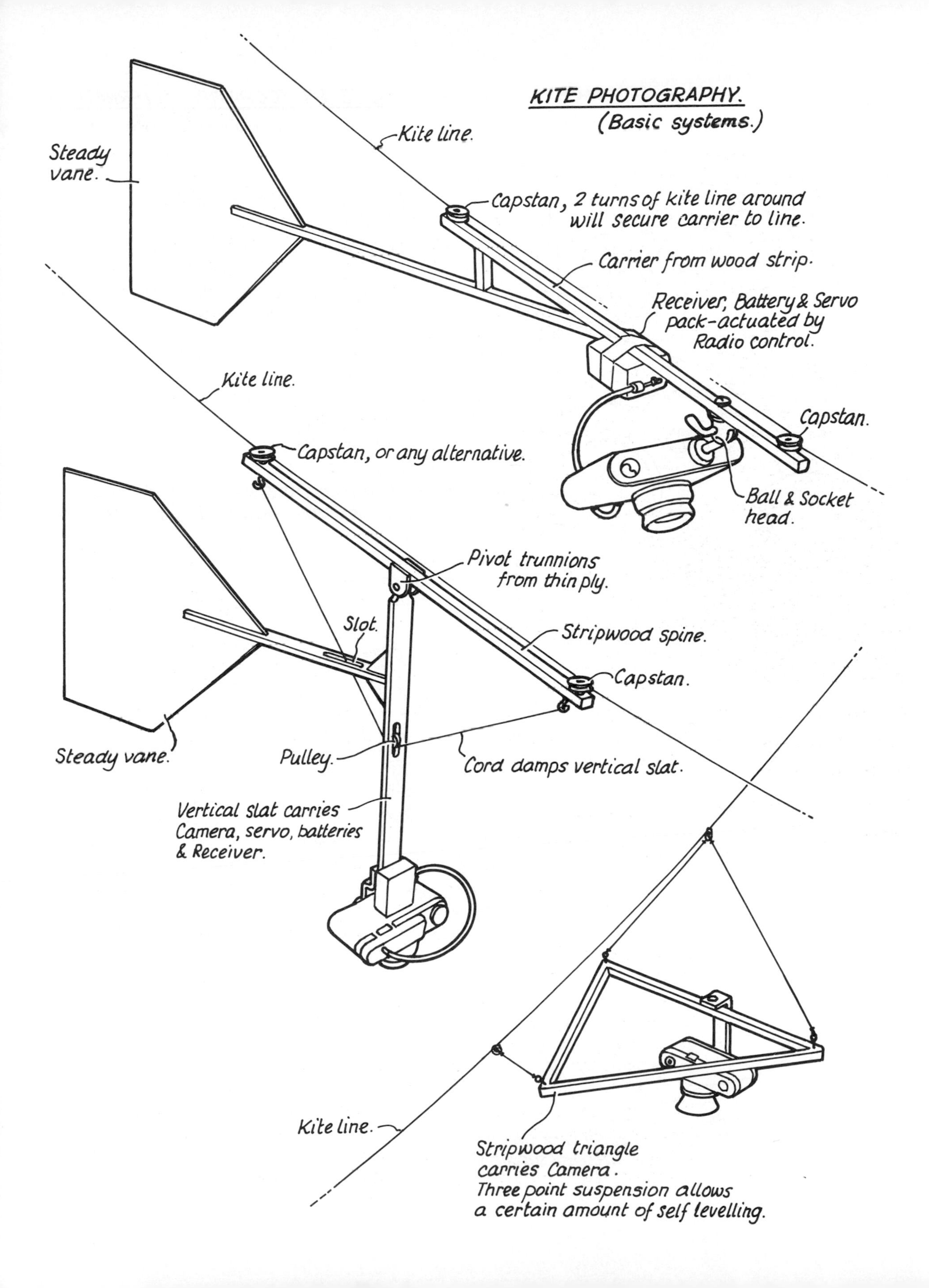
KITE PHOTOGRAPHY.
(Basic systems.)
Kite line.
Steady vane.
Capstan, 2 turns of kite line around will secure carrier to line.
Carrier from wood strip.
Receiver, Battery & Servo pack-actuated by Radio control.
Capstan.
Ball & Socket head.
Kite line.
Capstan, or any alternative.
Pivot trunnions from thin ply.
Slot.
Stripwood spine.
Capstan.
Steady vane.
Pulley.
Cord damps vertical slat.
Vertical slat carries Camera, servo, batteries & Receiver.
Kite line.
Stripwood triangle carries Camera.
Three point suspension allows a certain amount of self levelling.

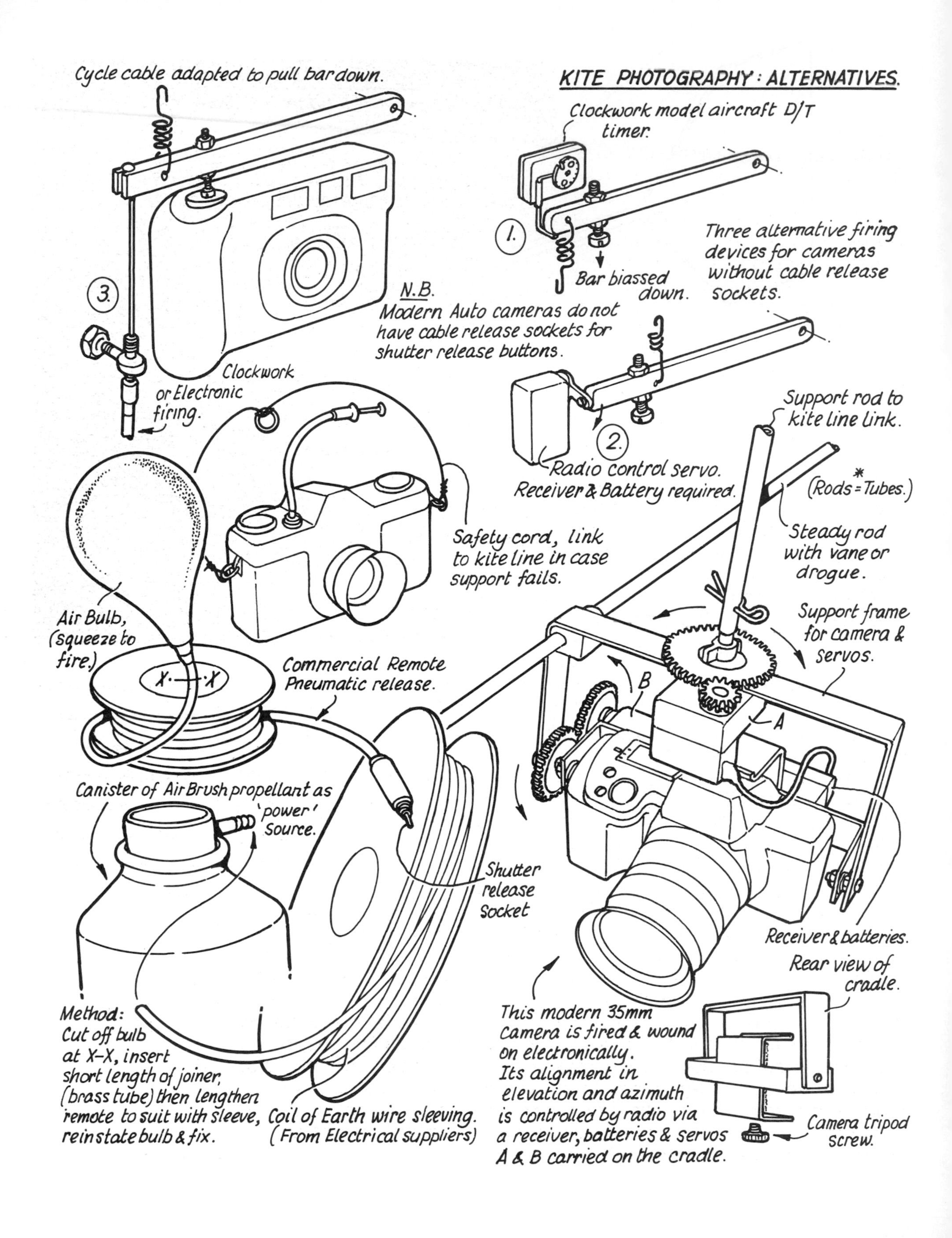
KITE PHOTOGRAPHY: ALTERNATIVES.
Cycle cable adapted to pull bar down.
Clockwork model aircraft D/T timer.
1.
Bar biassed down.
Three alternative firing devices for cameras without cable release sockets.
N.B.
Modern Auto cameras do not have cable release sockets for shutter release buttons.
3.
Clockwork or Electronic firing.
2.
Radio control servo. Receiver & Battery required.
Support rod to kite line link.
(Rods* = Tubes.)
Steady rod with vane or drogue.
Safety cord, link to kite line in case support fails.
Air Bulb, (squeeze to fire.)
X X
Commercial Remote Pneumatic release.
Support frame for camera & servos.
B
A
Canister of Air Brush propellant as 'power' source.
Shutter release Socket
Receiver & batteries.
Rear view of cradle.
Method: Cut off bulb at X–X, insert short length of joiner, (brass tube) then lengthen remote to suit with sleeve, reinstate bulb & fix.
Coil of Earth wire sleeving. (From Electrical suppliers)
This modern 35mm camera is fired & wound on electronically. Its alignment in elevation and azimuth is controlled by radio via a receiver, batteries & servos A & B carried on the cradle.
Camera tripod screw.

Tom Pratt's stack of four deltas lofts his Pentax unit. The secondary (vertical) line is used to fire the shutter and also exercises a measure of control over the kite train.

Right now we are dealing with basics. To take this further, you might consider the card pocket camera, particularly the 'Stretch' or panoramic, ready-loaded holiday camera as a suitable starter. They produce most satisfactory results, are relatively inexpensive, and can be located on the basic carrier systems shown by little more than a 'twinstick' mounting pad. Firing the shutter can be equally basic, by mounting a lever on the card camera box and even pulling it from the ground by separate line.

The serious aerial photographer will be looking for adjustment through vertical and horizontal axes so that the field of view is precise. Here the option is either to purchase a readymade carrier or to devise one's own. In the UK, Greens Kites of Burnley have a well-tried AP system which weighs 360g (13oz) and will accept most SLR cameras or smaller compacts. Aerial Innovations in the USA produce the 'Sky Pod' for compacts or cameras of similar size and weight. This is a 170g (6oz) unit, also made of flat strip aluminium, and arranged for two axis presetting. Instructions come with each; but the shutter operation has to be devised.

Otherwise, the prospective aerial photographer must construct the carrier using alloy strip or channel. The objective must be to keep it simple, yet allow adjustment for up and down inclination of the camera and rotary motion. These can be preset on the ground, or operated by radio control as in Belgian Raoul Fosset's system. This is a highly developed carrier for a sophisticated Ricoh XR-X electronic shutter, auto-wind, auto-exposure, auto-focus camera. Leaving aside these complexities, Raoul's carrier is adaptable for any 35mm camera, SLR or compact, with or without radio control.

A rod or tube, which drops down from the retainer on the kite line, engages a spindle on the main frame and is locked by a wire split pin. This enables the unit to be detachable. The main frame has a secondary 'U-shaped' sub-frame which can pivot by means of bolt bearings. In Raoul's case, there is a further frame attached to the sub-frame, and this is to hold both camera and radio equipment because, as you can see, he has utilised each of the pivot points as centres for cogwheels. These are driven direct by gears on the radio-controlled servo output spindles. Standard modelling parts are used but, for the non-radio kite flyer, the same framework can be adapted but with wingnuts to secure any preset adjustment.

That third frame is unnecessary if radio is not to be carried. Instead, the sub-frame should be drilled to accept a tripod mount screw so that the SLR or compact is fitted directly onto it. The purpose of the third frame is to carry the receiver and its battery, which in turn will balance the usually heavy auto-zoom lens.

Tom Pratt favours another form of carrier, made from alloy tube of square section, to hold his Mamiya Press camera. Tom is very professional in his approach. Any radial axis setting is by friction fit on the suspension tube that holds the radio-controlled shutter release system, and the camera angle is similarly preset for each large format exposure before sending aloft. A final touch is Tom's use of an independent flash unit to indicate that the radio has actually fired the shutter.

Raoul Fosset's fully articulated carrier, actuated by radio control. The Ricoh SLR XR-X is a sophisticated camera, another is the Canon EOS 5000 (only 330 gm). Lighter, and with electronic shutter release for direct link to a receiver, is the Compact Ricoh FF-9.

The shutter

There are dozens of ways you can fire the shutter, which is pressure sensitive to the finger in normal application; but it is not always easy to arrange a mechanical device to operate at distance. A lever with an adjustable contact point formed by simple bolt and nuts is a satisfactory method, and it can be actuated by clockwork timer for single shots or radio servo for auto-wind sequences, or by direct pull from the ground via a section of cycle cable anchored somewhere on the carrier. We've seen so many variations, from a radio servo driving an eccentric cam straight onto the release button, to the remarkable 'air' method, suited only to those cameras with cable release sockets.

Ever clever Mark Cottrell of the Kite Store in London suggests using 2mm I.D. electrical sleeving, which is both flexible and reasonably light, to adapt the photo studio pneumatic remote release system to a 'canned air' aerosol. Mark's self-produced book on aerial photography includes this method as used by Mike Miller of Essex KFG among many other gems of his experience.

Radio control is clearly the most attractive means of operating the remote camera, whether as a single channel on the shutter or, as Raoul Fosset uses, three channels for full control to orbit the field of view and fire the shutter. Many auto-winders have a remote control socket which accepts a polarised 'jack' for electrical contact to fire the shutter and transport the film. It is essential, though, that the correct frequency is used for your country, i.e. 27,35,40 or 72Mcs as applied to control of models *and* that, when the radio is operated, it is not within interference range of flying models or cars which might run dangerously amok if there is a clash of spot frequencies.

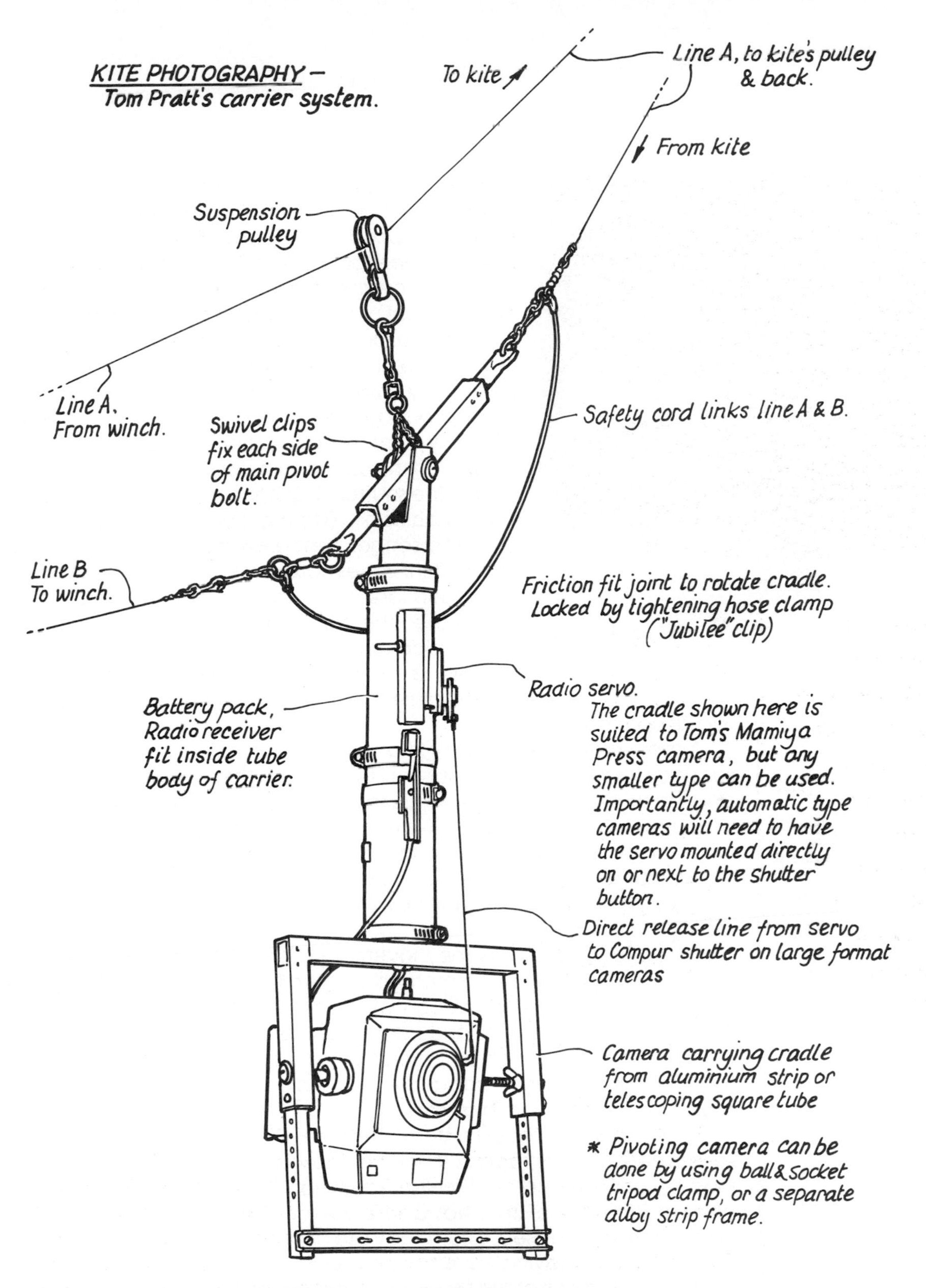

Lines A & B continue down to "double" pulley winch enabling line A to be paid out, as line B draws the camera back down to the operator, & vice versa.

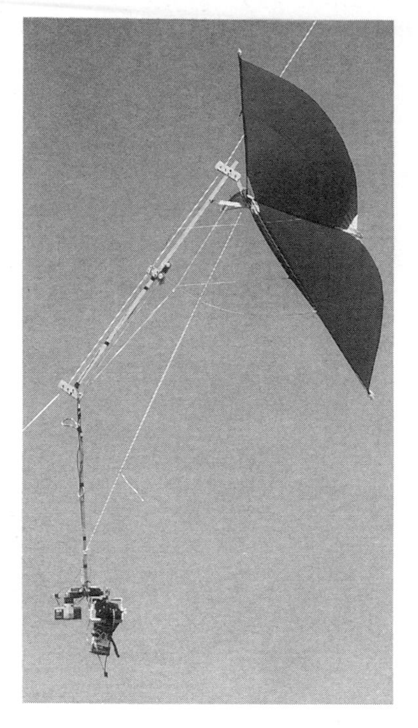

A Delta messenger hoisting a Polaroid camera for crowd shots. As the camera descends, processing is completed so that onlookers can see themselves from on-high within minutes of launch.

Getting it up there

We have already mentioned the simplest method is to tie a stripwood header to the kite line after the kite has been launched and is then able to lift the weight; that is to say, after about 20m or 65ft of line has been paid out, and the kite is free of ground level turbulence.

A variant of this basic system is to attach a pulley to the kiteline, and to hoist the camera unit by a second line via the pulley. This offers the opportunity to fire the shutter by having a trip rod to act as a plunger on the shutter button when contact is made with the pulley. Similarly a 'messenger' or transporter (see Chapter Six) could loft the unit up to a stop on the line which is used to fire the shutter. This is fine for a lightweight camera unit, such as a 110 format Disc camera, or a Polaroid. The messenger descends quickly enough when its propellent vanes collapse, so provision must be made for a soft landing of any weighty camera unit. We have seen the transported Polaroid used this way for beachside crowd shots, with results good enough to create a demand for continuous operation, and steady sales of the prints, each of which displayed a sea of upturned faces.

A more practical method of transporting the unit up the line is to utilise a second kiteline. This is paid out until the unit reaches a stop on the main line and the shutter is fired, either single shot or sequential if the camera is a motor wind type. To return the unit for reloading, or resetting the film by manual wind-on, the second kite is brought down. Such a system calls for kites with similar stability characteristics, otherwise there is a risk of tangling the two kitelines, as the shared weight draws them together when the unit nears the stop.

Far better control over the movement of the camera is obtained by the pulley and continuous loop method devised by Tom Pratt. The main kiteline terminates with a pulley through which the loop passes. A ground-based winch then winds the loop and camera unit to which it is attached, up to the pulley if that is to be used as a trip or, as in most instances, remote control of the shutter is used when adequate height is achieved. Here, the essential is a good lifting kite or train of kites.

Tom Pratt's developed system offers the ultimate approach for the really serious aerial photographer. It doubles the winch reels so that the continuous loop is wound in and out simultaneously, and the camera held firm at any altitude. The carrier itself has a safety link across the suspension points (a parachute can be used, also, to obviate disaster) and the pulley on the main line carries a steadying vane to retain loop separation, instead of having it twist with variations of line tension - braid line of generous breaking strain is recommended.

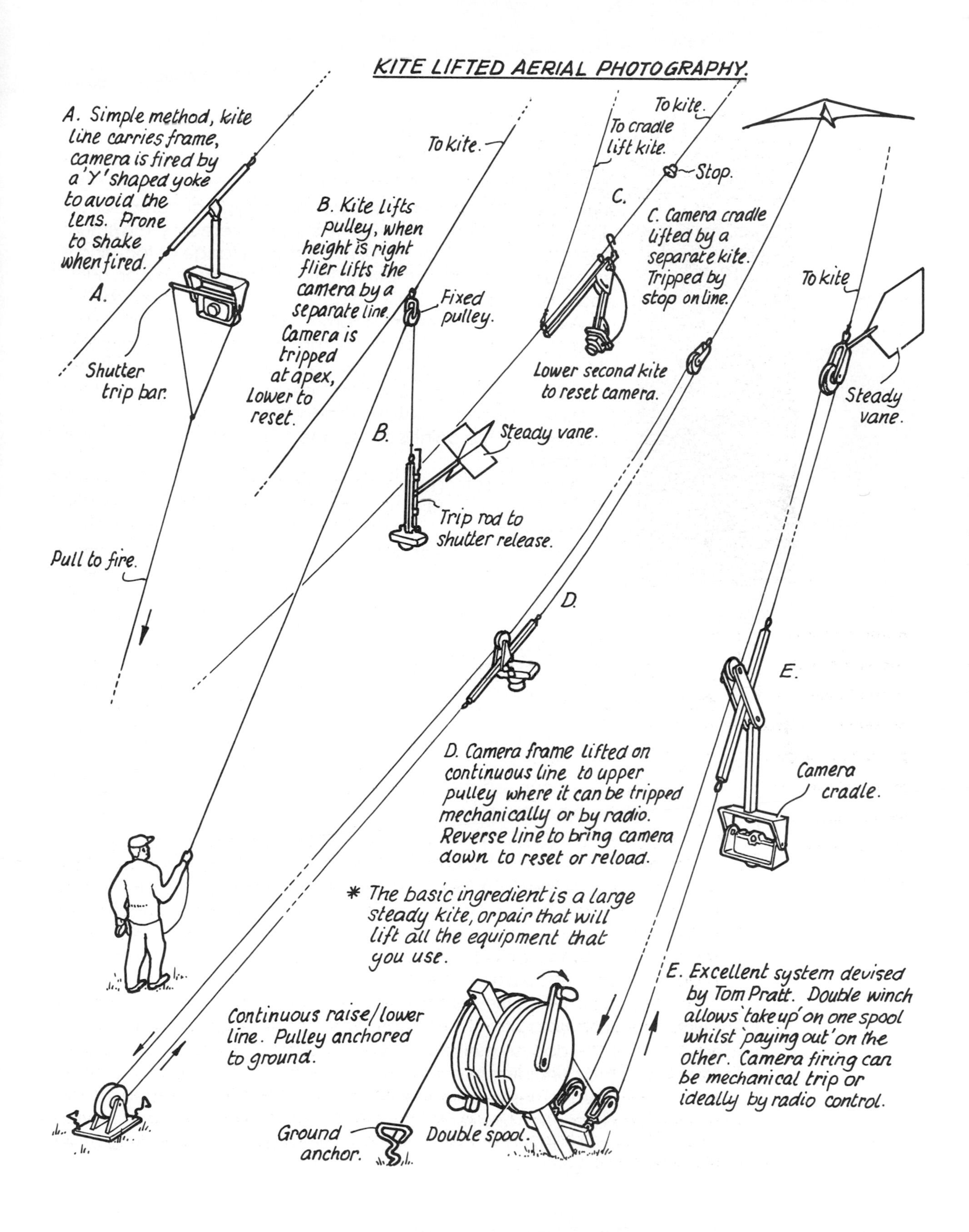
KITE LIFTED AERIAL PHOTOGRAPHY.
A. Simple method, kite line carries frame, camera is fired by a 'Y' shaped yoke to avoid the lens. Prone to shake when fired.
A.
Shutter trip bar.
Pull to fire.
To kite.
B. Kite lifts pulley, when height is right flier lifts the camera by a separate line.
Camera is tripped at apex, Lower to reset.
Fixed pulley.
B.
Steady vane.
Trip rod to shutter release.
To kite.
To cradle lift kite.
Stop.
C.
C. Camera cradle lifted by a separate kite. Tripped by stop on line.
Lower second kite to reset camera.
To kite
Steady vane.
D.
E.
Camera cradle.
D. Camera frame lifted on continuous line to upper pulley where it can be tripped mechanically or by radio. Reverse line to bring camera down to reset or reload.
* The basic ingredient is a large steady kite, or pair that will lift all the equipment that you use.
Continuous raise/lower line. Pulley anchored to ground.
E. Excellent system devised by Tom Pratt. Double winch allows 'take up' on one spool whilst 'paying out' on the other. Camera firing can be mechanical trip or ideally by radio control.
Ground anchor.
Double spool.

Getting the view

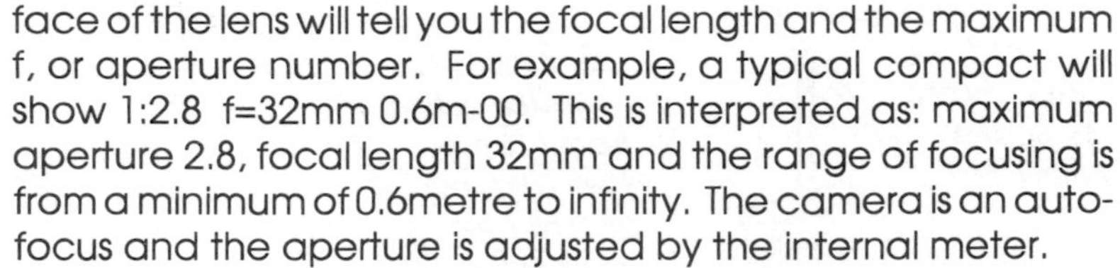

Wide-angle lenses are advised. A glance at the inscription around the front face of the lens will tell you the focal length and the maximum f, or aperture number. For example, a typical compact will show 1:2.8 f=32mm 0.6m-00. This is interpreted as: maximum aperture 2.8, focal length 32mm and the range of focusing is from a minimum of 0.6metre to infinity. The camera is an auto-focus and the aperture is adjusted by the internal meter.

A normal focal length is 50mm, so 32mm is a wide angle, and conversely 80,135 or 200mm lenses are progressively narrower angle or long focus lenses. Aside from 32mm; 35mm and 28mm lenses are found in single lens compacts, so all are suitable. SLR cameras with detachable lenses will have similar inscriptions. The point is to use one marked f=35 or f=28mm, if available.

Aperture setting on any aperture priority auto camera should be at least f8. The higher the number, then the slower the shutter speed. We should not want to go slower than 1/200th second because of camera movement and, for a bright day, f8 at 1/200th with ASA 200 rated film will give a result. Exactly how *good* a result will depend on many other factors. Haze, position of the sun, ground reflection and ranging of the field of view are the most critical.

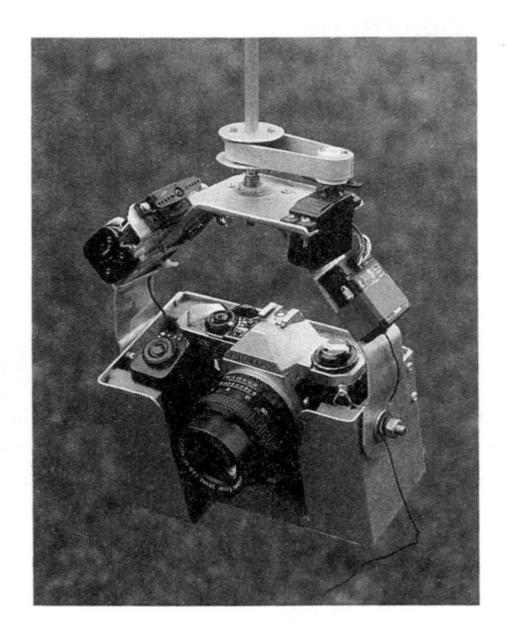

Simon Kidd's system for a Pentax unit under his Parafoil. The camera can be ranged around by radio control.

The all-auto compact offers only one means of control through the DX film cassette coding. The bars on the cassette will adjust the camera to the film speed. So use ASA 200 or faster film.

Other than the camera technique, which is a subject quite apart from this book on kites, we are now left with that vital question of setting the field of view. Thus far, we have reached a 'hit or miss' situation and, by using the wide-angle lens, have minimised the risk of having a print full of the cows in the next field or a horizon just above the baseline.

Some would say that it is the greatest part of the fun in aerial photography to achieve positive results by covering the target area with some precision. Adjustment of the two axes, even by radio control, in the air will not guarantee precision unless the flyer takes time to consider the camera angles. Even when using binoculars, it is difficult to sight the lens angle in reverse, but there are general rules that provide a start towards success through experience.

The direct downward view is dull unless it contains another subject at an interim level, such as a kite. At 10 degrees off the vertical, shadows and a degree of 'depth' convey a bird's eye view impression, and the surrounding activity begins to contribute interest to the photograph. Mix a lateral rotation with the downward angle, and inevitably the horizon will tilt due to the divergent wind direction and its effect on the kiteline. This adds further interest in the picture, but you should be careful not to have any more than a corner of sky in the final print.

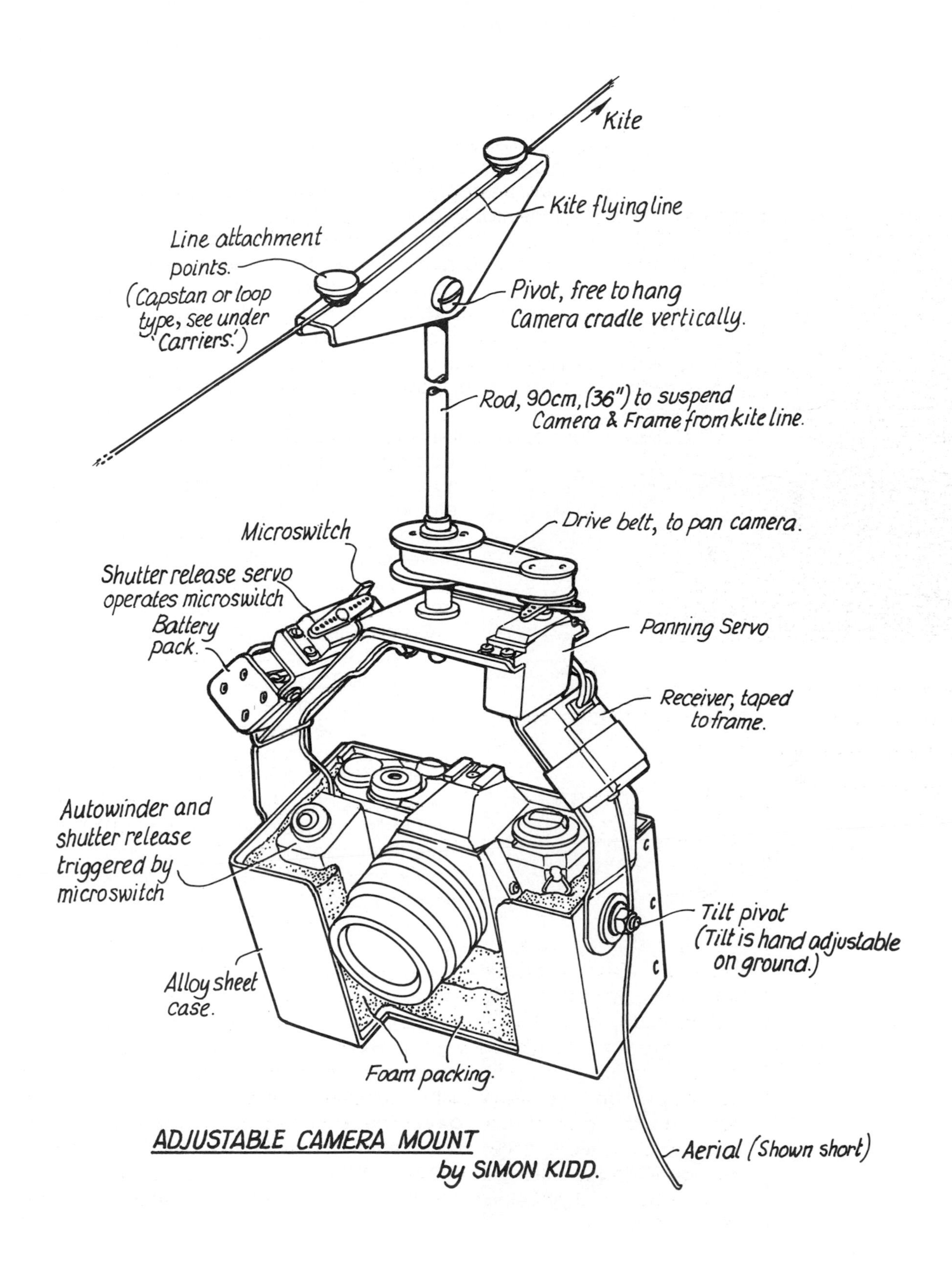

ADJUSTABLE CAMERA MOUNT
by SIMON KIDD.

What kite?

Fortunately, we can draw on all the experience of professional photographers and archaeologists in any decision on the ideal kite or kites.

The first requirement is stability with adequate lifting capacity. So it will be a large kite and will inevitably fall into the category of a Delta, a Parafoil, a Stratoscoop or a Box kite. The Delta is the easiest to make and it will need to have a base span of 8ft (2500mm) or more. Both the late Don Dunford and Tom Pratt have used delta shapes, Don with a smaller 'pilot' delta about 30 metres above the main lifter, and Tom with a stack of up to four deltas closely stacked barely a semi-span apart. Each has had magnificent results. Tom called his the 'Quadriplane' and, when the individual kites were only 40cm (16in) apart and flying as one, they would lift 10kg (221b) in a moderate-to-fresh breeze.

Simon Kidd's Parafoil being tested on the Isle of Wight.

The Jalbert Parafoil is an air-inflated aerofoil. Fully flexible or 'soft', it has excellent lifting power, taking air in at the leading edge to assume a wing section which is retained by multiple shroud lines. There are several commercial sources, of which Hagaman of Westport, Washington State has the highest reputation. It is not an easy kite to reproduce! Simon Kidd uses a kite with 6sq.m (60 sq.ft) area.

Greens Kites of Burnley, Lancashire, produce their 'Stratoscoop' which is similar but not the same as the Parafoil. These have been used for surveys of ruined areas in North Africa with, in the view of the archaeologists, a success rate better than could be achieved by any other aerial platform. The Stratoscoop we advise is the 180 x 150cm (70 x 59in) ST3, which is measured to lift an average of 4.5kg (101b). This is more than adequate for an SLR unit.

Our final recommendation is a kite that is classified as a 'Box' type but, due to its multi-triangulated panels, is not always recognised as such. Designed by the internationally famous Peter Lynn of New Zealand (whose Octopus appears on the cover of this book), it is popularly known as the Peter Lynn Tri-D and it can be made using square cells based on 90-degree Isosceles triangles as Pat has drawn, or with equilateral triangles. With the dimensions given, the result is a 1.52 x 1.84m (5 x 6ft) Tri-D capable of lifting 2.5kg (5 ½ 1b) dependent on wind strength.

The Tri-D offers excellent stability, as innumerable constructors around the world would verify, and it has the advantage of the traditional Box kite in that it does not need a bridle, and flies very well from a directly attached line. Quite definitely the ideal mate for your new camera unit.

Left: Lifting power of the Peter Lynn triangular cell 'Tri-D' is as impressive as its stability. Here one is supporting two large windsocks.

The line, and other precautions

It almost goes without saying that the camera lifter does not have to soar to great heights, and it will perforce have to fly in a steady breeze of 10 knots or more. So play safe - use a braided line of limited stretch, say Dacron, and with an added safety factor of high breaking strain. Greens suggest 170 kilo (3751b) line for their ST3 and this is sound advice for a 2.75sq.m (30sq.ft) kite and attendant dangling weights.

Then, take care that the photography does not invade the privacy of the neighbourhood. We're not referring to discreet overviews of the local naturists either! Some people object to having their property scanned by an airborne eye, so ensure that proper permission is obtained before snapping at random.

Lastly, do observe national limits. Many air regulations stipulate a maximum flying height of 60m (196ft), which can be exceeded by permit (Chapter Ten). Overall, exercise common sense. Flying in open spaces is ideal, but among buildings can be horrific. Always be sure that the lifting kite is stable *before* thinking about lofting the camera.

Below: The Nomansland flying site near St. Albans as photographed by Simon Kidd with Bob Piron's Tri-D type kite close to the camera.

TWIN KEEL DELTA by Tom Pratt.

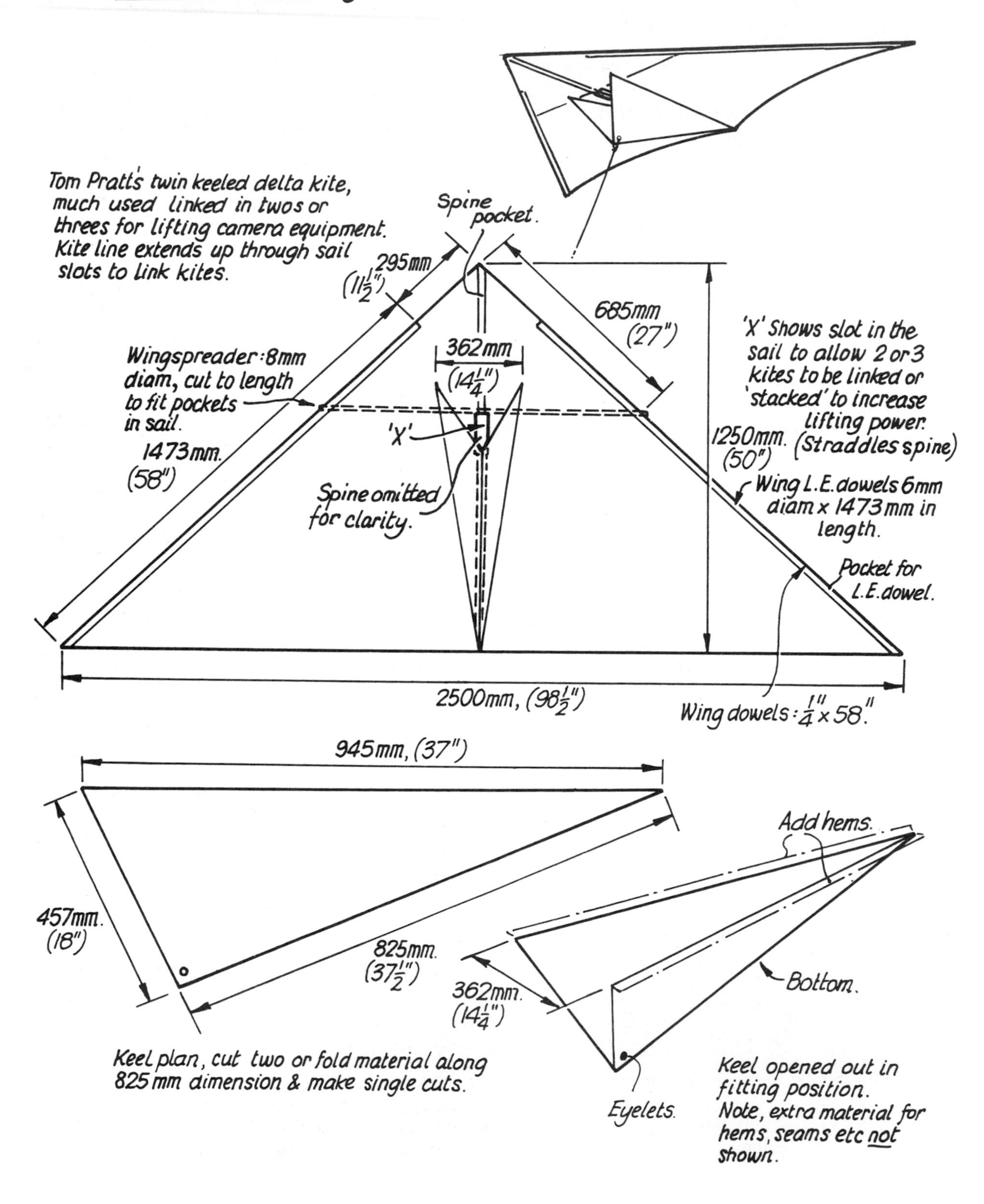

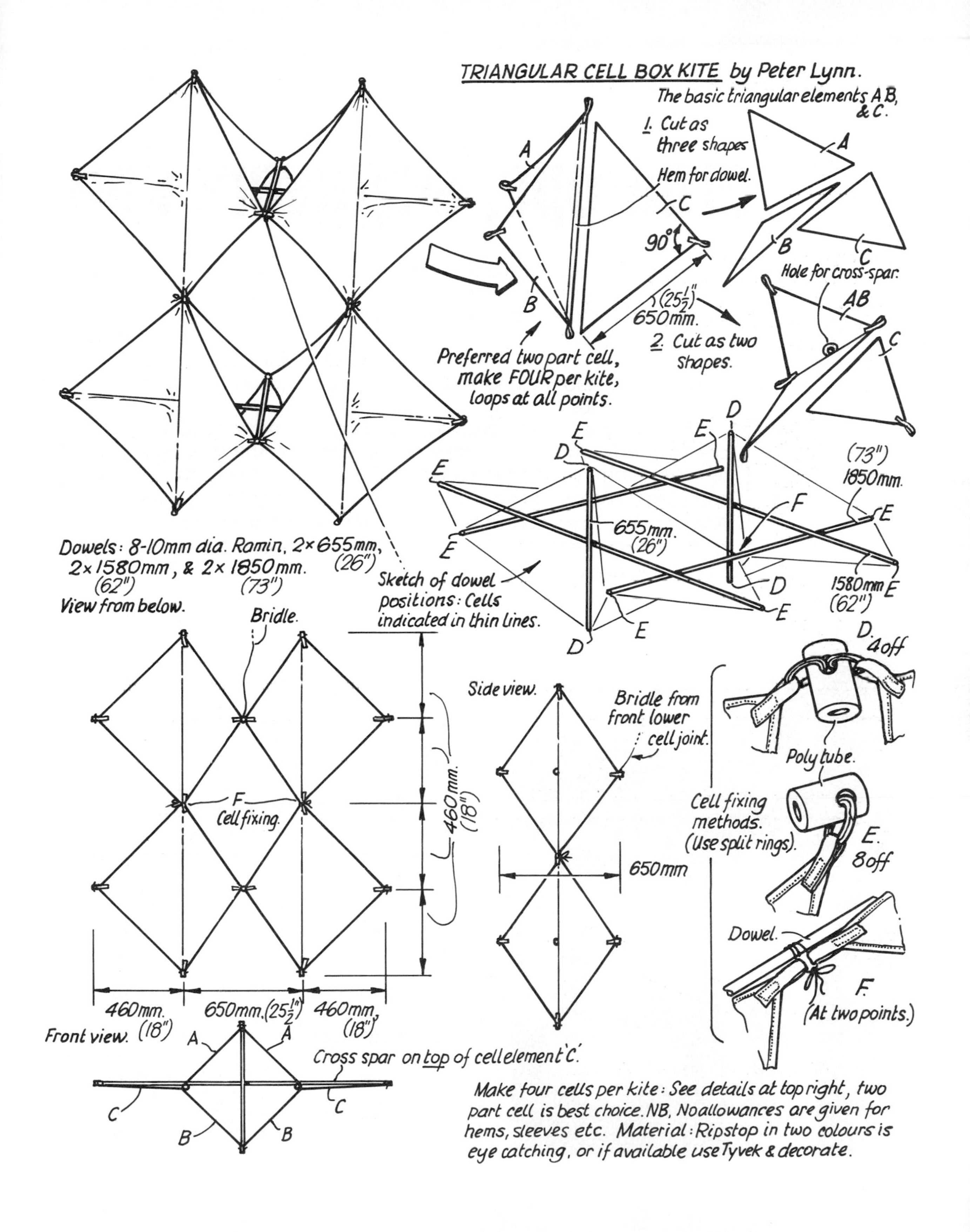

TRIANGULAR CELL BOX KITE by Peter Lynn.
The basic triangular elements A, B, & C.
1. Cut as three shapes
Hem for dowel.
A
C
90°
B
(25½")
650mm.
2. Cut as two shapes.
Hole for cross-spar.
AB
C
Preferred two part cell, make FOUR per kite, loops at all points.
D
E
(73")
1850mm.
F
655mm.
(26")
1580mm
(62")
Dowels: 8-10mm dia. Ramin, 2×655mm, (26"), 2×1580mm (62"), & 2×1850mm (73")
View from below.
Sketch of dowel positions: Cells indicated in thin lines.
Bridle.
F
Cell fixing.
460mm.
(18")
460mm.
(18")
650mm.(25½")
460mm.
(18")
Front view.
Side view.
Bridle from front lower cell joint.
650mm
D. 4 off
Poly tube.
Cell fixing methods. (Use split rings).
E. 8 off
Dowel.
F. (At two points.)
A
B
C
Cross spar on top of cell element 'C'.
Make four cells per kite: See details at top right, two part cell is best choice. NB, No allowances are given for hems, sleeves etc. Material: Ripstop in two colours is eye catching, or if available use Tyvek & decorate.

Chapter Six

Paradrops, skydivers and

As kite rallies began to get into the full swing of a tightly packed annual diary of events from the early 1980s, so a new trend developed. In reality, it was a resumption of the age-old practice of sending up 'messengers' along the kiteline which carried a parachute. On hitting a stop, the messenger released the parachute which was made of lightweight tissue paper, ballasted by a piece of card. That was the 'Atalanta' accessory, popular among British kite flyers of the 1930s.

However the 'new' paradroppers weighed many times that of their predecessors. Additionally, they had character. The passenger was far removed from an anonymous piece of card, and the parachutes correspondingly larger to slow the descent speed. Bob Ingraham, founder of the AKA and 'Kite Tales' coined the term 'Dropniks' and described in Will Yolen's book *Kites and Kite Flying*, his manikins with names like 'Fearless Fosdick' and 'One-jump Waldo'. The idea caught on.

Edward O'Bear

The principle was one thing, the passenger quite another element in the rapid growth of what has become an international craze. For John Barker, a founding member of the British kite flying movement, and a prime mover among the paradropping enthusiasts like Jim Whitehouse and Greg Locke before him, there was no other possible passenger to harness into a parachute than that childhood and perennial favourite, the 'Teddy Bear'.

The universal appeal of the 'Teddy' knows no bounds. We are, as we write, immersed in Bearaphanalia. The Top-selling hardback in the British book trade leading up to Christmas 1991 was Pauline Cockrill's *The Ultimate Teddy Bear Book* and it had held its position among the top ten hardbacks for over two months. It is but one of a plethora of bearable books on the ever lovable cuddly toy which has imprinted itself in the imagination of young and old alike, and always with a name by which to identify the original.

bearly made-it squads

Bearing in mind the Roosevelt connection, Theo was an obvious appellation far too obvious for John Barker's active creativity as he launched the dropniks' own newsletter, complete with its own unique cockney bear vernacular.

The skydive squad

Opposite: Chief 'Hume' of BMISS, John Barker and Fred Bear the fearless one! Left: Harnessed and on his way, Ted is hauled aloft, in this instance without a carrier. On reaching the stop on the kite line pulley, his ripcord is tripped and bingo - the chute opens for a perfect descent.

Roman Candle became the official publication of the UK Ted Devils. It was the mouthpiece of the Chairbear of the Bearly-Made-It Skydive Squad, otherwise known as BMISS with the password 'Geronimo' exhorted by an excited Fredbear who appeared in various guises near the masthead. As the bearer of good tidings reached foreign parts, enthusiasm expanded rapidly and the International Brotherhood of Parachuting Fauna (IBPF) brought to bear many fellow dropniks. The Dutch formed a 'Spring Team KNDST'. In Chicago, they even included a lookalike 'Bear and Hume' event.

Helped enormously by the continuous flow of Teddies from the toy trade, which even includes ready-made flyers in mock leather overalls, complete to the goggles and neck scarf, it was no wonder that BMISS should take-off (and drop down) so rapidly. Ethnic minorities crept into the scene, and fauna variety extended to pandas and gorillas. Bearmail brought together a broad range of bear shapes and sizes, each individually christened with the same affection as bestowed on the childhood comforter.

So 'Rasta', 'Rusty' and 'Camley' featured in the first *Roman Candle*. Their adventures were related in the first person, so giving a 'hume' character to the otherwise mute and very much stuffed fluffy playthings. Avoiding the copyright-protected names of bears that have been British National newspaper features for decades was quite easy for innovative owners of the new skydiving brood.

Just as Teddy itself was a diminutive of Theodore as in Roosevelt (or was it Edward, King of England?), so we began to hear of Charles Lindbear and Wilbear Wright the pioneer bearaviators. In the beargardens that became a feature of almost every major kite flying meeting, a parade of well-dressed bears took on the role of chairbears, observing in silence the antics of crazy 'humes' as they lounged in their designer suits, awaiting the call to elevation.

Up the line

So much for the way in which our furry fauna have captivated so many kite flyers; now to the practical aspects of how to take the skydiver up to a safe release height.

The previous Chapter described the means of approaching a similar task; but now we do not have any intention of launching a camera! The fauna can be elevated and released with far greater ease. Simplest of the systems is to attach the bear to the kiteline using a soft wire pipe cleaner. When the kite is launched on a fully paid-out line, it will soar to height where a few diligent shakes ought to detach our little friend, not forgetting the golden rule of anchoring the static line somehow so that the parachute deploys!

Such a 'Hi-Start' is however, making the launch a random 'will-it, won't it' affair. It is far better to determine the moment of release by positive control from the ground. The kite can be sent up with a carrier attached. A separate pull line is joined to a release pin or spring-loaded wire hanger so that the fauna can be dropped at will. An option is to use a clockwork timer, as for a camera or model aircraft dethermaliser, which can be set for specific time delay in minutes and seconds. Either way, this is a single drop system, and the kite will have to be walked down using a pulley, or similar, each time a reloading is wanted.

Much more satisfactory Is the method of lofting a pulley at the end of the main kiteline and passing a continuous loop of secondary line through the pulley. This loop is then paid out fully, and spread at the groundbase to form a triangle which is securely anchored at two points. A carrier is fixed by knots or capstan windings to the line, and it can be raised or lowered simply by rotating the secondary kiteline around the triangle. The system uses a lot of line, and pulleys plus tough gloves are suggested for operation of the 'loop'. As the lifting kite is going to be large enough to carry the weight of line, carrier and passenger(s), there will be quite a pull, so it pays to use good quality braided Dacron of adequate breaking strain.

John Barker prepares a lightweight Ted for a ride on the messenger.

Carriers

Ideally, the carrier should be easy to attach to the kiteline, either by capstan fitting, as shown for the camera carrier on page 75, or held by a stop on the line with pigs' tail wire wound links that allow the line to be worked into the central core. Remember that *knots weaken line*, so avoid undue stress.

The easiest convertible material for a carrier is an alloy tube. The sketches on pages 108 and 109 demonstrate two popular types, one with a spring-loaded hanger which is retained internally by a

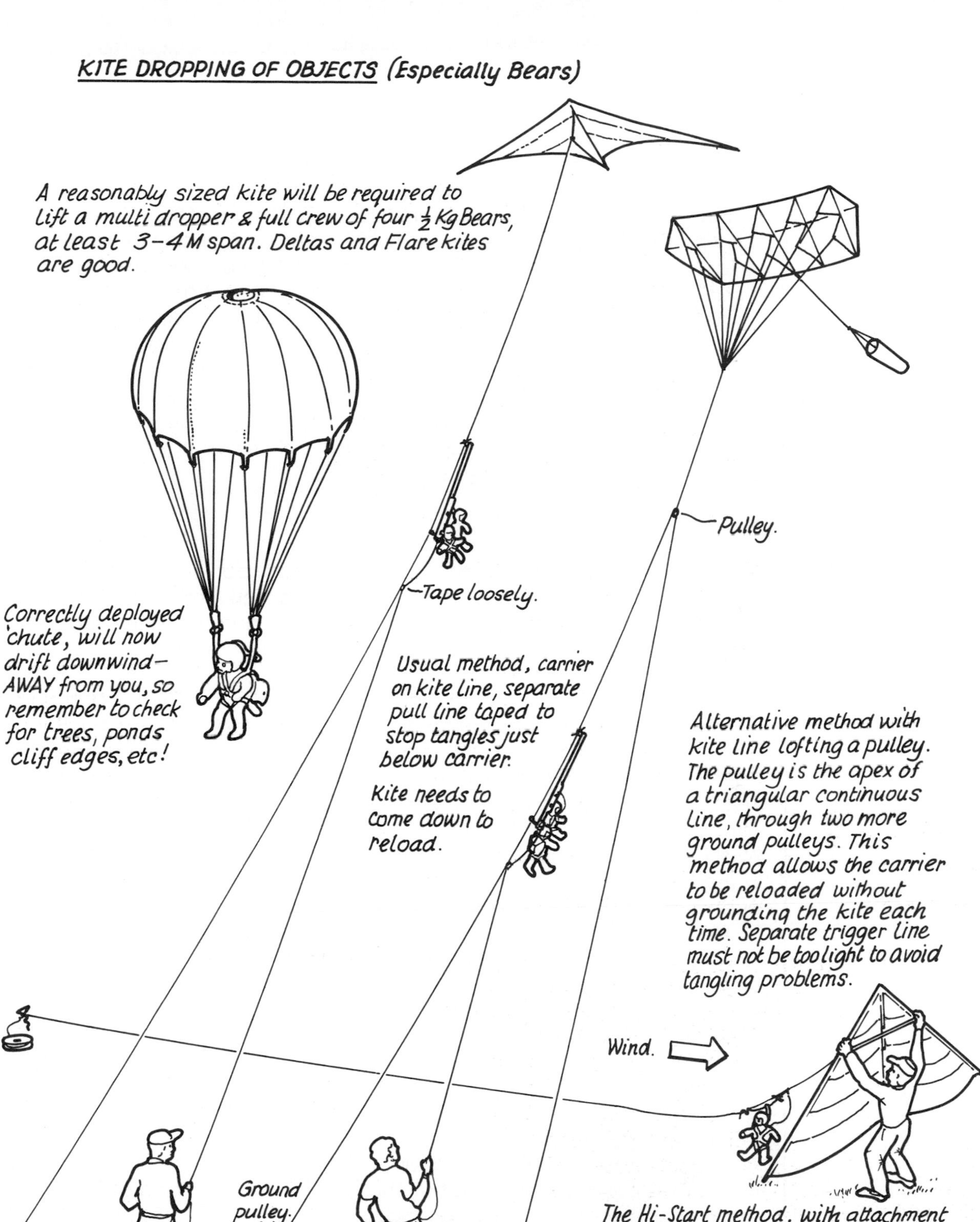
KITE DROPPING OF OBJECTS (Especially Bears)
A reasonably sized kite will be required to lift a multi dropper & full crew of four ½ Kg Bears, at least 3–4 M span. Deltas and Flare kites are good.
Correctly deployed 'chute, will now drift downwind– AWAY from you, so remember to check for trees, ponds cliff edges, etc!
Tape loosely.
Pulley.
Usual method, carrier on kite line, separate pull line taped to stop tangles just below carrier.
Kite needs to come down to reload.
Alternative method with kite line lofting a pulley. The pulley is the apex of a triangular continuous line, through two more ground pulleys. This method allows the carrier to be reloaded without grounding the kite each time. Separate trigger line must not be too light to avoid tangling problems.
Wind.
Ground pulley.
Pulley
The Hi-Start method, with attachment by pipe cleaner. Line anchored up wind, deploy kite and as sail fills out, step back to tauten line & release smartly. Climb to height should shake Bear loose.

split pin anchor, the other with a pull-through which is not spring-loaded but comes right through the tube after releasing a bear squad in any multiple drop. Remember always to attach that static line!

Multi-drops are spectacular. The most we've heard of thus far is a drop of thirty-nine at once (Greg Locke, May 1987) but just four can be exciting, as so far we have not mentioned the influence of the weather on our skydivers. It's an anomaly that a) we want the parabear to descend safely and not too far distant, while b) we have to rely on a fair breeze to elevate the lifting kite and carry the droppable load. Life is all compromise, so a first consideration when siting the launch pad is to ensure that the DZ (drop zone) is clear of hazards like forests, cliffs and rivers likely to cause much grief to the fauna. In actual fact, some well-worn pioneer parabears have survived tramlines, television aerials and town markets, though not without the concern of fire brigades and other uniformed services!

Transporters

The arrival of lightweight, plastic, foam-filled bears sparked off the use of ferry transporters to carry the fauna along the kiteline. This eliminates many of the previously described tasks of arranging a pulley and triangular secondary kiteline system and, in one single action, offers automatic release and return down the line for the next lift.

A complete parabear outfit comprising ferry, 25cm (10in) Teddy and a 64cm (25in) diameter parachute was soon marketed by Greens of Burnley (should that now be Bearnley?) to satisfy demand from those who preferred not to make their own, and to lift the unit they suggested their 2 metre span (79in) Delta and upwards. For the more serious, a twin-sailed transporter had already been devised that was capable of taking loads of around a kilo (2.2lb). Page 93 illustrates the general principle as used by both John Barker in the UK and Jan Fischer, the leading light among bear (also panda and even certain Disney characters) drops, in Holland.

As in other carriers, a tube incorporates a wire hanger, only in this case the release of the bear takes place when the wire hits a line stop with its forward end, and is pushed clear of a slotted tube at the aft end. The main tube has to run freely while suspended on the kiteline, so pulleys are used at each of its ends. To propel the transporter up the line, two sails are suspended from the tube. These are tensioned in the 'down' position by an elastic or bungee, one end of which is hooked on the bowsprit, and the other linked to a trip line which runs through the two sails to the slotted tube at the rear, where it is retained by the release pin.

The sails have mainspars in their upper edges, and these act as pivots where they pass through the tube. Thus, when tensioned and at 90 degrees to the kiteline, the sails catch the wind and will quickly carry the load up to the line stop. On hitting the stop, the wire trip is forced back in relation to the tube and, at the other end of a dowel pushrod, the release pin allows the parabear to fall away and its static line is pulled from the parachute pack. At the same time, the release ring falls free of the slotted tube and the sails collapse

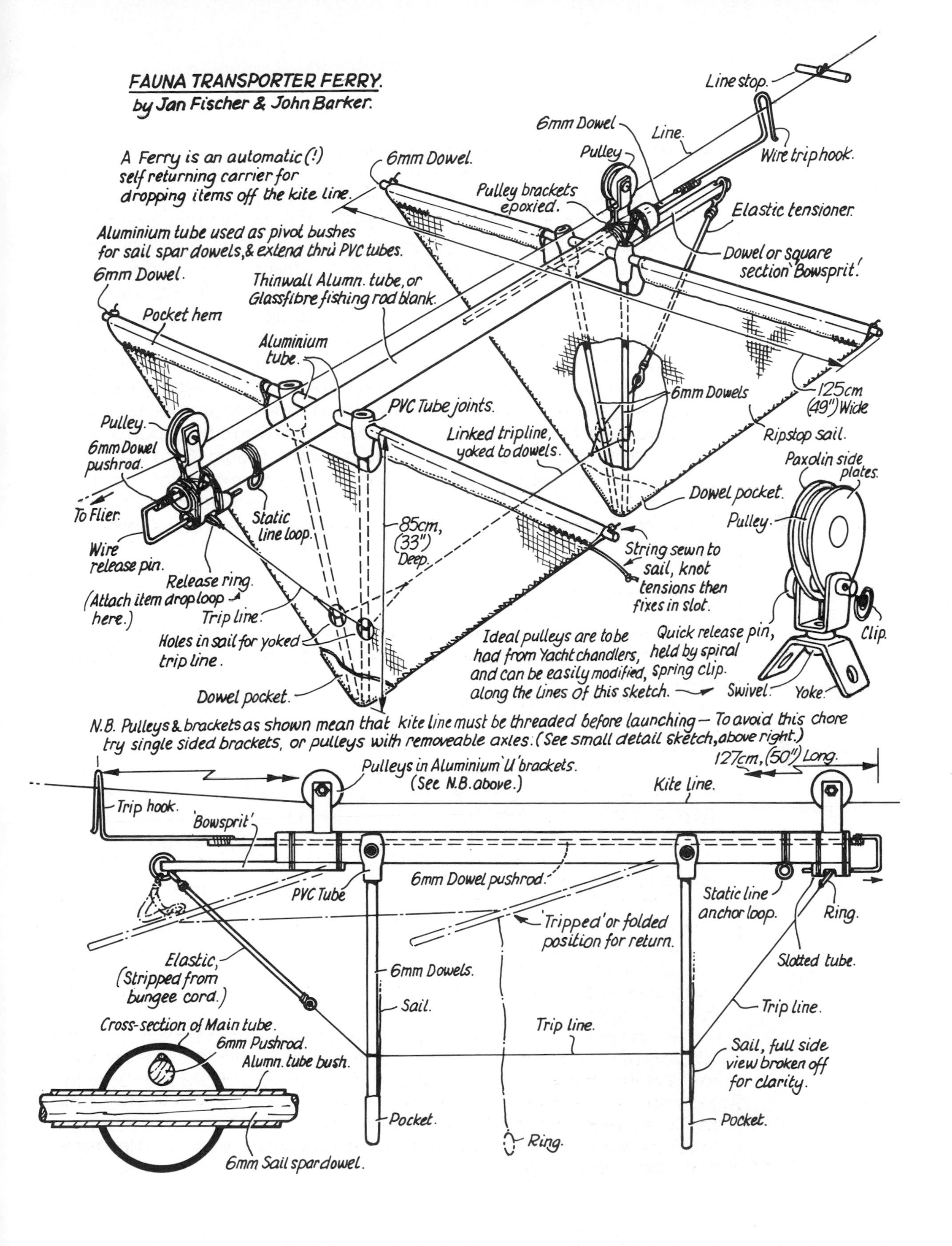

FAUNA TRANSPORTER FERRY.
by Jan Fischer & John Barker.
A Ferry is an automatic (!) self returning carrier for dropping items off the kite line.
Aluminium tube used as pivot bushes for sail spar dowels, & extend thru PVC tubes.
6mm Dowel.
Line stop.
6mm Dowel
Line.
Pulley
Wire trip hook.
Pulley brackets epoxied.
Elastic tensioner.
Dowel or square section 'Bowsprit'.
Thinwall Alumn. tube, or Glassfibre fishing rod blank.
Pocket hem
Aluminium tube.
PVC Tube joints.
6mm Dowels
125cm (49") Wide
Ripstop sail.
Pulley.
6mm Dowel pushrod.
Linked tripline, yoked to dowels.
Paxolin side plates.
Dowel pocket.
Pulley.
To Flier.
Static line loop.
85cm, (33") Deep.
Wire release pin.
Release ring.
(Attach item drop loop here.)
Trip line.
String sewn to sail, knot tensions then fixes in slot.
Holes in sail for yoked trip line.
Ideal pulleys are to be had from Yacht chandlers, and can be easily modified, along the lines of this sketch.
Quick release pin, held by spiral spring clip.
Clip.
Swivel.
Yoke.
Dowel pocket.
N.B. Pulleys & brackets as shown mean that kite line must be threaded before launching — To avoid this chore try single sided brackets, or pulleys with removeable axles: (See small detail sketch, above right.)
127cm, (50") Long.
Pulleys in Aluminium 'U' brackets. (See N.B. above.)
Kite line.
Trip hook.
'Bowsprit'.
6mm Dowel pushrod.
PVC Tube
Static line anchor loop.
Ring.
'Tripped' or folded position for return.
Slotted tube.
Elastic, (Stripped from bungee cord.)
6mm Dowels.
Sail.
Trip line.
Trip line.
Cross-section of Main tube.
6mm Pushrod.
Alumn. tube bush.
Sail, full side view broken off for clarity.
Pocket.
Pocket.
Ring.
6mm Sail spar dowel.

forward, so permitting the transporter to dump its effective lift, and then descend to the ground along the kiteline.

As well as the need to make the transporter as light as possible, commensurate with the strength required in the mainspars of the sails and the rigidity of the tube, you can employ a wide variety of materials in the transporter. John Barker based his on a discarded first joint of a fishing rod, stripping off the cork handle and ferrules for the main tube, and trying both wooden dowels and glass fibre for the sail spars. Simplest of all is the Brookite 'Paraman' unit which comes in pairs for a small outlay and is adaptable for dropping light loads. Jan Fischer produced a transporter that might well have come from a professional machine shop. They each work extremely well. A visit to the DIY store will suggest everyday accessories that will be easily adapted to the tube and its pulleys, while the sailspars are best made from carbon fibre tube.

Typical sail dimensions to lift up to one kilo (2.21b) would be a span or width across the mainspar of 125cm (49in) and a depth of 85cm (33in) so that the triangle is an included angle of 73 degrees approximately at the apex; however, this is not critical. Much depends on the wind strength of the day. These dimensions will offer just over a square metre or 11sq ft of sail, more than enough for a kilobear.

Best-dressed Ted

Like his big brother paratroopers, our favourite Teddy deserves a decent flying outfit, if only to give extra comfort on the occasion of that terrifying first drop into the void.

The parachute has to be packed into a realistic pack - no self-respecting dropnik could be expected to cast itself off with a bundle of nylon in its arms. Then the pack should have a harness, with over-arm and under-crotch straps linked to a waistband with quick release fastener.

A white scarf, preferably ex-real parachute, will show firm intent and act as a morale booster, and goggles are a must if our Teddy is at all myopic and wears specs or contact lenses. Even so, a perfectly-sighted Ted can suffer with streaming eyes for several reasons during the fall, so best play safe and make him some goggles from a pop bottle, as shown opposite. They could be ready tinted! Cheap swimming goggles fit Whopperbears but don't let it be tempted to go sub-aqua!

Then there is that essential helmet. Take a suitably sized plastic beach ball, cutaway for the face from nape of neck to forehead with nice curvy lines, and carefully estimate where the fluffy ears will have to project and trim the earholes tight. They're going to help keep the helmet in place, at the same time exposing the aurii to the full. They will act as stabilisers, beside being handy handles for recovery from bushes, marshland or similar enemy territory outside the DZ.

Ted devils on parade after a spectacular descent. 'Orace has yet to recover, and his companion's Pilot's wings have slipped to tummy level!

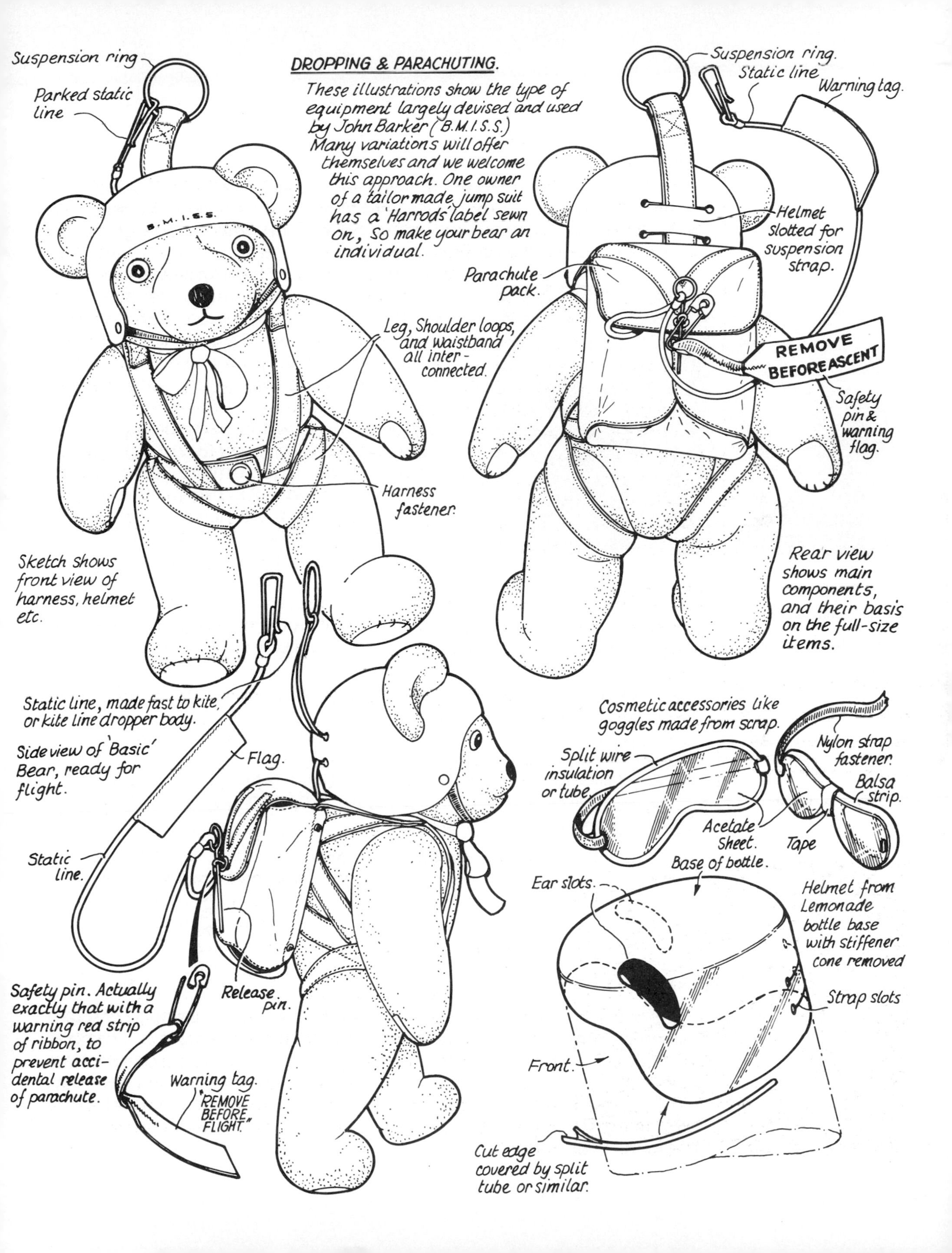
DROPPING & PARACHUTING.
These illustrations show the type of equipment largely devised and used by John Barker (B.M.I.S.S.) Many variations will offer themselves and we welcome this approach. One owner of a tailor made jump suit has a 'Harrods' label sewn on, So make your bear an individual.
Suspension ring
Parked static line
B.M.I.S.S.
Leg, Shoulder loops, and waistband all inter-connected.
Harness fastener.
Sketch shows front view of harness, helmet etc.
Suspension ring.
Static line
Warning tag.
Helmet slotted for suspension strap.
Parachute pack.
REMOVE BEFORE ASCENT
Safety pin & warning flag.
Rear view shows main components, and their basis on the full-size items.
Static line, made fast to kite, or kite line dropper body.
Side view of 'Basic' Bear, ready for flight.
Flag.
Static line.
Safety pin. Actually exactly that with a warning red strip of ribbon, to prevent accidental release of parachute.
Release pin.
Warning tag.
"REMOVE BEFORE FLIGHT."
Cosmetic accessories like goggles made from scrap.
Split wire insulation or tube
Nylon strap fastener.
Balsa strip.
Acetate Sheet.
Tape
Base of bottle.
Ear slots.
Helmet from Lemonade bottle base with stiffener cone removed
Strap slots
Front.
Cut edge covered by split tube or similar.

The `chute

We will want our fauna to enjoy his fall (well, at least *after* that first moment of terror!) and eventual arrival on terra firma must be reasonably safe and soft. So the size of the parachute will have to be determined against the weight of Ted and the attendant equipment.

They luv their bears! Two intrepid enthusiasts after recovery from a drop at Old Warden.

We can draw upon the experience of Hugh Andrew to work out the theoretical 'flat' size of the `chute. We all know that parachutes form themselves into a hemispherical shape as they fill with air; but, for simplicity, it is better if we follow Hugh's example and assume that we are dealing with a flat disc.

Our concern will be the disc-loading. Experience shows that a 230g fauna (8oz) should have a parachute of about 1 square metre (10 ¾ sq ft). The disc-loading is, therefore, 230g/sq m or 0.75oz/sq ft. To obtain this loading, we need a 'flat' size of 113cm diameter (44 ½ in). This is for a moderate rate of descent. By increasing the parachute diameter, the disc-loading is reduced, and the rate of descent is slower. Conversely, for faster descent, the area is reduced. We emphasise that the figures quoted are not in any way critical. It would take a lot of variation to show serious differences in performance, in fact one could `round down' the basic diameter to one metre (39in) and, with a given 230g or 8oz load, the loading will only vary to 294g/sq m or 0.95oz/sq ft which in any circumstances is very light loading indeed.

Similarly, the nominal 230g or 8 ounces can vary by 30g or an ounce or so, and only affect the loading by one-tenth. So it will be obvious that we have considerable flexibility in determining the size of parachute.

As a general guide for other weights, using tables produced by Hugh Andrew, we arrive at the following:

Bear Weight		Rates of descent / diameters					
Grams	Oz	Slow(cm)	Slow(in)	Medium(cm)	Medium(in)	Fast(cm)	Fast(in)
100	3.5	80	31 1/2	71	28	65	25
250	8.8	126	50	112	44	102	40
350	12.3	150	59	133	52	120	47
500	17.6	170	67	160	63	144	56
750	26.5	218	85	194	76	177	70
1 Kilo	35.3	252	99	224	88	204	80

All of these sizes are theoretical and err on the generous side to ensure a safe and floating descent, but they cannot take into account any climatic influence. Take them as a sea-level baseline. In hot and high areas where air density is much lower, performance will be different, with faster descents. On the other hand, thermal conditions can apply, and at light loadings as prescribed, a powerful thermal can take over and give Teddy an extra thrill of going up instead of down.

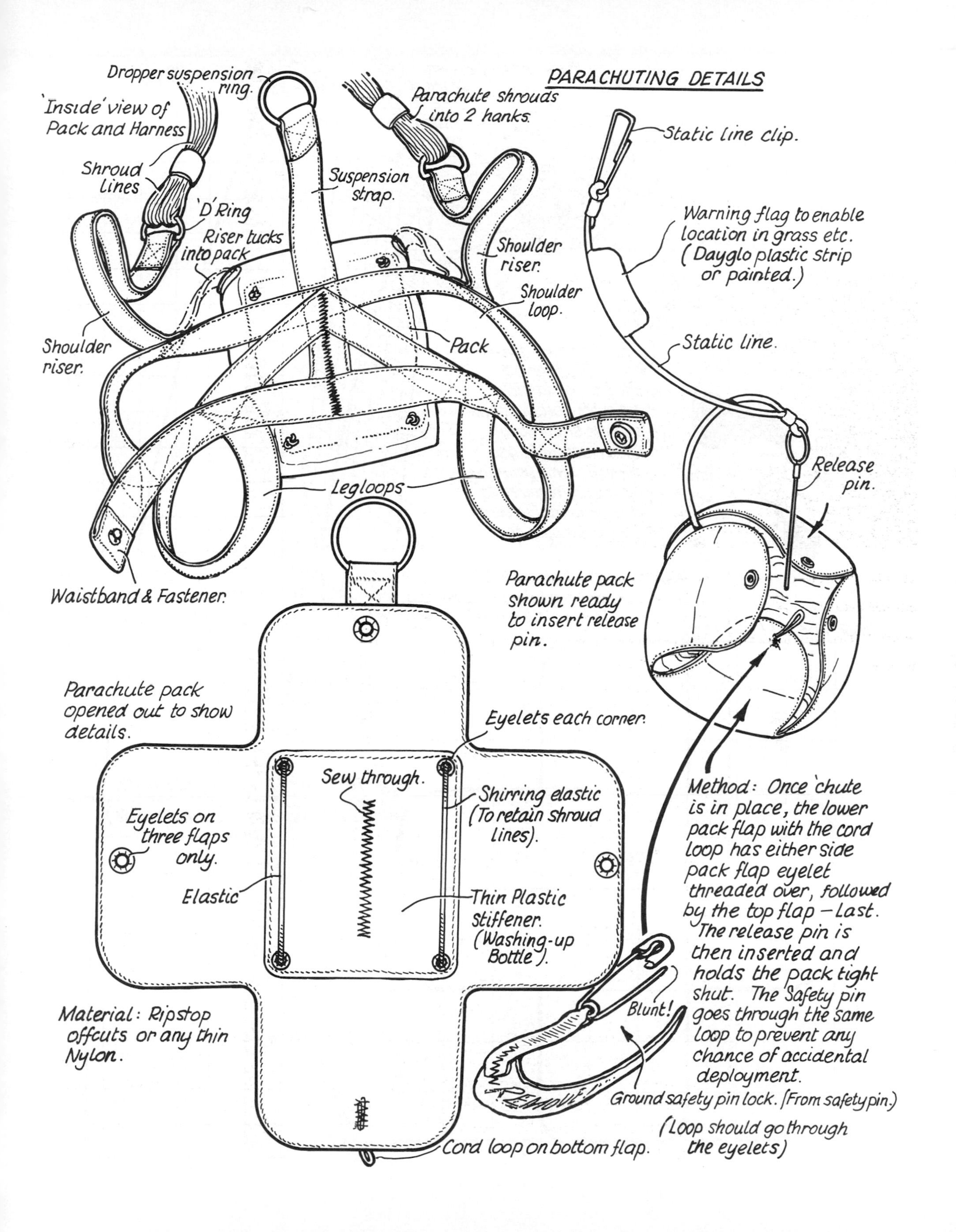

PARACHUTING DETAILS
Dropper suspension ring.
'Inside' view of Pack and Harness
Parachute shrouds into 2 hanks.
Static line clip.
Shroud lines
Suspension strap.
'D' Ring
Riser tucks into pack
Shoulder riser.
Warning flag to enable location in grass etc. (Dayglo plastic strip or painted.)
Shoulder loop.
Shoulder riser.
Pack
Static line.
Release pin.
Legloops
Waistband & Fastener.
Parachute pack shown ready to insert release pin.
Parachute pack opened out to show details.
Eyelets each corner.
Sew through.
Shirring elastic (To retain shroud lines).
Eyelets on three flaps only.
Method: Once 'chute is in place, the lower pack flap with the cord loop has either side pack flap eyelet threaded over, followed by the top flap – Last. The release pin is then inserted and holds the pack tight shut. The Safety pin goes through the same loop to prevent any chance of accidental deployment.
Elastic
Thin Plastic Stiffener. (Washing-up Bottle).
Blunt!
Material: Ripstop offcuts or any thin Nylon.
REMOVE!
Ground safety pin lock. (From safety pin.)
(Loop should go through the eyelets)
Cord loop on bottom flap.

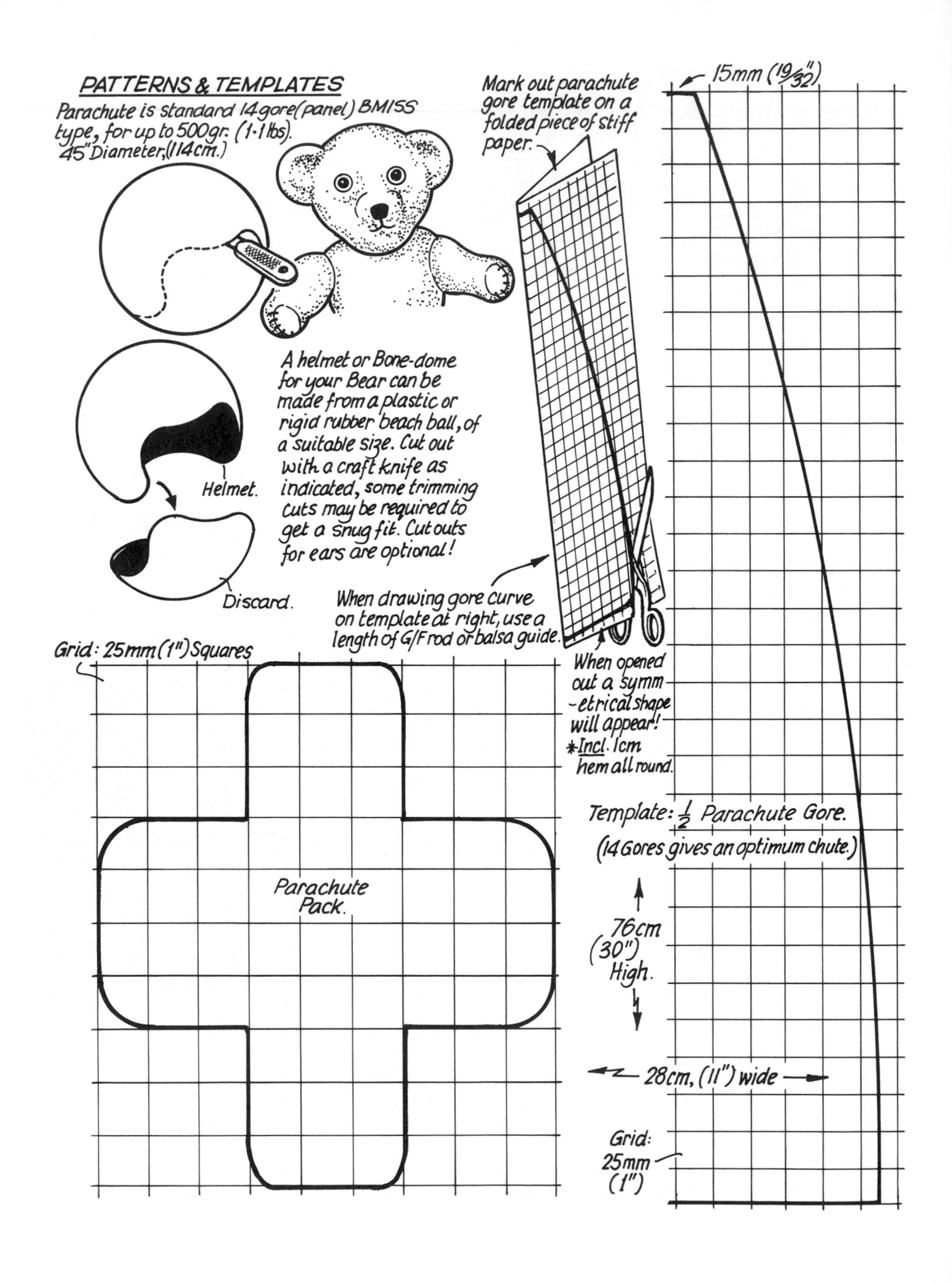
PATTERNS & TEMPLATES
Parachute is standard 14 gore (panel) BMISS type, for up to 500gr. (1·1 lbs). 45" Diameter, (114cm.)
Mark out parachute gore template on a folded piece of stiff paper.
15mm (19/32")
A helmet or Bone-dome for your Bear can be made from a plastic or rigid rubber beach ball, of a suitable size. Cut out with a craft knife as indicated, some trimming cuts may be required to get a snug fit. Cut outs for ears are optional!
Helmet.
Discard.
When drawing gore curve on template at right, use a length of G/F rod or balsa guide.
Grid: 25mm (1") Squares
When opened out a symm-etrical shape will appear!
*Incl. 1cm hem all round.
Template: 1/2 Parachute Gore.
(14 Gores gives an optimum chute.)
Parachute Pack.
76cm (30") High.
28cm, (11") wide
Grid: 25mm (1")

The considerable flexibility mentioned earlier could not be better illustrated than through one of John Barker's much-used parachutes. This results in a 'chute with 14 'gores' (see below) and a diameter of between 115cm (45in) and 123cm (48in). We say 'between' because there will always be a variation in the hem widths of the 14 gores when they are sewn together.

Reference to Hugh Andrew's tables will show that such a size of 'chute will give a fast descent when carrying 350g or 12.3oz, yet John Barker regularly operates with 450g (15.8oz) fauna, arriving at his parachute size by good old eyeball engineering.

So for starters, John's standard BMISS canopy is a fine practical size that suits a wide range of parabear weights. Scaling up or down on the 'chute diameter now raises another question, the number of gores to use.

Gores

To form the hemispherical shape of the traditional cupped parachute, the surface is divided into what are known as gores. Each will be of equal width and length and the shape is created by a curved taper which will be directly related to the circumference of the 'chute at each stage as it progresses from the open mouth to the crown.

To obtain the cupped or hemispherical shape with fairly smooth lines, the number of gores should be increased with the diameter. Starting with the smallest at say 65cm (25½ in) diameter we should have at least 10 gores, and for the huge 204cm(80in) 'chute 18 or even 20 gores. Alternating the colour of the gores is an added attraction; but beware having too many hems, as each overlap and its stitching will add weight.

So far, we have carefully avoided any suggestion of calculation, and we can still stay away from arithmetic by using tables for gore heights as they apply to a range of 'chute diameters.

Diameter		Gore height	
cm	in	cm	in
65	25 ½	47	18 ½
100	39	76	30
150	59	115	45
200	78	152	60

They do not *have* to be bears! Jan Fischer in Holland drops cartoon characters - even that Turtle!

To obtain the base width of the gores (minus any hem), the circumference (3.1412 x diameter) must be divided by the number of gores. So, a 1 metre (39in) parachute has a circumference of 314cm (123in) and, if we are to use 14 gores, then each must be 22.4cm (8 ¾in) wide at the base *plus* the hem which will be about 1cm or 0.4in.

There are ways of plotting the profile of the gore, and the mathematically minded will no doubt enjoy a computer exercise of setting up the progressively diminishing diameters, and the calculation of gore width at 10% stages from base to apex. As long as one allows a snub-ended top of 2cm (0.8in) on smaller and 3cm (1.2in) for larger 'chutes so that an air bleed hole appears in the top, then it is simple to sketch a curved profile.

The easiest approach is to plan the gore as one half only on a folded sheet of paper, then to cut out the pattern through the two leaves, open out flat and transfer to a piece of stout card or hardboard, With an average of 14 to cut, did you remember to include the hem width? It will pay to have a tough template for cutting by scalpel, hot iron or roller-cutter. Material is best if unproofed nylon or silk.

Shrouds

No parachute will deploy satisfactorily unless it has a set of equal length shroud lines of adequate length, and divided into two hanks for attachment to the harness.

When the gores have been sewn together, shroud loops are added at each hem, that is, one for each hem. On the BMISS standard 'chute, the lines are trimmed to 119cm (47in) length before sewing, or gluing with epoxy to the harness. On a smaller, 65cm (25½ in) parachute the shrouds can be 90cm (36in), and for a 2metre (78in) 'chute, up to the same dimensions, 2m or 78in.

The central vent is important, so make it neat by trimming to a circle and reinforce with stitching which will help the overall strength.

Pack and harness

Khaki, or at least dark brown ripstop offcuts, makes a nostalgic parachute pack and harness. This is the 'scale' colour, as first used by 'hume' paras; in fact, the whole pack is a scaled-down version of the real thing, even to the eyeletted flaps for the quick release pin to secure. Nowadays, parapacks come in all colours.

There are a couple of changes made necessary because our fauna have to be tagged to a carrier and use a static line for opening the pack, unlike 'hume' paras who can cast themselves off and tug at their own ripcords. So we have an extra strap on the harness, running vertically upwards, with a suspension ring at the end. For several reasons, one being to make sure our Ted hangs vertically from the carrier, it is best to pass this dropper strap through a couple of slots, in and out of the helmet.

The other parafauna necessity is the static line which is clipped to the carrier or transporter *and stays there*. This carries the release pin and, from that, a second line which we will explain later. Study of Pat's sketches will readily show how the lower flap has a loop permanently attached to it. The release pin holds the pack closed as it passes through this loop after securing the eyelets. On the ground, while Ted awaits the launch, a safety pin with warning flag is passed through the loop to prevent any premature opening.

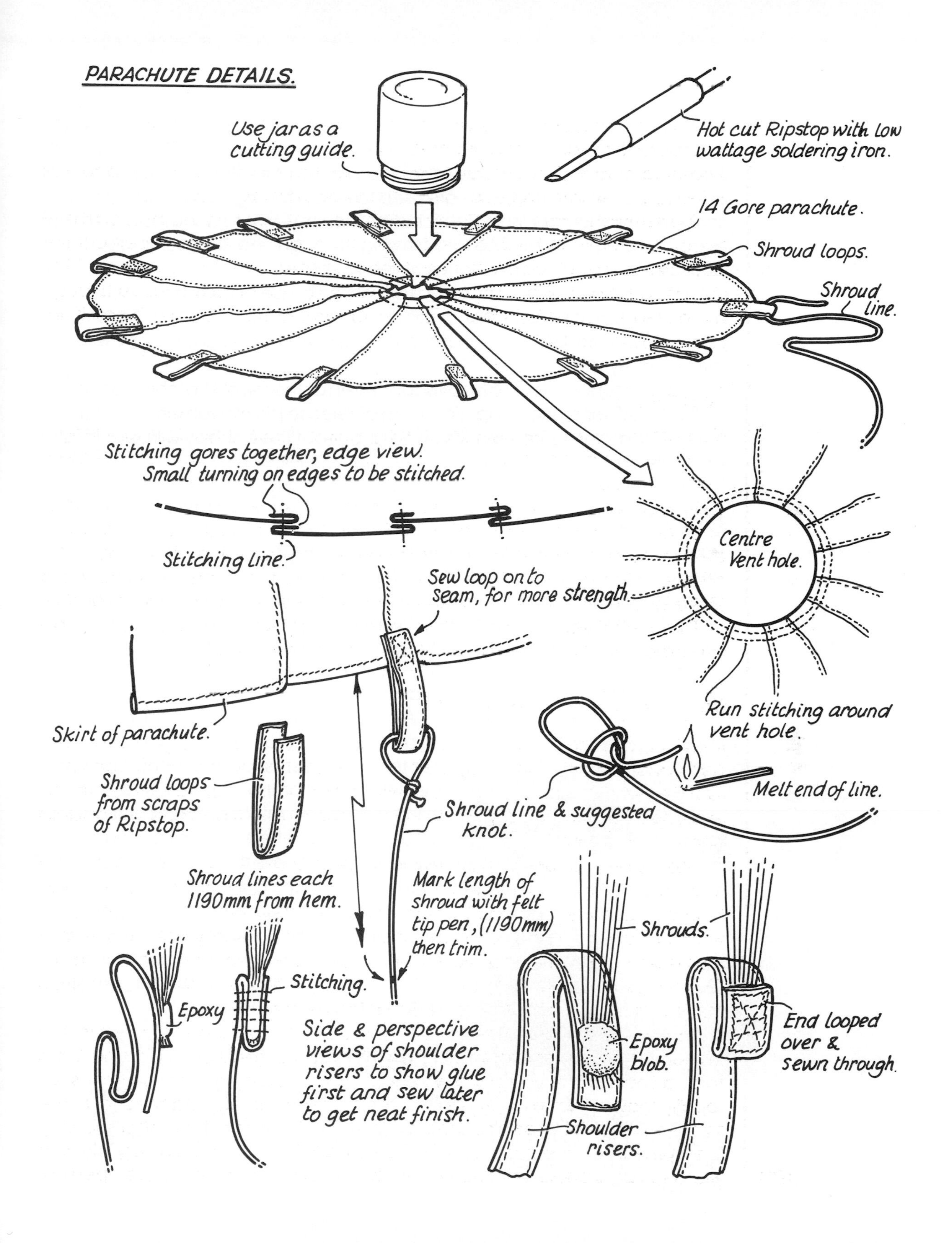

PARACHUTE DETAILS.
Use jar as a cutting guide.
Hot cut Ripstop with low wattage soldering iron.
14 Gore parachute.
Shroud loops.
Shroud line.
Stitching gores together, edge view. Small turning on edges to be stitched.
Stitching line.
Centre Vent hole.
Sew loop on to seam, for more strength.
Run stitching around vent hole.
Skirt of parachute.
Shroud loops from scraps of Ripstop.
Melt end of line.
Shroud line & suggested knot.
Shroud lines each 1190mm from hem.
Mark length of shroud with felt tip pen, (1190mm) then trim.
Shrouds.
Stitching.
Epoxy
Side & perspective views of shoulder risers to show glue first and sew later to get neat finish.
Epoxy blob.
End looped over & sewn through.
Shoulder risers.

Parapacking

Two elasticated loops, made of shirring elastic or similar - but certainly not tight across the inner eyelets - are used to retain the zig-zag of folded shroud lines over the shoulder risers. The 'chute is then pulled into a sausage, with that extra line from the release pin passing through the vent and down the parachute, leaving a few inches of line to spare.

By folding the chute concertina fashion, it can be packed relatively tightly under the pack flaps in the order of bottom, sides, then top, before the release pin is secured through the cord loop, after passing through the other three flap eyelets.

A few ground tests will be advisable, even at the risk of stretching patience with consequent repacking. Freedom of release pin movement and careful fitting of that cord extension which helps pull out the canopy will benefit from practice.

Parafoils

Brookite's extremely simple plastic carrier for dropping small 'chutes that in turn act as a messenger to pull the unit up the line.

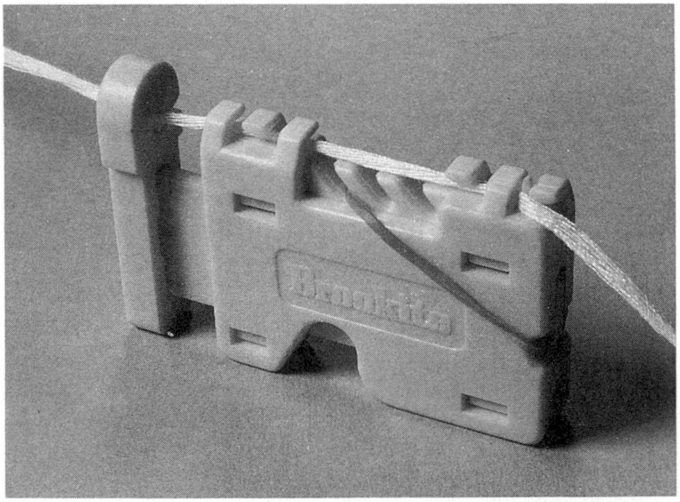

Round 'chutes have all but disappeared, except for supply drops and emergencies in the big world outside bearland, so it is only natural that the popular Parafoil should emerge as a fauna canopy. This is for the really keen seamstressers, as it calls for some tricky work assembling five risers or ribs within the top and bottom panels, and two combined riser/keels at either side. There's also the little matter of thirty-three shroud lines loops! The dimensions are correct to give a nose-down angle with anhedralled under surface.

All the details are shown on page 104. Again, we advise un-proofed material in order to obtain fastest deployment. Light silk is the best, as it compresses well and then expands rapidly on release. Packing a parafoil 'chute is a little different. It is best collapsed sideways so that is is the shape of one airfoil end. Then fold in a concertina from the rear, keeping all the shrouds straight. These are, in turn, packed as for a round parachute. Sometimes it is better to make the parafoil and fold it *before* the pack is made; then create a paper pattern around it.

The lifter

As with the camera units, we need a stable platform, so a Stratoscoop, Tri-D, or large Delta would be satisfactory. Ted Fleming, a founder of the Great Ouse Kite Flyers, centred on the river of that name, developed his concept of a super-stable Delta which he called a 'Double Ram Delta'.

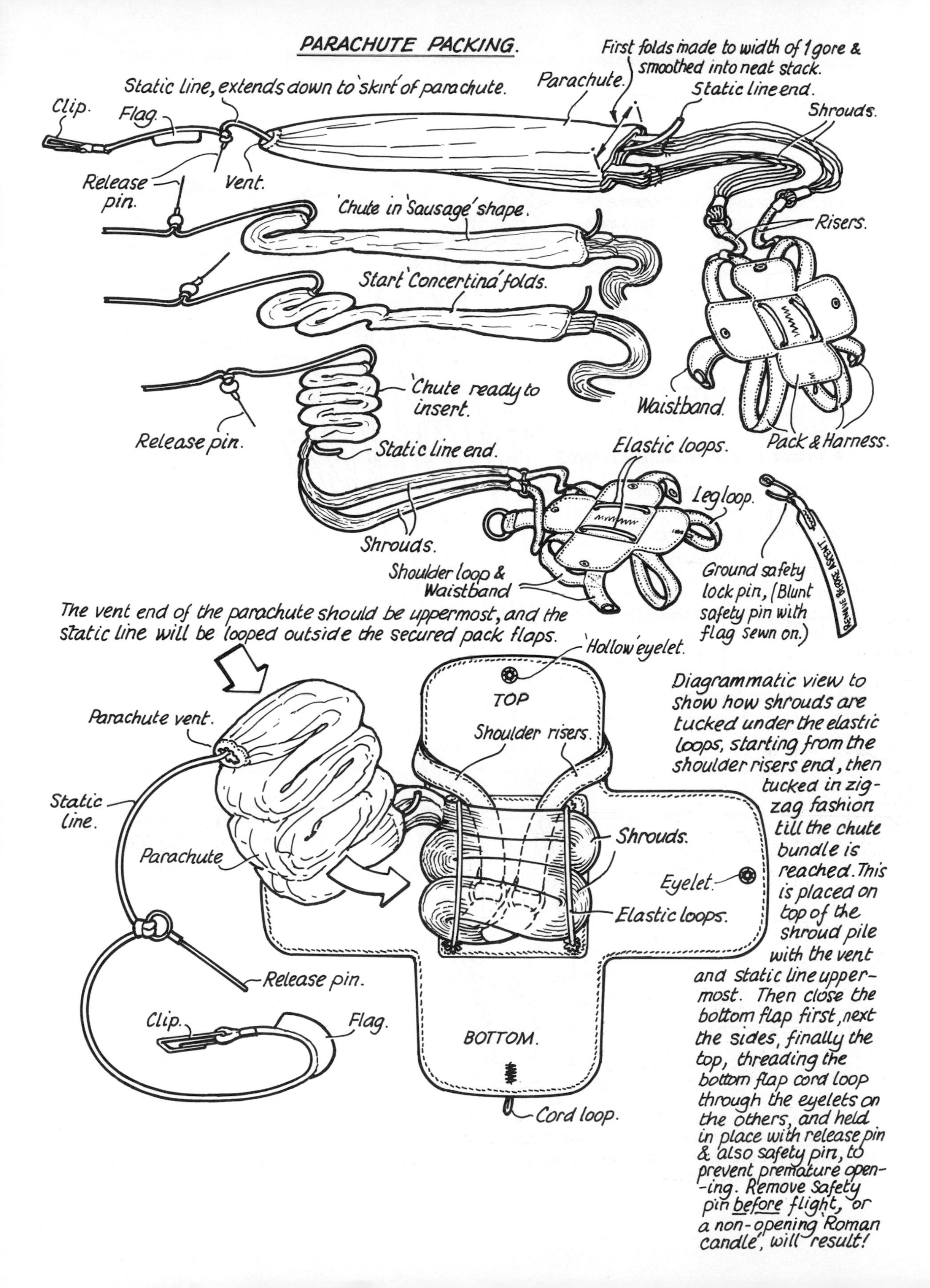
PARACHUTE PACKING.
First folds made to width of 1 gore & smoothed into neat stack.
Static line, extends down to 'skirt' of parachute.
Parachute.
Static line end.
Clip.
Flag.
Shrouds.
Release pin.
Vent.
'Chute in 'Sausage' shape.
Risers.
Start 'Concertina' folds.
'Chute ready to insert.
Waistband.
Release pin.
Static line end.
Elastic loops.
Pack & Harness.
Leg loop.
Shrouds.
Shoulder loop & Waistband
Ground safety lock pin, (Blunt safety pin with flag sewn on.)
The vent end of the parachute should be uppermost, and the static line will be looped outside the secured pack flaps.
'Hollow' eyelet.
TOP
Diagrammatic view to show how shrouds are tucked under the elastic loops, starting from the shoulder risers end, then tucked in zig-zag fashion till the chute bundle is reached. This is placed on top of the shroud pile with the vent and static line uppermost. Then close the bottom flap first, next the sides, finally the top, threading the bottom flap cord loop through the eyelets on the others, and held in place with release pin & also safety pin, to prevent premature opening. Remove Safety pin before flight, or a non-opening 'Roman candle', will result!
Parachute vent.
Shoulder risers.
Static line.
Shrouds.
Parachute
Eyelet.
Elastic loops.
Release pin.
Clip.
Flag.
BOTTOM.
Cord loop.

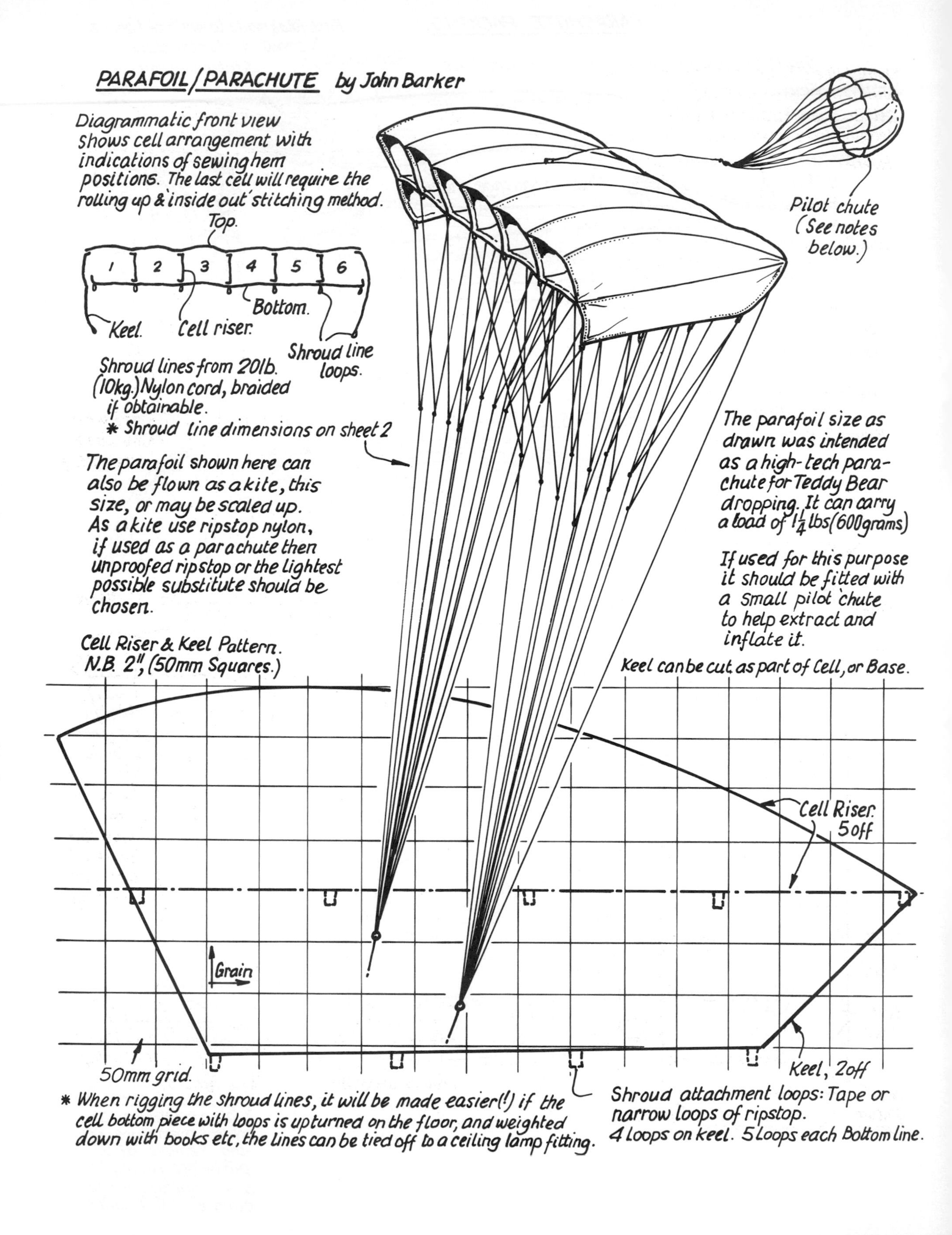
PARAFOIL/PARACHUTE by John Barker
Diagrammatic front view shows cell arrangement with indications of sewing hem positions. The last cell will require the rolling up & 'inside out' stitching method.
Top.
1
2
3
4
5
6
Bottom.
Keel.
Cell riser.
Shroud line loops.
Pilot chute (See notes below.)
Shroud lines from 20lb. (10kg.) Nylon cord, braided if obtainable.
* Shroud line dimensions on sheet 2
The parafoil shown here can also be flown as a kite, this size, or may be scaled up. As a kite use ripstop nylon, if used as a parachute then unproofed ripstop or the lightest possible substitute should be chosen.
The parafoil size as drawn was intended as a high-tech parachute for Teddy Bear dropping. It can carry a load of 1¼ lbs (600 grams)
If used for this purpose it should be fitted with a small pilot 'chute to help extract and inflate it.
Cell Riser & Keel Pattern. N.B. 2", (50mm Squares.)
Keel can be cut as part of Cell, or Base.
Cell Riser. 5 off
Grain
Keel, 2 off
50mm grid.
* When rigging the shroud lines, it will be made easier(!) if the cell bottom piece with loops is upturned on the floor, and weighted down with books etc, the lines can be tied off to a ceiling lamp fitting.
Shroud attachment loops: Tape or narrow loops of ripstop.
4 Loops on keel. 5 Loops each Bottom line.

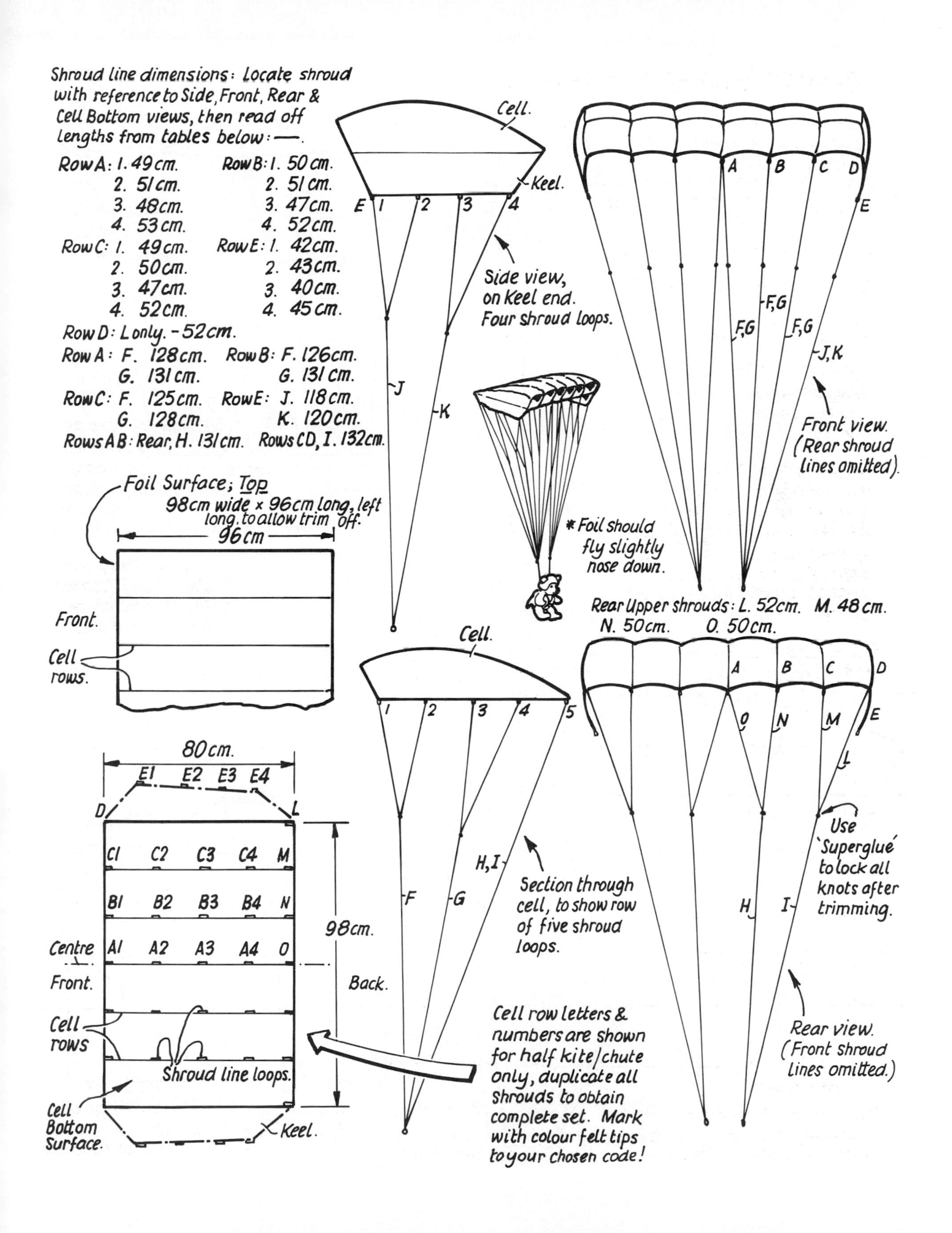

Shroud line dimensions: Locate shroud with reference to Side, Front, Rear & Cell Bottom views, then read off lengths from tables below:—.
Row A: 1. 49cm. 2. 51cm. 3. 48cm. 4. 53cm.
Row B: 1. 50cm. 2. 51cm. 3. 47cm. 4. 52cm.
Row C: 1. 49cm. 2. 50cm. 3. 47cm. 4. 52cm.
Row E: 1. 42cm. 2. 43cm. 3. 40cm. 4. 45cm.
Row D: L only. - 52cm.
Row A: F. 128cm. G. 131cm.
Row B: F. 126cm. G. 131cm.
Row C: F. 125cm. G. 128cm.
Row E: J. 118cm. K. 120cm.
Rows AB: Rear, H. 131cm.
Rows CD, I. 132cm.
Cell.
Keel.
E 1 2 3 4
J
K
Side view, on Keel end. Four shroud loops.
A B C D E
F,G
F,G
F,G
J,K
Front view. (Rear shroud lines omitted).
* Foil should fly slightly nose down.
Foil Surface; Top
98cm wide x 96cm long, left long to allow trim off.
96cm
Front.
Cell rows.
Rear Upper shrouds: L. 52cm. M. 48cm. N. 50cm. O. 50cm.
Cell.
1 2 3 4 5
F G H,I
Section through cell, to show row of five shroud loops.
A B C D
O N M E
L
H I
Use 'Superglue' to lock all knots after trimming.
Rear view. (Front shroud lines omitted.)
80cm.
E1 E2 E3 E4
D L
C1 C2 C3 C4 M
B1 B2 B3 B4 N
A1 A2 A3 A4 O
Centre
Front.
Back.
98cm.
Cell rows
Shroud line loops.
Cell Bottom Surface.
Keel.
Cell row letters & numbers are shown for half kite/chute only, duplicate all shrouds to obtain complete set. Mark with colour felt tips to your chosen code!

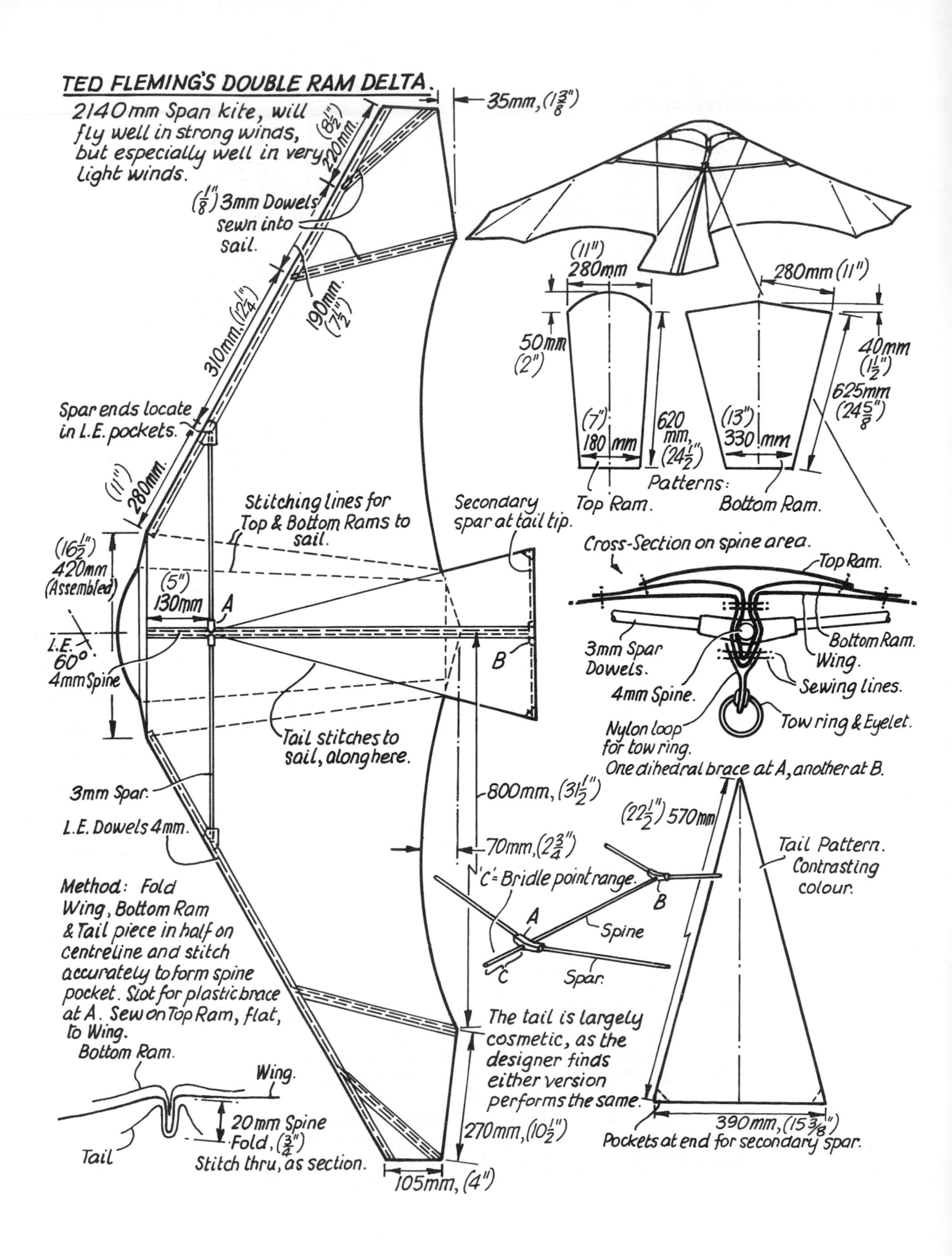
TED FLEMING'S DOUBLE RAM DELTA.
2140mm Span kite, will fly well in strong winds, but especially well in very light winds.
35mm, (1 3/8")
(8 1/2") 220mm.
(1/8") 3mm Dowels sewn into sail.
190mm. (7 1/2")
310mm. (12 1/4")
Spar ends locate in L.E. pockets.
(11") 280mm.
(16 1/2") 420mm (Assembled)
(5") 130mm
A
L.E. 60°
4mm Spine
Stitching lines for Top & Bottom Rams to sail.
Secondary spar at tail tip.
B
Tail stitches to sail, along here.
3mm Spar.
L.E. Dowels 4mm.
800mm, (31 1/2")
70mm, (2 3/4")
Method: Fold Wing, Bottom Ram & Tail piece in half on centreline and stitch accurately to form spine pocket. Slot for plastic brace at A. Sew on Top Ram, flat, to Wing.
Bottom Ram.
Wing.
Tail
20mm Spine Fold, (3/4")
Stitch thru, as section.
105mm, (4")
270mm, (10 1/2")
The tail is largely cosmetic, as the designer finds either version performs the same.
(11") 280mm
50mm (2")
(7") 180 mm
620 mm, (24 1/2")
280mm (11")
40mm (1 1/2")
625mm (24 5/8")
(13") 330 mm
Patterns:
Top Ram.
Bottom Ram.
Cross-Section on spine area.
Top Ram.
3mm Spar Dowels.
Bottom Ram.
Wing.
4mm Spine.
Sewing lines.
Nylon loop for tow ring.
Tow ring & Eyelet.
One dihedral brace at A, another at B.
'C'= Bridle point range.
B
A
Spine
C
Spar.
(22 1/2") 570mm
Tail Pattern. Contrasting colour.
390mm, (15 3/8")
Pockets at end for secondary spar.

Its real advantage is performance through a range of wind speeds, the battened tips on its 214cm (84¼ in) span contributing to the lateral stability while the traditionally flexing leading edges of the Delta shape pivot about the cross-spar. A tail adds to the bird-like configuration and, to effect the double ram feature, Ted introduced two additional layers of ripstop nylon at the centre section. These form open-fronted pockets, held by a run of stitches on their outer sides to a symmetrical pattern. The bottom ram is part of the spine assembly, so that it creates two 'nostrils' beneath the top ram. Ted Fleming's original has been copied many times, always with appreciated success, so we have no hesitation in recommending it as a lifter for any parafauna that is within reasonable weight limits.

Which brings us to one vital and final point on the matter of taking our fauna aloft and then casting them off to the winds not knowing exactly where they are going to land. If there is enough wind strength to take the lifting kite to almost any height, there will be a temptation to make a high release. *Don't.* There's no reason for launches higher than 200 metres (650ft) and then only for a parafoil type which needs the initial fall to take in air and inflate. Any standard round canopy will drift at wind speed as soon as that release pin opens the pack, and that means a chase if you take the carrier to any great height. If the wind at a meteorological height of 10 metres (33ft) is gusting up to 32kmh (20mph), it will be 48kmh (30mph) at 60 metres (200ft) and pro-rata, so be satisfied with drops from moderate heights.

Above: Ted Fleming of Great Ouse Kite Flyers, Bedfordshire, devised the double ram Delta, originally designed as a thermal soaring kite.

Left: The Parafoils made by Doug Hagaman of Westport, Washington State, are very large and can lift considerable weights. This one is used by the White Horse Group, here at Weymouth.

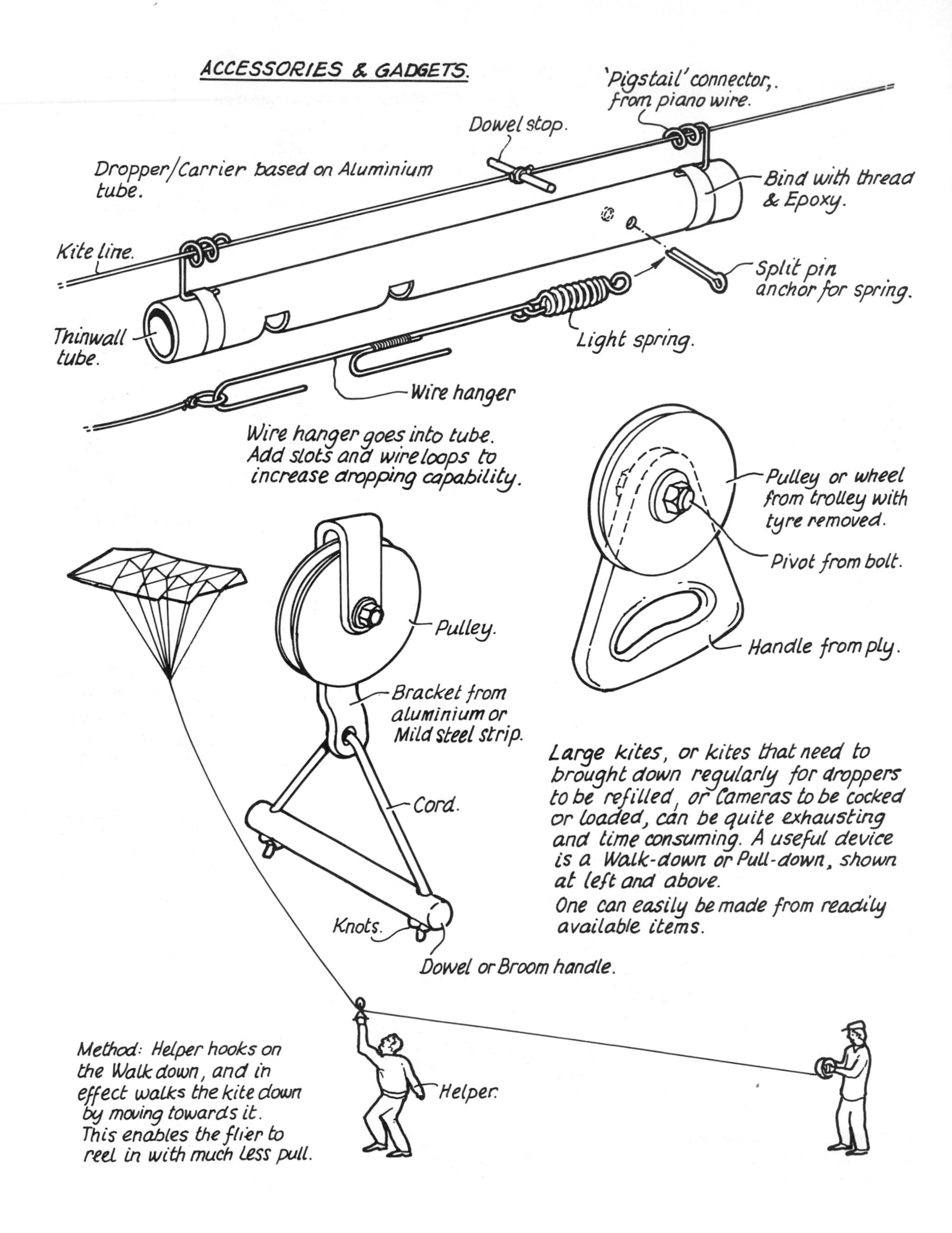
ACCESSORIES & GADGETS.
'Pigstail' connector, from piano wire.
Dowel stop.
Dropper/Carrier based on Aluminium tube.
Bind with thread & Epoxy.
Kite line.
Split pin anchor for spring.
Thinwall tube.
Light spring.
Wire hanger
Wire hanger goes into tube. Add slots and wire loops to increase dropping capability.
Pulley or wheel from trolley with tyre removed.
Pivot from bolt.
Pulley.
Handle from ply.
Bracket from aluminium or Mild steel strip.
Large kites, or kites that need to brought down regularly for droppers to be refilled, or Cameras to be cocked or loaded, can be quite exhausting and time consuming. A useful device is a Walk-down or Pull-down, shown at left and above.
One can easily be made from readily available items.
Cord.
Knots.
Dowel or Broom handle.
Method: Helper hooks on the Walk down, and in effect walks the kite down by moving towards it. This enables the flier to reel in with much less pull.
Helper.

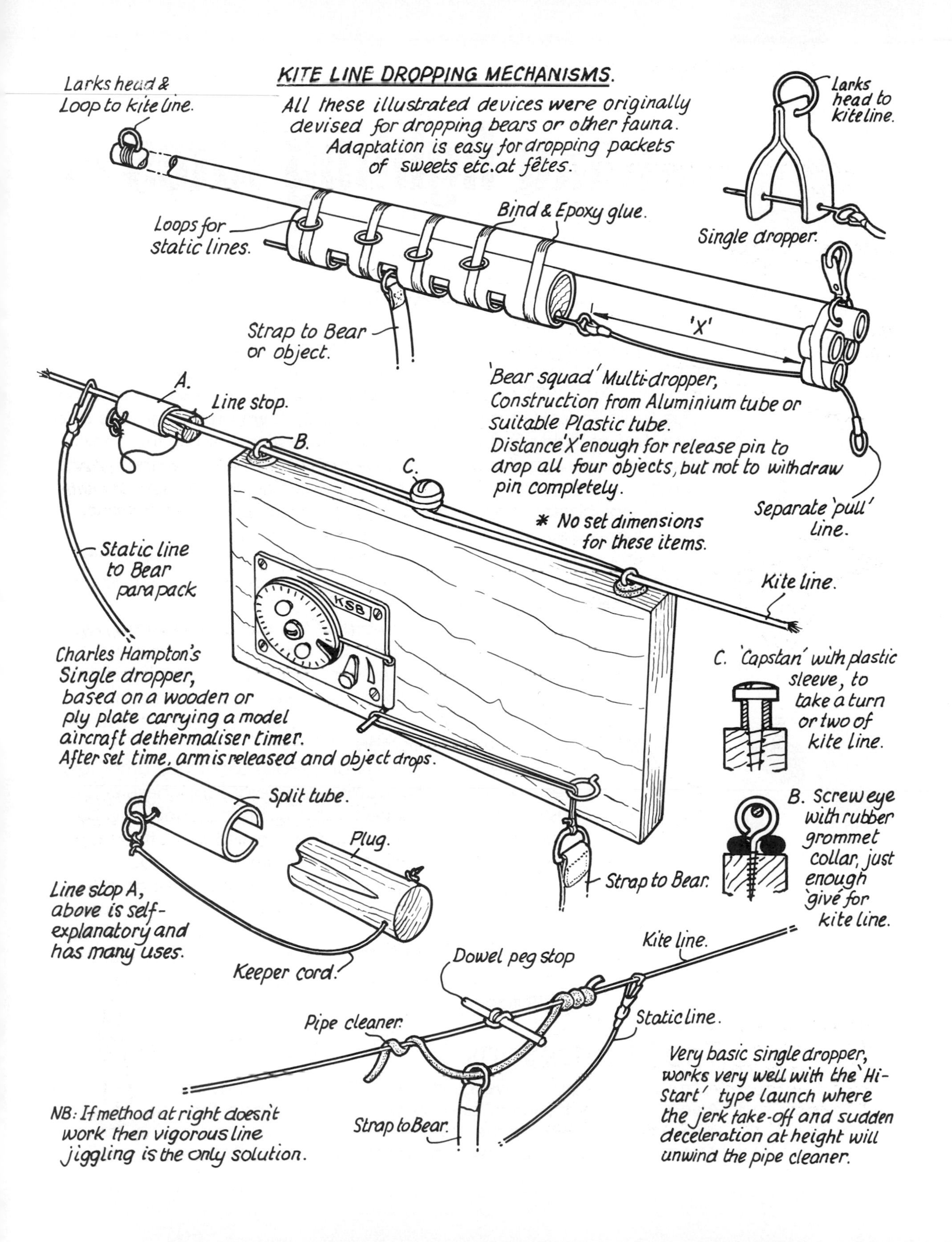
KITE LINE DROPPING MECHANISMS.
All these illustrated devices were originally devised for dropping bears or other fauna. Adaptation is easy for dropping packets of sweets etc. at fêtes.
Larks head & Loop to kite line.
Larks head to kite line.
Single dropper.
Loops for static lines.
Bind & Epoxy glue.
Strap to Bear or object.
'X'
'Bear squad' Multi-dropper, Construction from Aluminium tube or suitable Plastic tube.
Distance 'X' enough for release pin to drop all four objects, but not to withdraw pin completely.
Separate 'pull' line.
A.
Line stop.
B.
C.
* No set dimensions for these items.
Kite line.
Static line to Bear para pack
KSB
Charles Hampton's Single dropper, based on a wooden or ply plate carrying a model aircraft dethermaliser timer. After set time, arm is released and object drops.
C. 'Capstan' with plastic sleeve, to take a turn or two of kite line.
Split tube.
Plug.
B. Screw eye with rubber grommet collar, just enough 'give' for kite line.
Line stop A, above is self-explanatory and has many uses.
Strap to Bear.
Keeper cord.
Kite line.
Dowel peg stop
Pipe cleaner.
Static line.
Very basic single dropper, works very well with the 'Hi-Start' type launch where the jerk take-off and sudden deceleration at height will unwind the pipe cleaner.
NB: If method at right doesn't work then vigorous line jiggling is the only solution.
Strap to Bear.

Chapter Seven

Rokkaku fighting kites -

This Chapter provides an exception to our deliberate omission of the historical aspects of kite flying, because we are about to describe an incredible revival of an age-old practice. For, with the use of modern materials and construction techniques, the long established Sanjo Rokkaku fighting kite from Shirone, Japan, has, since 1983, become adopted as a cult by the Western kite world.

The Sanjo Rokkaku is a hexagonal 'flat' kite, flown on a single line as a *kenka-dako* or 'fighting kite' at the annual kite festival held on the banks of the Nakanokuchi canal near the small town of Shirone. It is flown there as the alternative to the massive O-dako or rectangular Edo kites, largely because its three-spar construction enables it to be rolled up when the spine is removed. It is also infinitely more manageable in the air and, unlike the diamond-shaped Nagasaki hata fighting kite (similar to those found in India, Malayasia and many other Eastern nations), it is reasonably stable and large enough to carry distinctive decoration.

Dedicated kite flyers had been aware of the Sanjo Rokkaku for ages, largely through Tal Streeter's excellent book *The Art of the Japanese Kite* (Weatherhill, New York 1974) in which he vividly described a visit to Toranosuke Watanabe, the Shirone kite maker who specialised, as did generations of his family before him, in making these hexagonal Rokkaku fighters.

However, the giant O-dako Edo kites of the Hamamatsu festival, as flown by teams of exuberant Japanese enthusiasts in their emblazoned *happi* coats, had gained greater publicity through international coverage in the media. These massive constructions of bamboo and paper, often passed down from year to year and re-covered for the next May festival by a new team, were far beyond the capacity of any individual flyer. Yet the spirit of excitement which came through each TV showing of the battles between happi-coated teams was an inspiration that called out for adoption.

The Mama-Sans

A group of veteran kite flyers, led by past-President of the American Kitefliers Association Bevan Brown and his wife Margo, made the first move. A challenge was issued to the AKA membership in their February 1983 issue of *AKA News*. It gave dimensions for an 80 x 96in (203 x 244cm) Rokkaku and

into battle!

described how teams of two or more could fly against each other as in the traditional Japanese events. Forecasting 'more of a mêlée than a contest' Bevan stated that awards may be given to recognise outstanding kites, team uniforms and accessories, team spirit etc.

The challenge was taken up by no less than 23 teams during the AKA Convention that year at Columbus, Ohio, all of them fulfilling the specification of being colourful, distinctive and exciting. With names like 'Moon dance', 'Black dragon', 'Plain white delight' and 'Grand butterfly', they provided a boisterous outlet for the conventioneers. Eileen Kinnaird, whose husband Rick was a prime mover of the event, led an all-girl team soon to be known as the 'Mama-Sans' which made its mark by beating the men on a disqualification. This, and other attention-getting highlights, was to spark off widespread interest.

In 1984 and 1985, the Rokkaku challenge, as it was to become known, was a feature of the AKA conventions; though it was regarded more as a participant than a spectator event. This was to change as the 'Mama-Sans' made an international impact in Europe in June 1985. At the beach sites of Scheveningen (Netherlands) and Cervia (Italy), the girls demonstrated and fought with valour. They were narrowly beaten by a four-nation, eight-man team of elite males in Italy in a spectacle that proved you didn't have to be one of the team to enjoy the fun. The die was cast. Within the year, both the International and the UK Rokkaku Challenges became established.

Above: Traditional Rokkaku using nine bridle lines. Note the curved spine and cross-spars under wind pressure.

Left: Jostling for position, a group of Rokkaku kites at Blackheath, London. Designs are similar but the decorations are distinctive.

Rules

Martin Lester had seen the American all-girls team in action and recognised the potential of Rokkaku fighting, both as a sport and a spectator event. Through the Kite Society of Great Britain and its *Kiteflier* quarterly newsletter, Martin proposed standard rules to apply for his International Challenge. They have been adopted by the Kite Society who undertook to organise a Rokkaku League for 1987 in the UK, coupled with a supplementary challenge for solo flyers using a smaller Sanjo.

The simple rules are:

Kite	Sanjo Rokkaku shape and style. Int'l: Minimum sail height 2metres (78¾ in) Solo: Minimum 1m (39.37in) Maximum 1.5m (59in)
Line	A suggested maximum length 60metres (200ft) was made but is not adhered to in practice. Kevlar, abrasive and cutting lines are strictly prohibited.
Team	Minimum of two persons.
Objective	Teams or Solo flyers will endeavour to cause other Rokkaku kites being flown simultaneously to make contact with the ground. This may be by direct knockdown or by severing the opponents' kiteline.
Procedure	All teams and kites are spread across an arena with lines extended downwind. They are then launched at a signal. After one minute, a second signal is given to start fighting. As soon as any kite touches the ground it is eliminated. Judges will note the sequence of landings. A third signal to terminate the round is given after twenty minutes have elapsed since the second signal. There shall be no intentional physical contact between teams. There shall be no catching of kites.
Scoring	Each event will consist of 3 rounds. The last kite flying in each round is considered the winner. Points will be awarded in descending order for duration of survival in the round, to 5th place that is:

1st place	6 points
2nd place	4 points
3rd place	3 points
4th place	2 points
5th place	1 point

On the conclusion of the third round, scores will be added. In the case of a draw, the team with the highest number of cutdowns as distinct from knockdowns will be considered the winner. For the UK League, the highest total from any two events will count.

Since the British challenge made its start in 1987 at four locations, it has blossomed into a major feature of up to six kite meetings each year. The international challenge did not meet Martin Lester's target in full, largely due to the fact that, wherever international representatives attend the major rallies, they do so for the very purpose of demonstrating their national speciality and no other.

Nevertheless, the Rokkaku cult spread from the USA and UK into Australia, Tasmania, New Zealand, Italy, Belgium and Germany, such that it became the ideal barrier-breaker that brought flyers together. At the same tirne it offered great entertainment for the onlookers.

Setting-up

Ideally, the arena should be marked by a rope, or similar. All flyers should be encouraged to wear a team identity, such as a coloured jacket, cap or sash. They should wear gloves and soft shoes, avoid physical contact and use whistles to express their fervour. The line should be at least 100kg (2201b) braided and low-stretch, although there is no rule for this.

The organisers should brief teams on safety. They should warn offenders guilty of so-called 'dirty-tricks' by loud hailer, and equally disqualify passive participants who hover outside the battle zone.

A commentator is the most important component of any major Rokkaku challenge. Kite and team identity must be relayed to the spectators with live-wire intensity. This is a cut-and-thrust spectacle where as many as 24 kites out of 25 have been cast off, their lines to float down in defeat within twenty minutes. The contest demands physical energy and not a little skill from the flyers. It should never be mute and matter of fact. This is the one kite event that benefits from noise!

Gordon Ellis of the Vectis Flyers, Isle of Wight, uses marker pens to apply his Dragon decor, though not always enough to ward off the evil intent of competitors!

Tactics

Imagine the scene. Lined along the windward side of the roped-off area are the teams. They are spaced only a few metres apart, each wearing the garish colours of their clan, two on the kiteline, others as reel carriers, coaches or vociferous fans. All are eager to do battle. The line stretches across to the leeward side where two other assistants are poised, their brightly-marked Sanjo Rokkaku supported vertically, all bridle lines in tension awaiting the signal to launch.

The airhorn blows. In seconds the air is filled with hexagons, their Dacron lines shining bright white against the azure sky. They sag again, some tilt momentarily before the slack line is hauled taut by the agile crew. Like oriental flowers on fragile weavlng stalks, the Rokkaku flock is impatiently ready for the next signal.

The airhorn blows again. To shouts of 'Wasshoi-Wasshoi' (Forward-Forward) the battle is 'on'. Darting teams zig-zag across the arena, their

Rokkakus turning, dropping, then soaring vertically as the line is tugged vigorously. 'Katsuru' yells the first team to make a cut. The victim settles horizontally, floating downwind as it waddles side to side, then gently descends to earth.

In the centre of the mêlée (Bevan Brown *was* right!) a trio of kites is twirling, their lines irretrievably wrapped, and the bunch comes down as one. Meanwhile, our heroes of that first encounter have dashed across to challenge another. Their kite is rocking as though on a pendulum as the line is relaxed, tugged, relaxed and tugged again. Now it swoops across and underneath its target. The lines contact and instantly our heroes turn and pull away upwind. The running line saws its way through the strands of the opposition and once again the *katsuru* cry announces victory.

By now the field is narrowed to the more experienced. The fight is less cut-and-thrust, much more a game of patient positioning. The first knockdown shows us how one Rokkaku can bear down on another, its cross-spars pushing at the other's sail in its vulnerable side where shape is vital. Distorted, the vanquished falls as the victor is pulled aloft at the last moment.

How long is twenty minutes? The judges must have forgotten their watches. Truth is, the battle is so absorbing it always extends until an obvious champion emerges; the last lonely Rokkaku deserves its 6 points as it descends unscathed. The victims are gathered for quick repairs before another round.

Control

When introducing the Sanjo Rokkaku we were exaggerating a little in stating that it was a 'flat' kite, and that it is reasonably stable. In truth, it is neither. The surface must be bowed into a curve by tensioning the two cross-spars and, while this contributes a lot to stability, the Rokkaku is still a single line *fighting* kite. By this, we mean that it will ascend perfectly when the sail is under pressure, and the line tension applies a driving force; but it will tumble or spin off to right or left if the line tension is relaxed. So this characteristic is put to use.

The initial ascent with a 'tuned' Rokkaku is straight up, following the line of the spine. When the line tension is relaxed, the kite will tilt. Pull hard, and it will resume its climb. Move laterally, and the manoeuvre can be converted to a sweeping move across the field. Make irregular pulls, and the kite will respond with similar jerky movements. With experience, the flyer soon learns how to steer by allowing the kite to determine its direction as it turns from the vertical, then, with a combination of ground movement and line pull, it is controlled with precision. You'll be working in three dimensions, in the air, and on the ground.

The difference between the lightweight, diamond-shaped fighter and the Sanjo Rokkaku is speed. Whereas the typical Indian fighter flits like a demented moth around a light bulb, the Rokkaku allows its flyers time to think. Given a good breeze, the 3¾ square metres (40 sq ft) of a standard Rokkaku can be allowed to hover close to ground effect where wind speed is lower. It is then in the ideal position to attack from below. As the kite is pulled up, so the line is fast moving as it contacts the victim's line and tension quickly severs the strands.

Selection of the line breaking strain and thickness is another tactical element. You can use a stronger line which is less likely to be cut; but, inevitably its extra weight and drag will put you at a disadvantage in terms of manoeuvrability. Thicker lines are no guarantee of invulnerability, as they can be weakened by a partial cut and snap at the next pull; equally, they are more expensive!

Knots are another taboo. Not only are they a natural weak point, they also entrap any intersecting line and form an unwelcome hazard. Hence the number of discarded lines after any bout - another very good reason for balancing cost against breaking strain and thickness!

Toronto Kitefliers team at the AKA convention, Columbus Ohio, flew as jolly sailors, led by the Hinchcliffe and Woodcock families.

The Rokkaku itself

There is but one restraint on design and that is that the sail height should be greater than 2 metres or 78 3/4 inches. From earliest days, the proportion of the hexagon was recommended to be 5 units deep, 4 units wide with the cross-spars 3 units apart and this simple formula with units at 500mm (19 3/4in) has sufficed adequately. Sometimes the proportions have to be modified by a small margin because materials for the spars and spine, or the central rectangle of the sail, are not available to the ideal dimension.

As kite-makers and everyday enthusiasts for kite flying, the Peacock family of St.Albans had also been the winners over the first two years of the Kite Society's Rokkaku Challenge. Who better then, to provide a well-proven Rokkaku design for this Chapter?

As we were to discover, Alan Peacock's aeromodelling experience had obviously influenced the detail in his fighter. Model aircraft demand preparation for many eventualities, most of them concerned with crashing. So Alan must have had well in mind the fact that any Rokkaku bout is going to offer short odds on a kite crash!

First, the frame size - 13mm (1/2in) Ramin dowel is used throughout for spars and spine. It comes in lengths that tend to vary around 2440mm (8ft). Allowing for rough ends on the dowel's random overall length, plus the short extensions at each end of the spar, the sail height is set at 2220mm (87 1/2in) and the spine cut to 2300 (90 1/2in). Similarly, the cross-spars have extensions, and the sail width is set at 1900mm (74 3/4in) so the cross-spars are cut to 1980mm (78in) lengths.

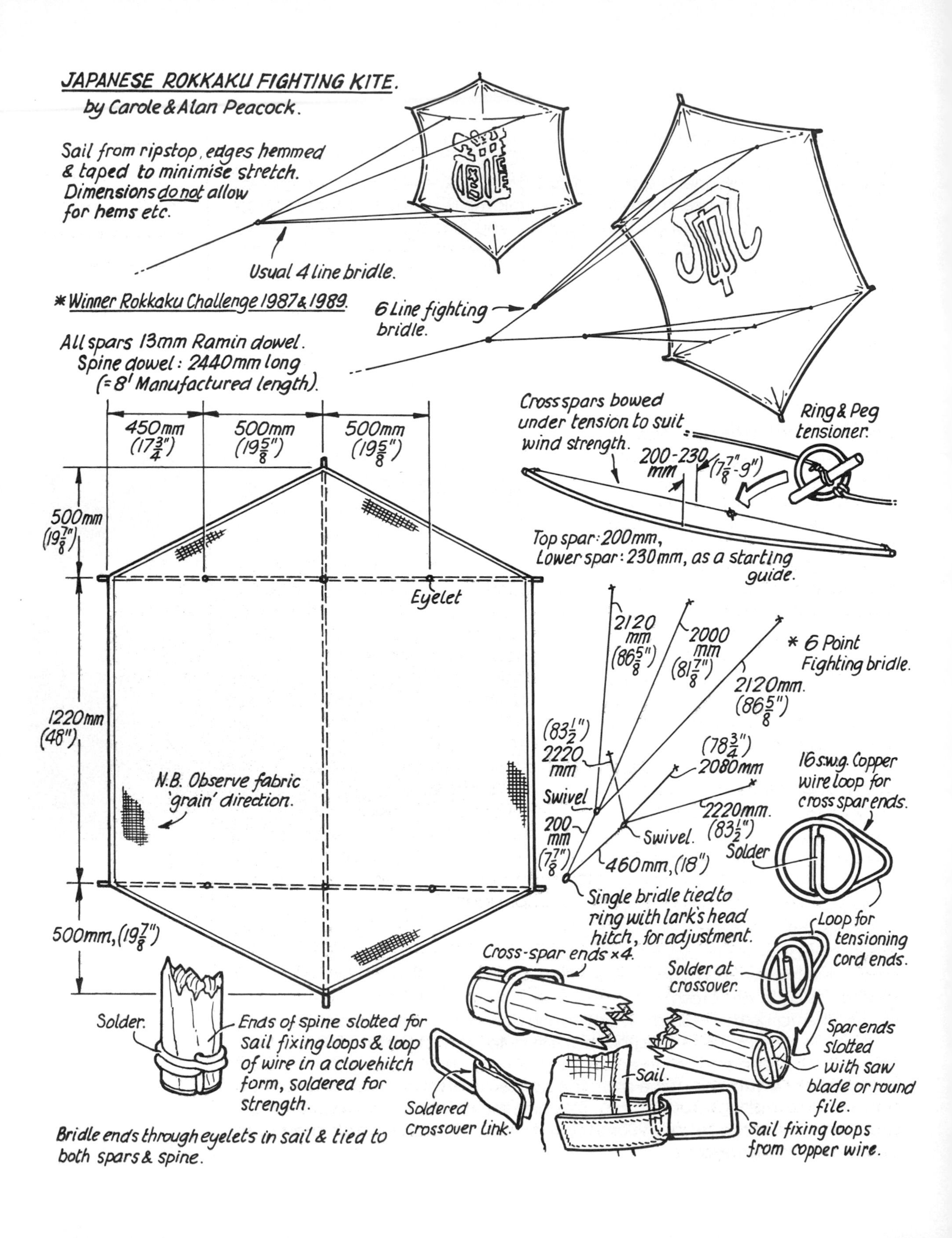

JAPANESE ROKKAKU FIGHTING KITE.
by Carole & Alan Peacock.
Sail from ripstop, edges hemmed & taped to minimise stretch. Dimensions do not allow for hems etc.
Usual 4 line bridle.
* Winner Rokkaku Challenge 1987 & 1989.
6 Line fighting bridle.
All spars 13mm Ramin dowel.
Spine dowel: 2440mm long (= 8' Manufactured length).
450mm (17¾")
500mm (19⅝")
500mm (19⅝")
500mm (19⅞")
1220mm (48")
500mm, (19⅞")
Eyelet
N.B. Observe fabric 'grain' direction.
Cross spars bowed under tension to suit wind strength.
200-230 mm
(7⅞"-9")
Ring & Peg tensioner.
Top spar: 200mm, Lower spar: 230mm, as a starting guide.
2120 mm (86⅝")
2000 mm (81⅞")
2120mm. (86⅝")
* 6 Point Fighting bridle.
(83½") 2220 mm
(78¾") 2080mm
2220mm. (83½")
Swivel
Swivel.
200 mm (7⅞")
460mm, (18")
Single bridle tied to ring with lark's head hitch, for adjustment.
16 s.w.g. Copper wire loop for cross spar ends.
Solder
Loop for tensioning cord ends.
Solder at crossover.
Cross-spar ends × 4.
Spar ends slotted with saw blade or round file.
Solder.
Ends of spine slotted for sail fixing loops & loop of wire in a clovehitch form, soldered for strength.
Sail.
Soldered crossover link.
Sail fixing loops from copper wire.
Bridle ends through eyelets in sail & tied to both spars & spine.

All six of the spine and cross-spar ends have 2mm or 1/16in saw cuts about 10mm or 3/8in deep across the centre. These have to accept the tensioner loops at each of the 'points' of the hexagonal shape, and should have a smooth interior. Use sand-paper to clean the sawcuts and also to radius off the edges of the spar ends.

Danish graphic artist, Jurgen Møller-Hansen, has a style of his own, using appliqué panels of random shapes.

Now comes the crash protection. We have to bind the spar and spine ends to prevent splitting on impact. Equally they have to sustain sail tension and, in the case of the cross-spars, must anchor the tensioners which create the bow in the sail. Copper wire is best for the job - 16 swg is ideal. This is near to 1.5mm, and it is soft enough to wrap tightly around the dowel, and accepts soft soldering for a neat fix. On the spine, these bindings are positioned at the bottom of the sawcuts and do not have any protrusions; but, on the cross-spars, an extra triangular form extension (it isn't critical though a pointed shape is best) for the bow tensioner has to be made. Make sure that these extensions are on the same side of the spar. That's the woodwork finished. Or is it? Better to make extras while in the mood. If you watch the Peacocks, you'll find they have spare parts for every eventuality!

Carole is the sail expert. She is also the appliqué decoration ace as those who have seen her 'Disney' Rokkaku would agree. So, for starters, we actually have a six-piece sail to take advantage of the ripstop nylon. While the centre rectangle has the 'grain' in-line with the edges, each of the triangles which will be joined to make the head and foot will have their 'grain' running at right angles (and parallel to) the edges. This is not much of a problem, as colours can be varied if a simple identification is sought.

As we have advised before, allow for the hems and, especially in the case of a Rokkaku, for reinforcement with a cotton tape all around the sail. It is very important that the sail edges are stressed taut on the spars and that the degree of tautness is equal from side-to-side. You will quickly discover on the first launch, if otherwise.

To obtain the tensions, we resort to the copper wire again and make up rectangular loops which are sealed by soldering. Six are needed, each to fit firmly into those sawcuts. They are then retained on the sail by tape made into double thickness tags. Tailor the sail over the frame for a fit, then sew the tags secure in their positions. With a trial fit, we now have a flat Rokkaku and must position the bridle eyelets.

Bridle

You can use four point or six point bridling, the latter bring preferred by the majority, and advised here. Reverting to that basic unit proportion which for convenience is 500mm (19 ¾ in), the outer eyelets are positioned this distance from the central eyelets and over the cross-spar in all cases. Symmetry is critical. The eyelets have to accept 100kg (2201b) breaking strain braided polyester line for the bridles. The bridle lengths must also be symmetrical, and by general standards are long. Pat's sketch offers the dimensions used on Carole and Alan's winning kite. Note that the lower bridles are longer to start with, and that each set of three is fixed from the cross-spar to a swivel so that these are permanent. Then the two sets are joined by a single line, swivel to swivel.

Using a Lark's Head knot for adjustment, the ring on this connecting bridle line then links to the flying line via a strong clip and another swivel. Spare connecting bridles are always carried!

The final part of assembly is the bow tensioning. The same line material is used for this. Four pieces are tied permanently to those triangular extensions on the cross-spar end bindings. Two will carry rings on their other ends; the other two must be fitted with wire or strong plastic (GRP or CF rod) pegs *after* trial and error fit to obtain the suggested bow. Note that the upper bow is 15% less than the lower.

Carole Peacock's Disney Special is an eye catcher of a Rokkaku, not flown in the mêlée of competition.

As the Rokkaku is a 'roll-up' kite, the spine should be detachable although, once the quick-release bow tensioners are disconnected, the cross-spars can be left in place on the sail. If this is the plan for carriage, then extra tapes will be needed to tie the spars to the spine. They can be left free or sewn to the back of the sail.

Ensure that there is no interference with the centre bridles. Ideally, the spine ought to be next to the sail and at the intersections, the loop on the centre bridle made large enough to permit the spine to slip through. A sleeve as used for the Aramid lines, Spectra and Dyneema would help, but don't think of using this material for the flying line!

Trimming

Most Rokkakus will soar straight off, holding a reasonably stable position as long as the flying line is under tension and the wind speed is steady. There is enough area there to call for a two-person crew, so launching is rarely a problem.

An out-of-balance Rokkaku is, however, a taskmaster. The balance must be aerodynamic and dynamic. The cross-spars should flex equally either side of centre and their weight be evenly disposed. It is not unknown to have ballast on a spar tip to even out the dynamic balance, and, if you are as well prepared as the Peacocks, you'll have spare spars handy which will have different flexibility to try against the breeze of the day.

Two adjustments are possible. The first is that, if the Rokkaku is a real handful and pitches over from the vertical even when a steady pull is made on the flying line, then extend the upper section of the single line bridle connector by slipping the Lark's Head knot *down*. You will then have lowered the towing point relative to the sail.

The other adjustment is in those tensioner bows. By increasing the curve or bow on the kite face, the dihedral angle is increased, the turbulence at the sail edges is reduced (but not eliminated by any means) and stability is improved. Between the curving cross-spars, the sail adopts a vee form not unlike that on a Malay or triangular kite. By keeping the lower vee steeper than the other, we introduce what might be called a 'tail-effect', so keep this in mind when you are practising for that first Rokkaku Challenge.

Painting in ripstop nylon has been taken to extremes by Tom Casselman. On this Rokkaku he combines Chinese traditional artform with a sepia arch depicting early flight. (G. Maurer)

CHAPTER EIGHT

The steerable revolution

Nothing new

Although a very great proportion of the revival in kite flying is attributed to the growth in popularity of the 'Stunt' kite , one should not run away with the idea that controlled flight is a new discovery. The traditional Diamond fighter kite as flown all over Eastern nations is remarkably controllable even on a single line, though its flight path is limited to a narrow sector of the downwind area.

Use of more than one line to steer a kite dates back as far as the 1820s when George Pocock pioneered use of kites for traction of carriages and a boat.He utilised variable tension on two lines to elevate or depress peg-top kites successfully enough to transport as many as 16 lads to a cricket match, including his grandson, W.G.Grace the famous cricketer. An amusing description of the system appeared in *Model Aeroplanes and Airships* circa 1910 ,with claims that the *Char-Volant* could travel at up to 42km/h or 26mph (in gale force winds!).

Pocock's method of 'shifting the belly band' or adjusting the bridle link position was not what we currently accept as a 2-line system.That described by J.Woodbridge Davis in *Aeronautics* of August 1894 for his method of steering a life-saving device undoubtedly was a progenitor of what we use today. Separated bridles linked to two lines enabled this American invention to be positioned within reasonable accuracy over any likely shipwreck victim.The kite was star shaped and would thus lack the refinement of current designs; but nevertheless, it must be credited as the first of the steerables.

Reproduction of a 1930's advertisement for the Air-o-bian controllable Box kite.

Refinements

One of the earliest patent applications for a stunt kite was that filed in December 1928 for H.De Haven in the USA and eventually sold as the 'Air-o-bian' fighting kite, with claims for 'loops dives and climbs. In reality it was a basic Box kite with a pulley at the normal line attachment, over which the two lines passed. As the lines were alternately pulled so the pulley actuated a semi-circular rudder hinged across the diagonal of the rearmost box.One does not need to have much imagination to see how the rudder deflection might send the Box kite into limited stunts, given a strong wind; but we doubt if the rigid stability of a Box kite could be persuaded to perform over a wide range of movement.

Paul E.Garber

Also using a rudder control but coupled to a spreader bar in the bridle, was the Target kite developed by Paul Garber assisted by Lloyd Reicher and Stanley Potter.This was a kite for the specific purpose of training small calibre gunners and on the Eddy-shaped sail, an outline of the Japanese Zero fighter appeared in silhouette.

Flying on 180m (200 yards) of line, the operator moved a control bar to send the kite through radial manoeuvres as the gunners took aim (or tried to!) in an effort to destroy it. Thousands were made for the US Navy and Paul Garber filed his patent in August 1944 to pre-date all the subsequent claims to be 'first' with the present day concept of the steerable kite. In fact 125,000 were made by the sports goods specialists, A.G.Spalding and although heavy and bulky even for its 1.5 x 1.5m (5 ft.sq.) dimensions, any survivors are treated with great reverence and are highly valued by their collectors.

Above:Paul E Garber at his roll-top desk in the Smithsonian Institute. No man ever did so much to popularise the kite as an educational instrument. Garber died, aged 93, on 23 August 1992.

Left: Paul Garber's Target kite, one of the first to use two-line control as known today.

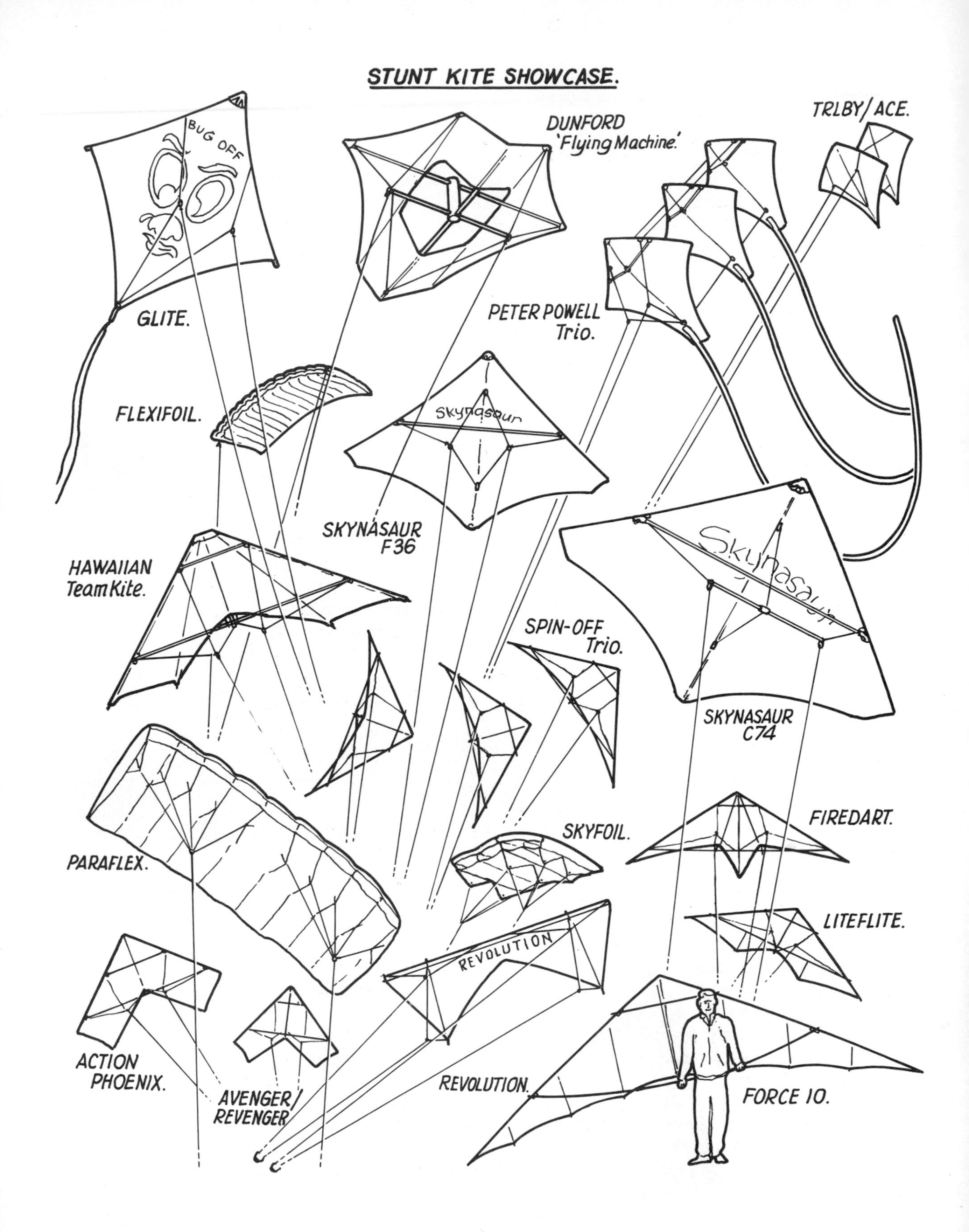
STUNT KITE SHOWCASE.
BUG OFF
GLITE.
DUNFORD
'Flying Machine.'
TRLBY/ ACE.
PETER POWELL
Trio.
FLEXIFOIL.
Skynasaur
SKYNASAUR
F36
HAWAIIAN
Team Kite.
Skynasaur
SPIN-OFF
Trio.
SKYNASAUR
C74
FIREDART.
SKYFOIL.
PARAFLEX.
LITEFLITE.
REVOLUTION
ACTION
PHOENIX.
AVENGER/
REVENGER
REVOLUTION.
FORCE 10.

Francis Melvin Rogallo

In 1948, Francis Rogallo and his wife Gertrude were granted their patent for the flexible kite that has been credited so often as the origin of the 'modern' species. It was a diamond shape, flown as the Flexi-kite for some years before the patent application was made, and it heralded the concept of a lifting surface with two cones which could be controlled as a parachute or kite.

Strangely we find that its diamond shape was less like the present day Delta with its rigid leading edge spars than the design by George D.Wanner of Dayton as filed 11 months prior in January 1948, as against November 1948 for Rogallo. But that is not to take anything away from Rogallo whose researches with NACA at the Langley Research Center into Parawings were well known, and had been picked up by several inventive minds in the kite world.

In our view it was Charles H. Cleveland's 'Glite' for the North Pacific Products Co. which took the conical cambered diamond into the mass marketplace and stimulated the two-line variation of its basic single line operation; but there was yet one other little known patent that pre-dated almost all these attempts to obtain control by rudder or lateral tension in the bridles.

Wilhelm H.Hatecke of Dornbusch filed a patent in Germany in July 1949 for an aeroplane-shaped kite to be flown on two lines.It was not so much the kite design, rather, the principle of moving the centre of lift as the lines were pulled that made this early application a forecast of today's steeringmethods. Hatecke had recognised that it was possible to obtain control even in diving flight by using the lift force itself instead of any mechanical rudder system.

Don Dunford and Peter Powell

If anyone is ever to be held responsible for the steerable revolution, it has to be one of these two eccentric personalities.

Working on entirely different design principles, each successfully publicised their stunters to capture public attention in the mid 1970s and open up a new wave of enthusiasm. The late Don Dunford's 'Flying Machine' was more like a Conyne winged Box kite and in consequence was relatively heavy with its four spars and a spreader plus stout plastic tube and cambric sail. It called for strong winds in order to execute the full range of manoeuvres; but it also had sufficient lifting power for Don to go on television and demonstrate its use as a crop sprayer, photo-ship or bird scarer! Made with glass fibre spars and using a ripstop nylon sail, the 'Flying Machine' has been improved in many ways for the 1990s.

Don had been experimenting for years before he took out the patent in 1970 and significantly stated that 'The difference in camber and relative incidence on the central fore and aft sections serves to stabilise the kite in pitch'. It is this same difference of angles that makes the current stunt kite so stable and controllable, though Don used the Conyne planform. The Delta employs the same principle with its reflex on the swept-back tips. Current

Dunford 'Flying Machines' have ripstop nylon, CF spars and unique venting to improve performance way above that of the originals. Its long life is a credit to the designer.

Peter Powell's adventures with kites had already put him into the headlines when he 'flew' his grandmother, long before he perfected the diamond-shaped stunter that made him a worldwide name from 1974.The plastic sail, stressed over an aluminium tube frame, with a deep bow formed into the spine was frequently misnamed a Rogallo. Still popular, and made in new materials with plastic or nylon sail, the Peter Powell kite has introduced millions to the sight of its long tube tails following the looping flight pattern, and still, after over 15 years of its production when all else around has seen so much change, is a firm recommendation for anyone wishing to take up stunt flying for the first time.

Flexifoils

Andrew Jones and Ray Merry experimented with air-filled kite shapes as students and after many frustrations, evolved the Flexifoil which like the Powell, has stood the test of time and is still in production now in five sizes, from 122cm(48in) up to 393cm(154.5 in)span and in many colours. It introduced entirely new aerodynamic principles to kiting which we shall describe under 'Foils' later, and because of its efficiency, it has been applied to power-kiting with some astonishing results, not least being measured speeds approaching 195km/h (120mph) in horizontal flight.

Irreverently called the flying mattress or airbed, the Flexifoils are unique. They have no bridles, the main and only spar is at the extreme leading edge and improves efficiency as it curves in flight. Moreover, the aerofoil section of the Flexifoil seems upside down!

Flexifoil principle explained - by the designers Ray Merry & Andrew Jones

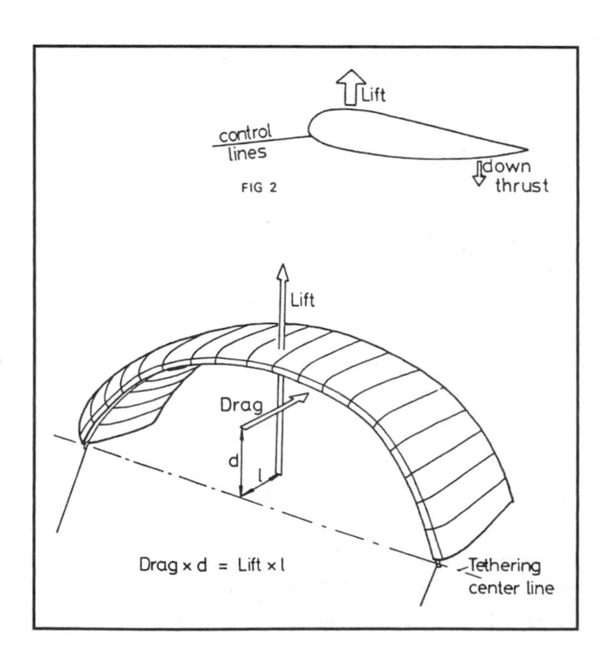

As a stunt kite it excels in smooth, sweeping manoeuvres and when 'stacked' in train they make a spectacular sight, as well as offering a considerable physical challenge for the operator.

These patent protected designs by Dunford,Powell,Merry and Jones created the initial wave and still survive. Other Diamond kites - the Ace, Trlby, Zig-Zag, Rainbow, Hyperkites Ghostie, Blazer, Brookite Stunters and Dunford Stingray - have carried on that first generation with low-cost products that perform impressively for any novice and have served over the years to bring so many into the kite fraternity. Similarly, the delta form Skynasaur F36, like the surviving Diamonds, now equipped with latest technology glass fibre rod spars and braces, comes into the first stage bracket of starter-stunters with honours.

Second generation - the narrow delta

A Hawaiian at low level, holding position at the edge of the window.

No sooner had 'Red' Braswell, then President of AKA, publicised in 1978 a conversion of the simple plastic-sailed Gayla 'Baby Bat' from single to two-line for what was then christened Figurekiting, than the inventive kiters in America applied new thought to open up a new phase.

Richard Radcliffe's re-bridled 'Bat' was a typical 90-degree Delta. Its leading edges were free to articulate, the central keel was retained simply because short of cutting it away,there was no choice, and there was but one forward cross-spar. But it flew well enough to confound the sceptics and won first place at the April 1978 Maryland Kite Festival.

Precisely who first devised the conical form of the sail on the special stunters that followed out of California and Hawaii is not recorded.The camber of the sail was natural anyway, the Radcliffe bridle had established the bridle points, but the new designers (Tony Cyphert and Don Tabor are often credited) added the rear cross-spar and this permitted the trihedron bridle which also absorbed the strain on glass fibre spars. In one move, there was a new generation and it adopted the generic term of 'Hawaiian'. The real Hawaiian, still a favourite for team aerobatics, is the indented and battened type, since regarded as relatively heavy, but very precise in control, with plenty of 'pull' in strong winds.

Rigid frames

Unlike the single line delta with its articulating leading edges that pivot about the cross-spar joint, the second generation stunters with cross-spars fore and aft introduced the rigid frame.As the spine retained the sail on the neutral datum centreline, the sail pattern was made larger than the actual frame plan in order to obtain the conical camber which is the characteristic of the Rogallo design.

The 'billow' or cone shape which develops in the sail under airflow couples with sweep back on the leading edges to offer a natural washout. In other words, the angle of attack relative to the airflow is steep at the centre of the sail, and neutral or even negative at the tips.

In the Hawaiian, glass fibre tube battens stretch the sail across its chord, front to rear and in flight they form an angular camber so that there is a deep vee centre body, while the outer panels could be described as flat. When

BRIDLE MODIFICATIONS.

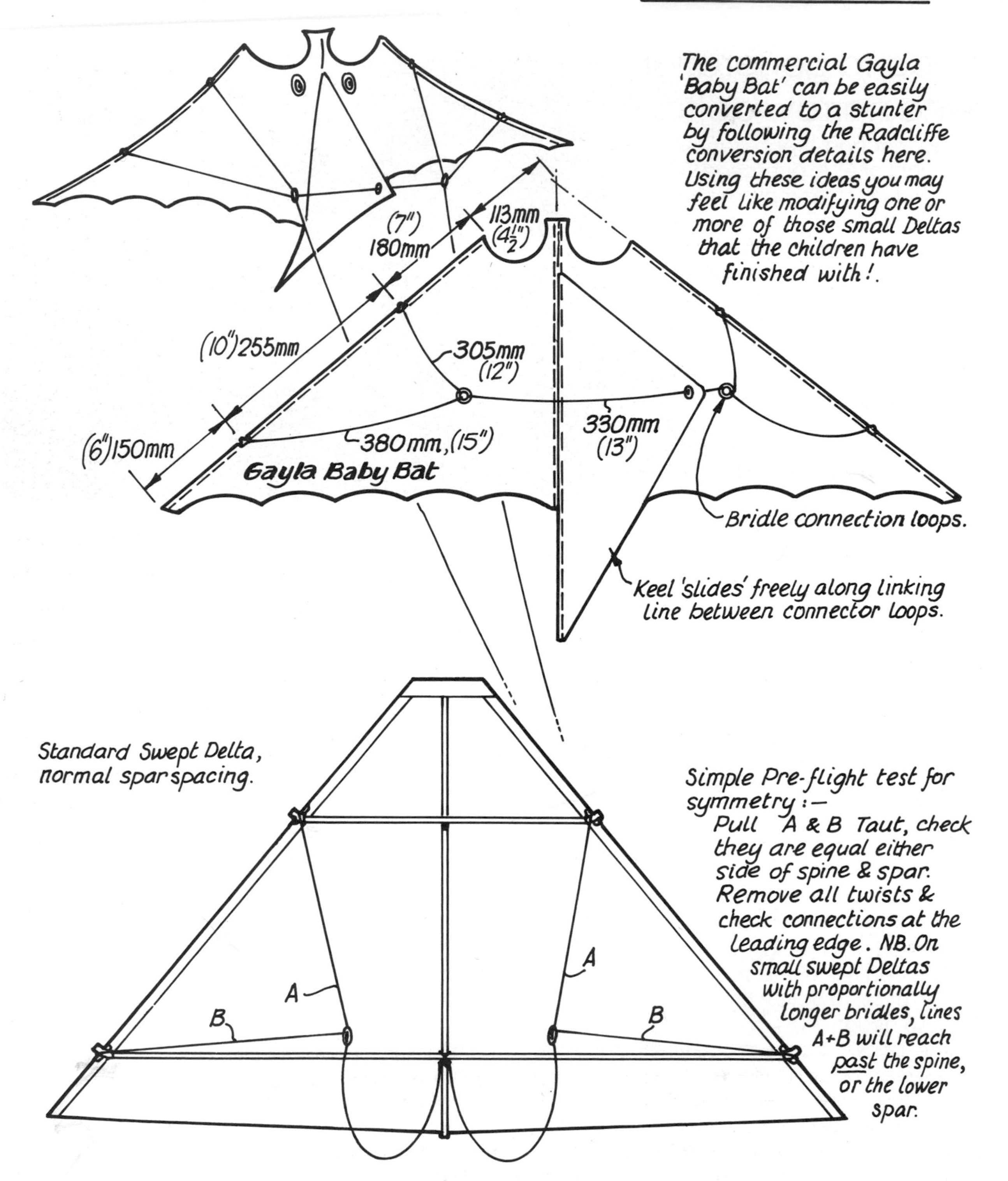

observed from the rear, the trailing edge is gull shaped and from the side, the angle of attack created by the bridle is very steep over the central delta and shallow over the outer panels. What had been devised was a swept wing with reflex (washout) at the tips as found in the majority of free-flying tailless aeroplanes.

The British Sport Kite Team AIRKRAFT that overtook American dominance in 1995 when they became World Cup Champions. Carl and James Robertshaw, Jeremy Boyce and Nick Boothby. From 1993 to 1996 they took 28 first places and 5 seconds in major British, European and World events. The kite is a Top of the Line.

Flying the Hawaiians in Californian events, the 'Top of The Line' demonstrators created a wave of interest that was not far short of being explosive. Many variations in shape quickly emerged to explore the potential. Extended tails as in the Dart series showed that by using indents in the outer trailing edge, the conical camber of the sail could be varied spanwise. This has since been employed in the wide variety of commercial designs that have arrived following the introduction of the Hawaiian; but for a simple and long-standing development one need look no further than Top of The Line's own 'Spin-Off'. This narrow angle swept wing with the leading edges at 105 degrees and corresponding sweepback on the trailing edge was a weight-saving exercise that established a standard by which most others are judged.

Originally 254cm (100in) and since made at 3/4 size 198cm (78in),the Spin-off also contributed to the arrival of what we might call third generation stunters by its eventual introduction of stand-off struts.

Use of carbon fibre reinforced tube and rod not only added to frame rigidity, it enabled lighter and smaller designs to be created such as described already in Chapter Two, and it led eventually to sail designs with multiple stand-offs for improved aerodynamic characteristics.

Third generation

Competition breeds improvement. In the case of steerable kites, skill of the operator is paramount; but anything that could make the kite fly to a tighter turning radius, open its radius of action in the 'wind window' or launch and land more easily was a critical element. Similarly the detail design of swept-wing kites contributes to a consistent performance. Improved handling has resulted from a series of innovations most of which have been related to aerodynamic efficiency. They are listed on page 130.

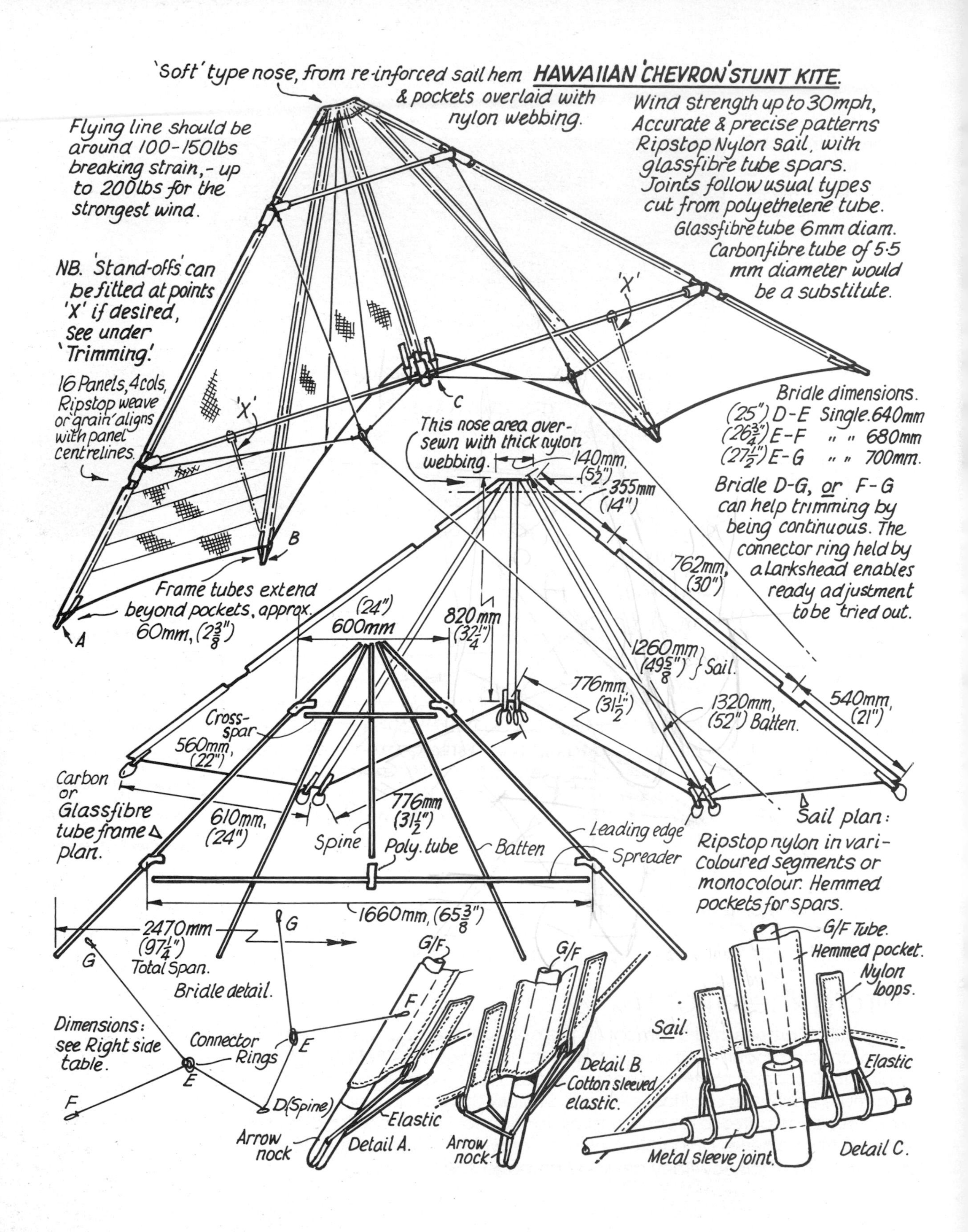
HAWAIIAN 'CHEVRON' STUNT KITE.
'Soft' type nose, from re-inforced sail hem & pockets overlaid with nylon webbing.
Wind strength up to 30mph, Accurate & precise patterns Ripstop Nylon sail, with glassfibre tube spars. Joints follow usual types cut from polyethelene tube.
Glassfibre tube 6mm diam. Carbonfibre tube of 5·5 mm diameter would be a substitute.
Flying line should be around 100-150lbs breaking strain, - up to 200lbs for the strongest wind.
NB. 'Stand-offs' can be fitted at points 'X' if desired, see under 'Trimming'.
'X'
'X'
16 Panels, 4 cols, Ripstop weave or 'grain' aligns with panel centrelines.
C
B
A
Frame tubes extend beyond pockets, approx. 60mm, (2 3/8")
This nose area over-sewn with thick nylon webbing.
140mm, (5 1/2")
355mm (14")
762mm, (30")
Bridle dimensions.
(25") D-E Single. 640mm
(26 3/4") E-F " " 680mm
(27 1/2") E-G " " 700mm.
Bridle D-G, or F-G can help trimming by being continuous. The connector ring held by a Larkshead enables ready adjustment to be tried out.
(24") 600mm
820mm (32 1/4")
1260mm (49 5/8") Sail.
776mm, (31 1/2")
1320mm, (52") Batten.
540mm, (21")
Cross-spar
560mm, (22")
Carbon or Glassfibre tube frame plan.
610mm, (24")
776mm (31 1/2")
Spine
Poly. tube
Batten
Leading edge
Spreader
Sail plan: Ripstop nylon in vari-coloured segments or monocolour. Hemmed pockets for spars.
1660mm, (65 3/8")
2470mm (97 1/4") Total Span.
Bridle detail.
G
G
Dimensions: see Right side table.
Connector Rings
E
E
F
F
D (Spine)
G/F
G/F
Elastic
Arrow nock
Detail A.
Detail B.
Cotton sleeved elastic.
Arrow nock.
G/F Tube.
Hemmed pocket.
Nylon loops.
Sail.
Elastic
Metal sleeve joint.
Detail C.

ichel Gressier's French artforms are internationally nowned. His pair of colourful flare kites fly head to ad as male and female with opposing bridles while gh above, the Dutch Oostveen/Schiefer Circoflex ngs hover almost stationary in space.

The multi-cell is a European phenomenon of the 90s, bringing flower petal patterns with the colours of a rainbow. The technique was first seen in German kites. This one is from Belgium.

This huge kite measures approximately 3.5 x 8 metres. Made by Gerard Rosenblatt it illustrates the classic stories of 1001 Arabian nights, painted on nylon. Such a large kite would only have been considered as a rectangular shape in the 80s, now all shapes are possible.

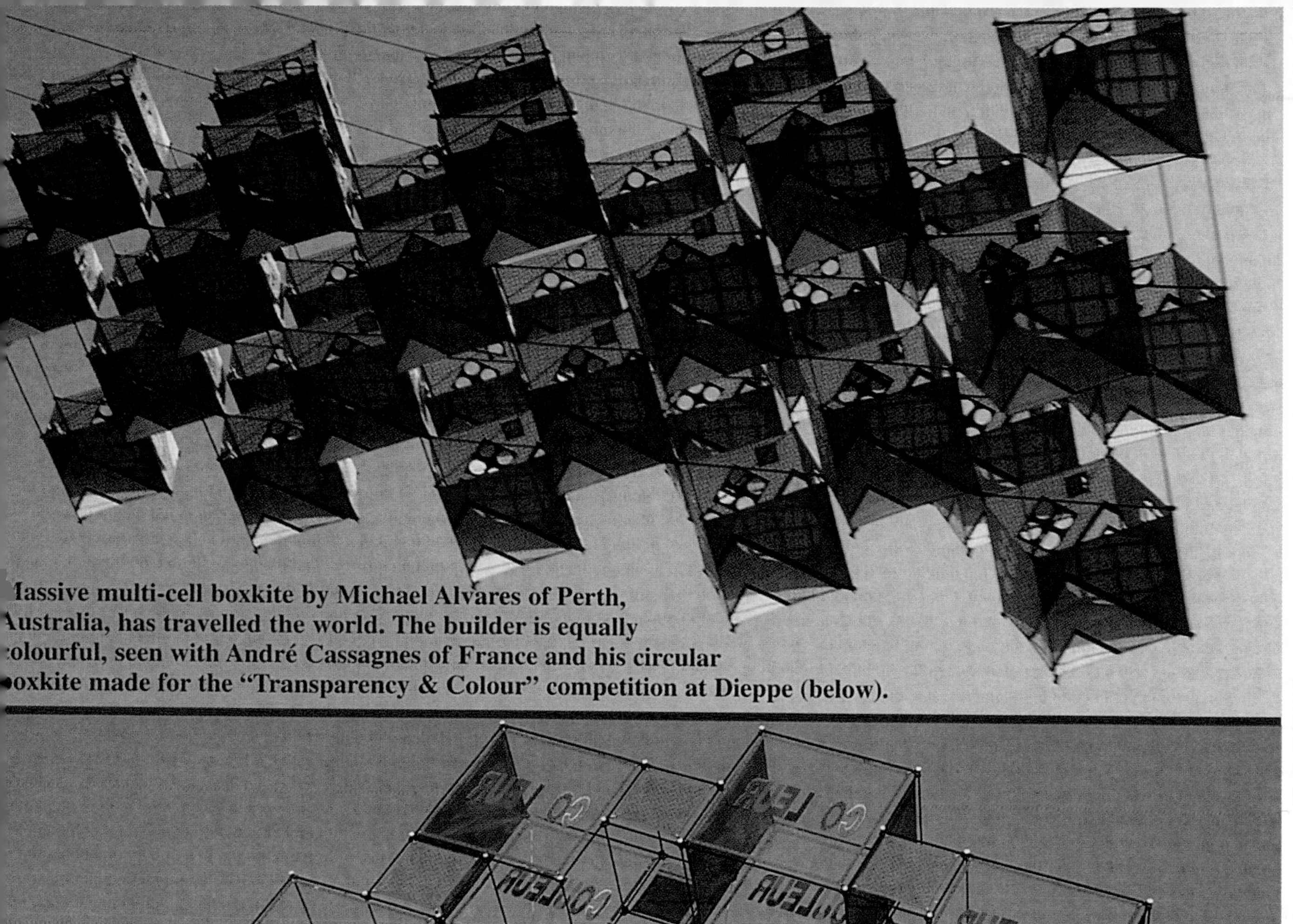

Massive multi-cell boxkite by Michael Alvares of Perth, Australia, has travelled the world. The builder is equally colourful, seen with André Cassagnes of France and his circular boxkite made for the "Transparency & Colour" competition at Dieppe (below).

Gigantic windsocks and rotating "ballons" are a feature of the Dieppe bi-annual festival. This one carries Michel Gressier's art in clever spiral panels.

The Japanese Edo shape offers a natural canvas for artistic expression, as in this Cherubic scene, signed by the artist.

Sails, wings, wind farms, tents and the booths for thirty nations to display their traditional kites are part of the scene which has qualified the claim that Dieppe is the "kite capital of the world" bringing East and West together.

Intricacy of the typically Italian approach to a multi-cell design calls for accuracy in machining panels to identical size. A cruciform frame supports the peripheral line to tension the rainbow toned kite body.

The popular Japanese Rokkaku, now adopted the world over.

Above, a tall Edo on a multiple line bridle by Danielle Theillac carries a symbolic picture story. The aeroplane kite is one of many by Eric de Montarnal from Bordeaux in his Virgule series. Made with bamboo frames and covered with paper from Nepal, they represent the pioneers of flight.

erre Fabre won the 94 Dieppe contest for presentation of the a with his startling uclear Swordfish. hough of remarkably gh aspect ratio, the eral fins stabilised e kite. At the other treme, Don Mock om Seattle specialises the art of American dians on the Pacific m. His kites are symetrical, as evident this topside view hen readied for unch. Kite shapes and lours progressed pidly in the 90s.

Kites are often designed to signify a gesture of goodwill, nation to nation. The twin doves of peace, five children of all continents holding hands around an Eiffel Tower picture, form one of the more elaborate symbols offered to the organisers of the Dieppe Festival, and it flies too! The many pointed star kite at left is from the Austrian "Drachenbau und Flugverien" in Vienna. It confirms the rapid development of kite design and international spread of enthusiasm for kite flying.

THE SPIN-OFF TYPE STUNT KITE.

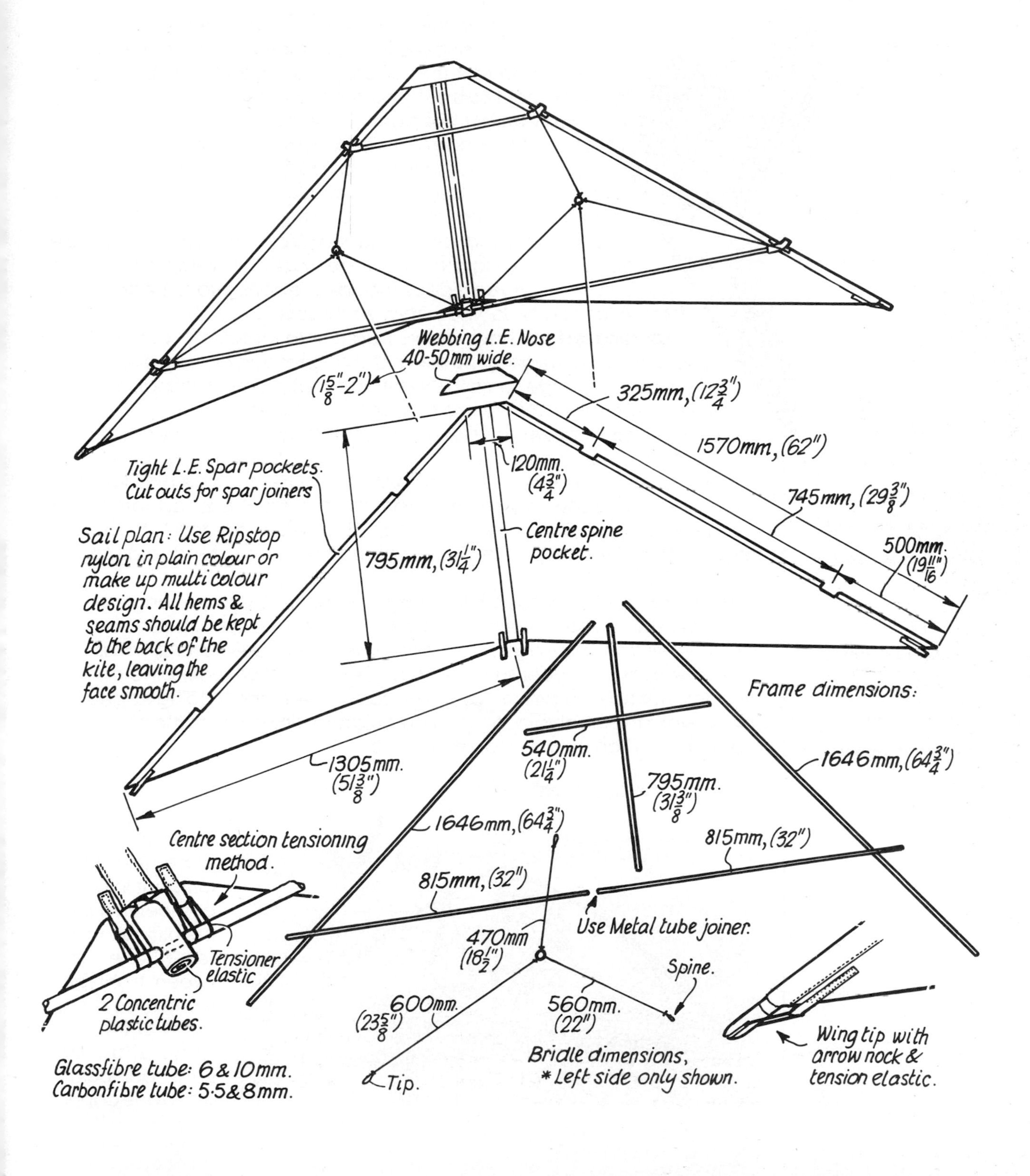

Leading edge profiles have been improved with the use of stiff Dacron tape sleeves which primarily reinforce the sail but equally important,they offer a 'Tear Drop' entry to the aerofoil as on yacht sails or hang glider wings.

Sail camber has been formed by use of battens, either flat or of round section in sleeves from leading to trailing edge.Tensioned into compression, the battens become curved and the degree of camber is controlled by vertical stand-offs from the forward cross-spar.

Sail curvature has been induced by clever tailoring of multiple panels. By changing direction of the bias in the ripstop and cutting patterns with curved edges,the sail balloons into a better cone form. A lot can be learned from yacht sails.

Leading edges have been pulled into parabolic curves by a combination of tailoring the sail and strong tension on long and unsupported leading edge spars.This is both cosmetic and practical, as it can convert the conical camber to cylindrical camber.

Trailing edge vibration has been dampened by reinforcement with thin glass fibre rod inserts or mylar overlays, so eliminating excessive noise.

Tail shapes have been added to extend the central vee and improve stability in pitch. These vary in shape from the curved 'Bustle' adopted from the full-size Burgess-Dunne flying wings of 1911 to the divided Tern shaped forked tails and the vented 'Air Fin' which converts the extended delta into twin stabilisers.

Wing tips can be warped into a negative or 'washed-out' angle by sail stretchers. Use of open mesh material here can reduce the turbulence at the finely tapered trailing edge,and improve performance in strong winds.

Airflow over the upper surface at the centre section where the chord is broadest has been improved by open mesh close to the leading edge spar. This converts the leading edge into a turbulator, behind which the airflow has a better transition over the top of the kite.

Above: Tim Benson's Swallow Tail has extra leading edges for the tail section. Very attractive in flight.

Below: The 'Rare-Air' King Cheetah C70 design by Zbys Kaczmarek of St Ives, Cornwall, using battens and a tensioned leading edge.

All of these design developments have been used to gain a commercial advantage by offering that 'extra something' to be different in a very competitive field. It is difficult to differentiate between flight performances of many of the third generation designs, particularly when one cannot

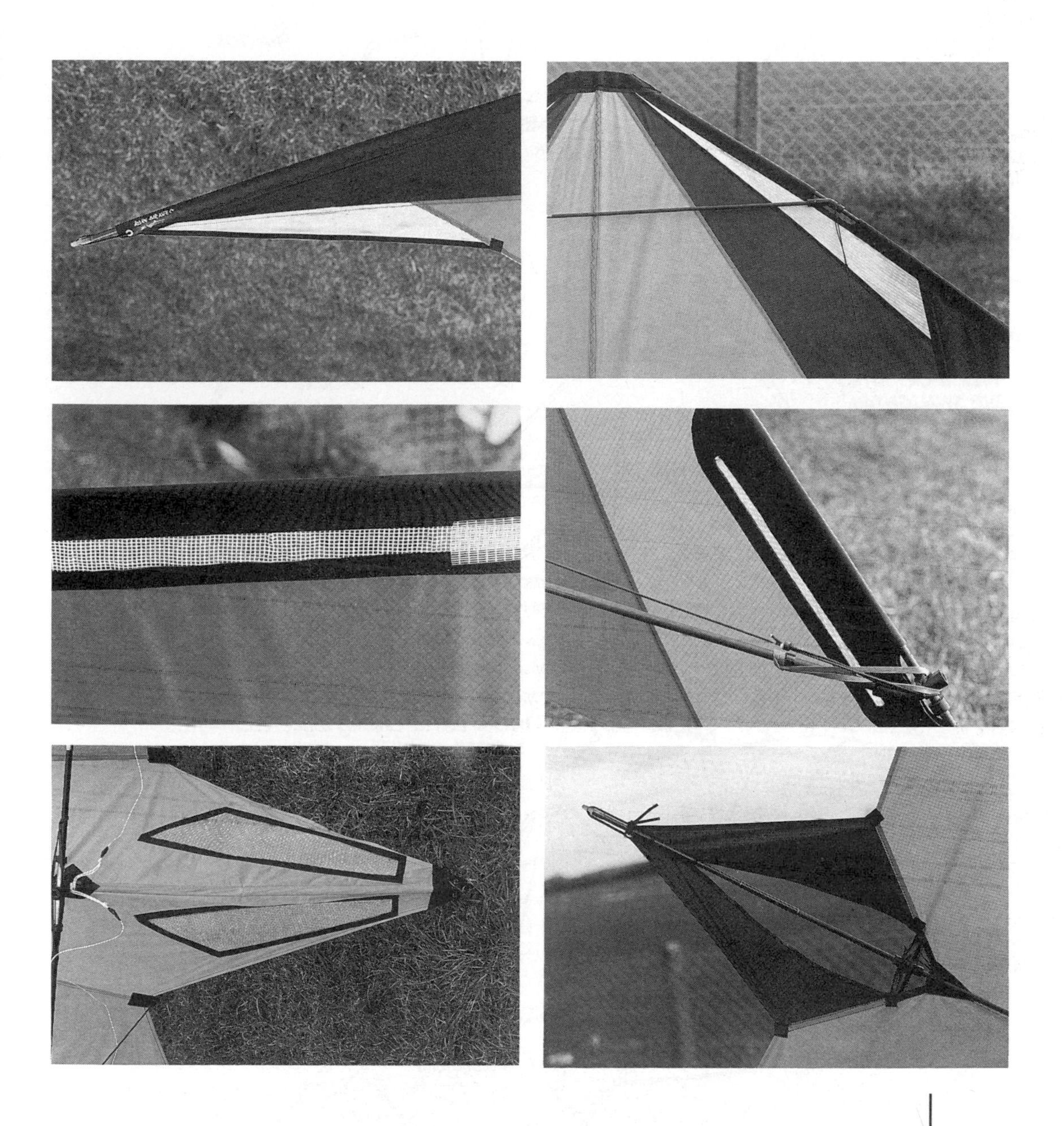

Top left: Mesh in the tip section of Rare-Air's Flash Angel enables fast turns.

Top right: Leading edge mesh on the innovative Flash Angel design by Zbys Kaczmarek.

Centre left: Open mesh (mosquito netting) provides a slot on this leading edge.

Centre right: Fine mesh slot on a Powerhouse kite - there are two slots on each leading edge.

Bottom left: Mosquito net on this tail is not quite as effective in the central open area as on Flash Angel (bottom right).

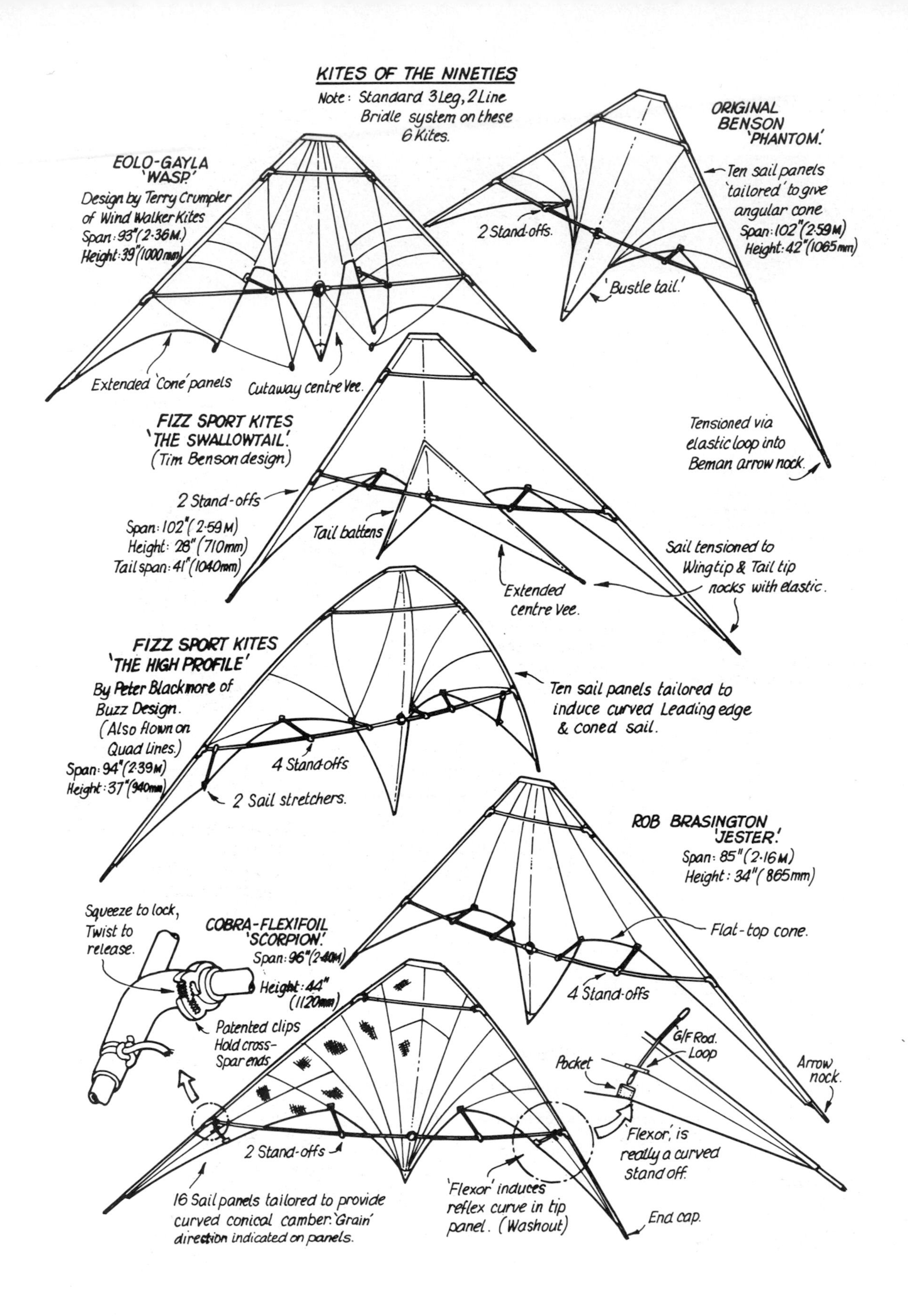
KITES OF THE NINETIES
Note: Standard 3 Leg, 2 Line Bridle system on these 6 Kites.
EOLO-GAYLA 'WASP.'
Design by Terry Crumpler of Wind Walker Kites
Span: 93" (2·36M.)
Height: 39" (1000mm)
Extended 'Cone' panels
Cutaway centre Vee.
ORIGINAL BENSON 'PHANTOM.'
Ten sail panels 'tailored' to give angular cone
Span: 102" (2·59M)
Height: 42" (1065mm)
2 Stand-offs.
'Bustle tail.'
Tensioned via elastic loop into Beman arrow nock.
FIZZ SPORT KITES 'THE SWALLOWTAIL.'
(Tim Benson design)
2 Stand-offs
Span: 102" (2·59M)
Height: 28" (710mm)
Tail span: 41" (1040mm)
Tail battens
Extended centre Vee.
Sail tensioned to Wingtip & Tail tip nocks with elastic.
FIZZ SPORT KITES 'THE HIGH PROFILE'
By Peter Blackmore of Buzz Design.
(Also flown on Quad Lines.)
Span: 94" (2·39M)
Height: 37" (940mm)
4 Stand-offs
2 Sail stretchers.
Ten sail panels tailored to induce curved Leading edge & coned sail.
ROB BRASINGTON 'JESTER.'
Span: 85" (2·16M)
Height: 34" (865mm)
Flat-top cone.
4 Stand-offs
Squeeze to lock, Twist to release.
COBRA-FLEXIFOIL 'SCORPION.'
Span: 96" (2·40M)
Height: 44" (1120mm)
Patented clips Hold cross-Spar ends
G/F Rod.
Loop
Pocket
Arrow nock.
'Flexor,' is really a curved stand off.
2 Stand-offs
16 Sail panels tailored to provide curved conical camber. 'Grain' direction indicated on panels.
'Flexor' induces reflex curve in tip panel. (Washout)
End cap.

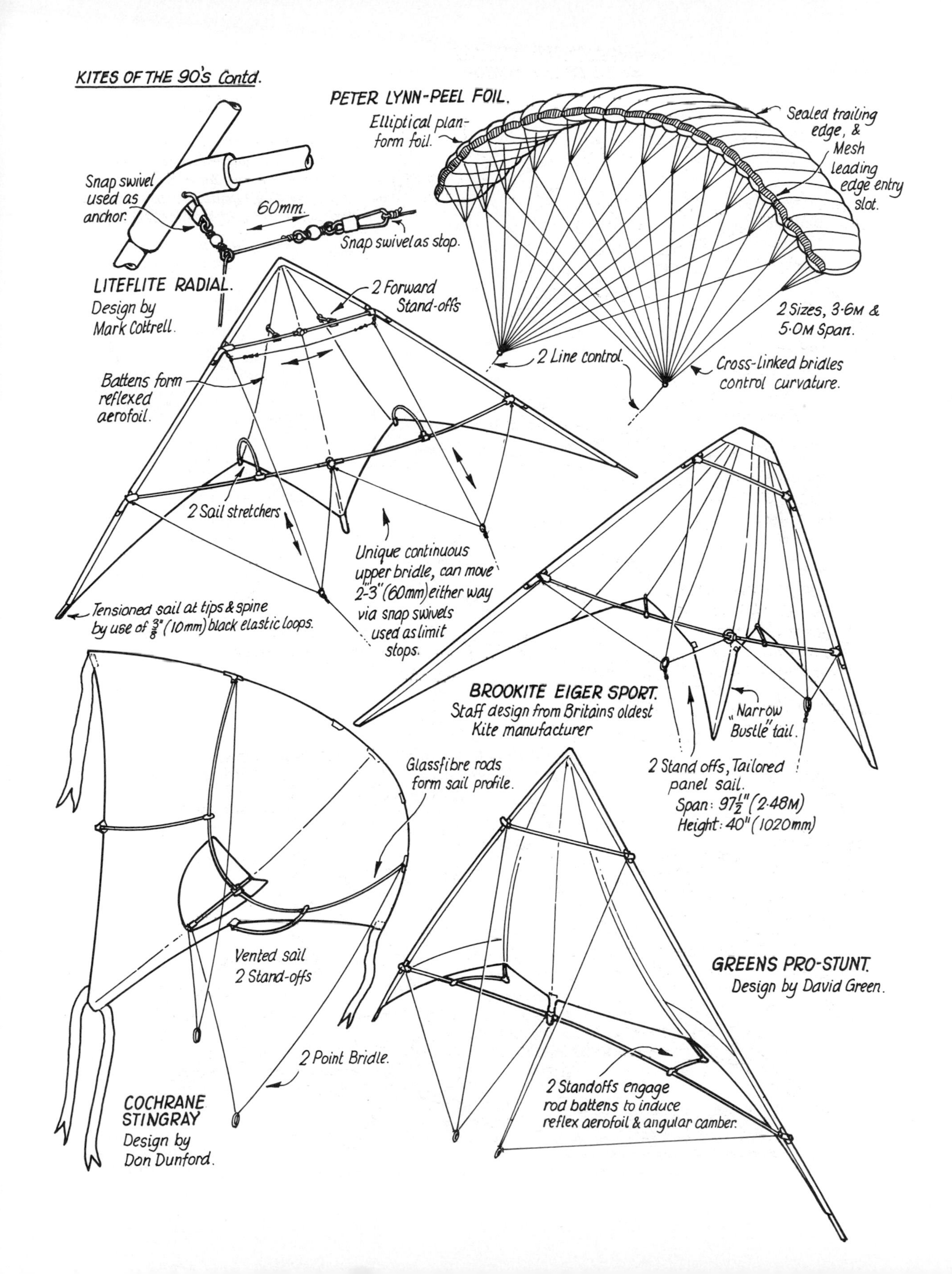
KITES OF THE 90's Contd.
PETER LYNN-PEEL FOIL.
Elliptical plan-form foil.
Sealed trailing edge, & Mesh leading edge entry slot.
2 Sizes, 3·6M & 5·0M Span.
2 Line control.
Cross-linked bridles control curvature.
Snap swivel used as anchor.
60mm.
Snap swivel as stop.
LITEFLITE RADIAL.
Design by Mark Cottrell.
2 Forward Stand-offs
Battens form reflexed aerofoil.
2 Sail stretchers
Unique continuous upper bridle, can move 2"-3" (60mm) either way via snap swivels used as limit stops.
Tensioned sail at tips & spine by use of 3/8" (10mm) black elastic loops.
BROOKITE EIGER SPORT.
Staff design from Britains oldest Kite manufacturer
"Narrow "Bustle" tail.
2 Stand offs, Tailored panel sail.
Span: 97½" (2·48M)
Height: 40" (1020mm)
Glassfibre rods form sail profile.
Vented sail
2 Stand-offs
2 Point Bridle.
COCHRANE STINGRAY
Design by Don Dunford.
GREENS PRO-STUNT.
Design by David Green.
2 Standoffs engage rod battens to induce reflex aerofoil & angular camber.

KITES OF THE NINETIES Continued

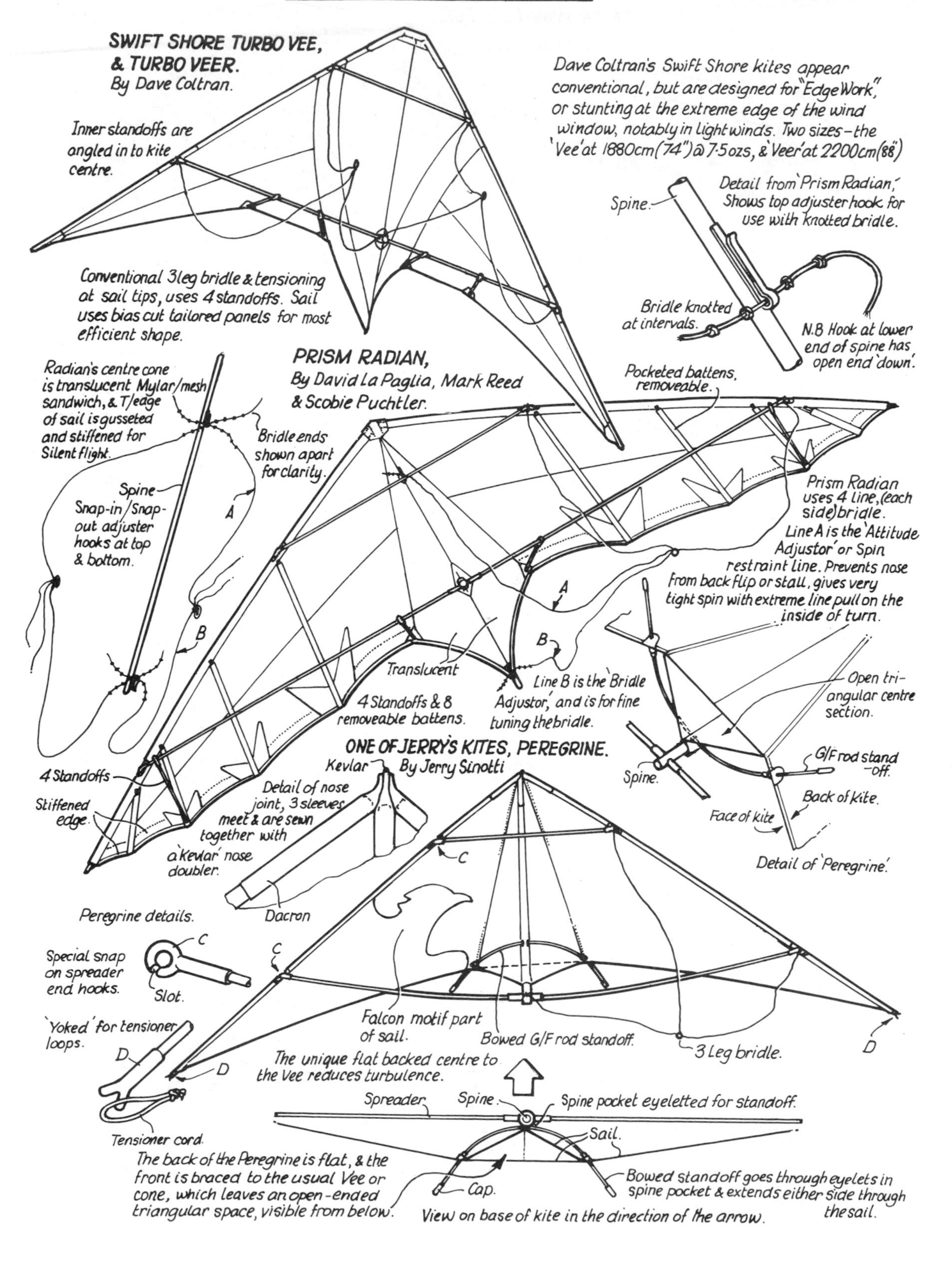

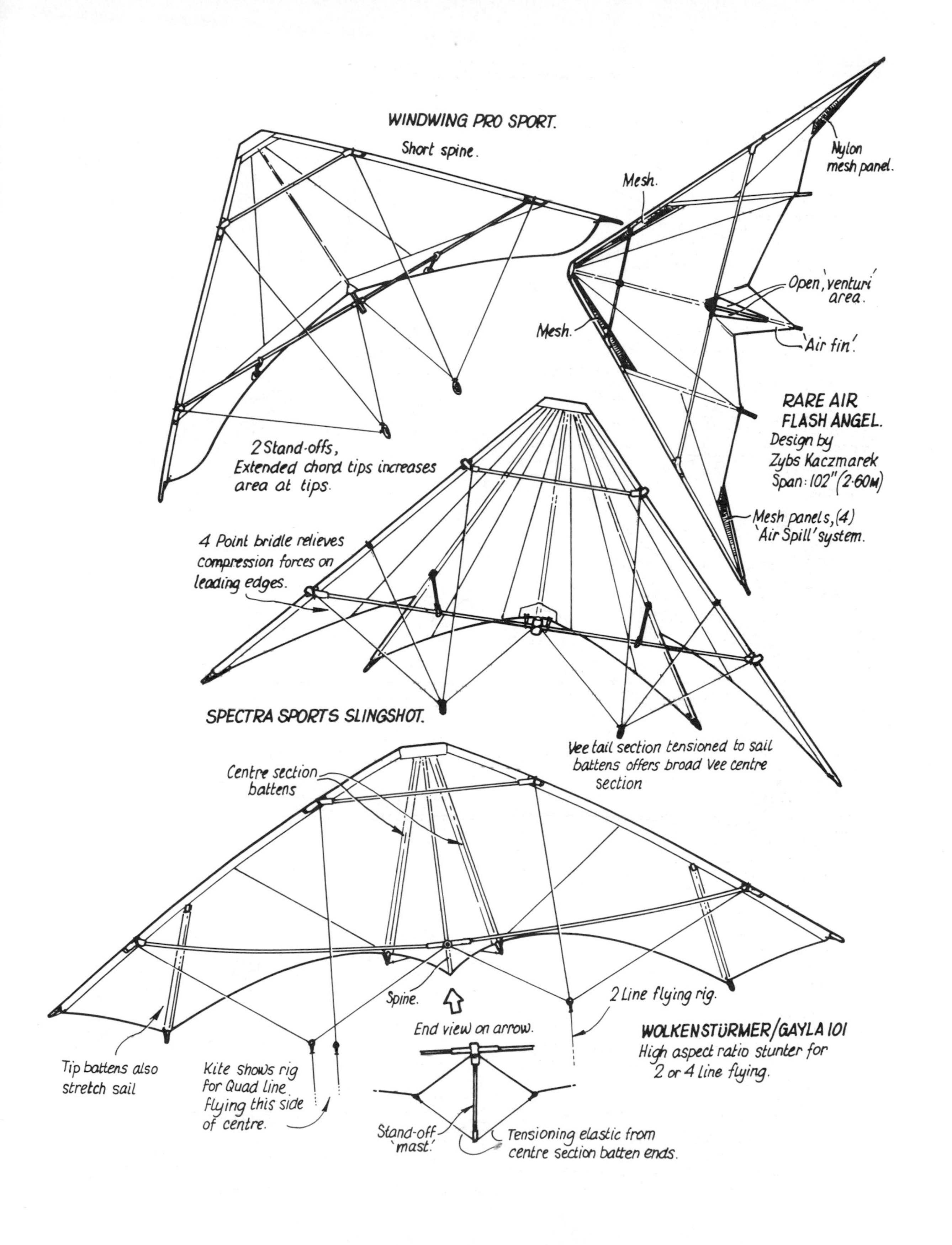

WINDWING PRO SPORT.
Short spine.
2 Stand-offs,
Extended chord tips increases
area at tips.
Mesh.
Nylon
mesh panel.
Open 'venturi'
area.
Mesh.
'Air fin'.
RARE AIR
FLASH ANGEL.
Design by
Zybs Kaczmarek
Span: 102" (2·60M)
Mesh panels, (4)
'Air Spill' system.
4 Point bridle relieves
compression forces on
leading edges.
SPECTRA SPORTS SLINGSHOT.
Vee tail section tensioned to sail
battens offers broad Vee centre
section
Centre section
battens
Spine.
2 Line flying rig.
End view on arrow.
WOLKENSTÜRMER/GAYLA 101
High aspect ratio stunter for
2 or 4 line flying.
Tip battens also
stretch sail
Kite shows rig
for Quad line
flying this side
of centre.
Stand-off
'mast.'
Tensioning elastic from
centre section batten ends.

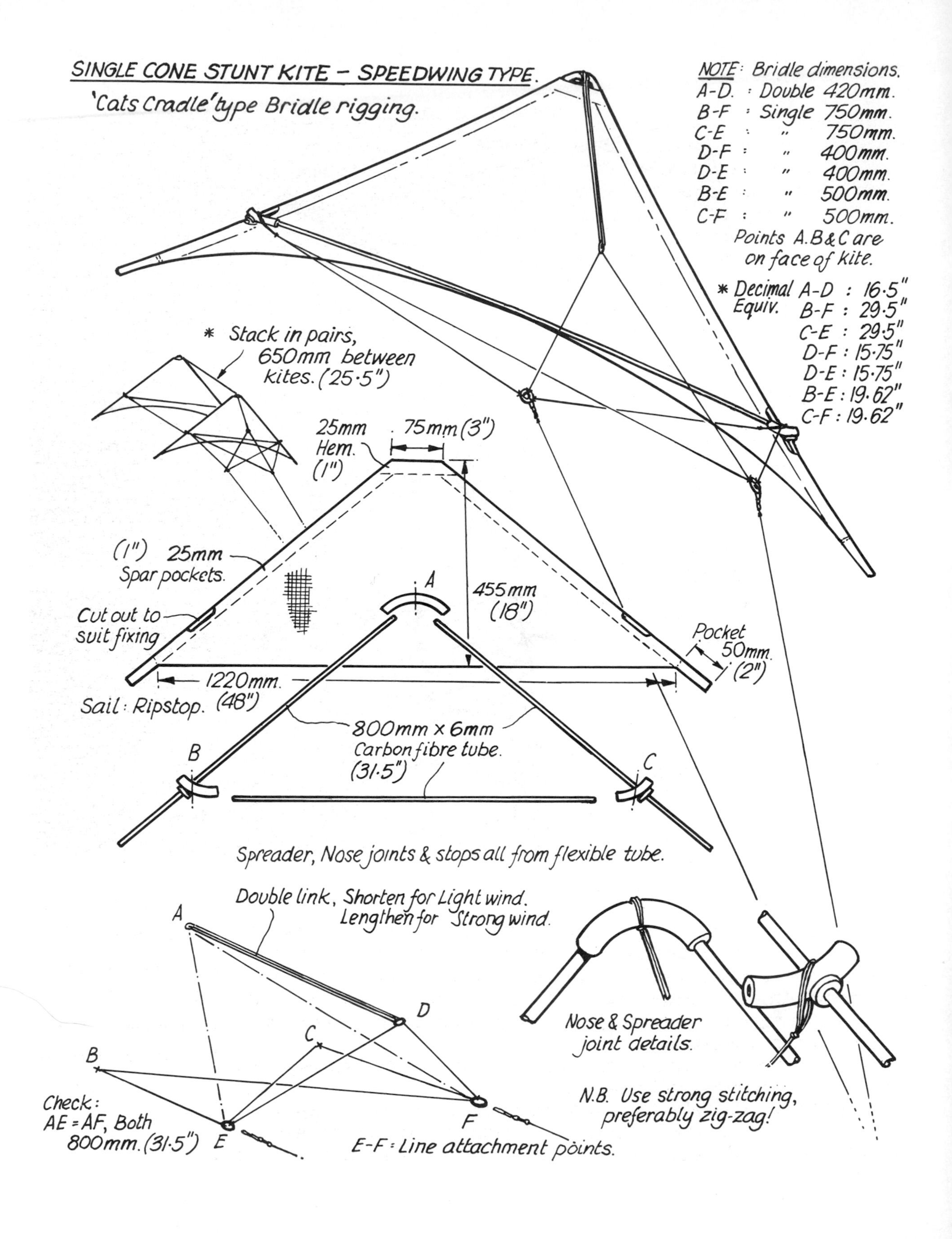
SINGLE CONE STUNT KITE – SPEEDWING TYPE.
'Cats Cradle' type Bridle rigging.
NOTE: Bridle dimensions.
A-D. : Double 420mm.
B-F : Single 750mm.
C-E : " 750mm.
D-F : " 400mm.
D-E : " 400mm.
B-E : " 500mm.
C-F : " 500mm.
Points A.B & C are on face of kite.
* Decimal Equiv.
A-D : 16·5"
B-F : 29·5"
C-E : 29·5"
D-F : 15·75"
D-E : 15·75"
B-E : 19·62"
C-F : 19·62"
* Stack in pairs, 650mm between kites. (25·5")
25mm Hem. (1")
75mm (3")
(1") 25mm Spar pockets.
Cut out to suit fixing
A
455mm (18")
Pocket 50mm. (2")
1220mm. (48")
Sail: Ripstop.
800mm x 6mm Carbon fibre tube. (31·5")
B
C
Spreader, Nose joints & stops all from flexible tube.
Double link, Shorten for Light wind. Lengthen for Strong wind.
A
D
C
B
Check: AE = AF, Both 800mm. (31·5")
E
F
E-F = Line attachment points.
Nose & Spreader joint details.
N.B. Use strong stitching, preferably zig-zag!

expect a constant windspeed. Selection of a purchase could well depend upon the standard of manufacture.

Neat sewing, balloon seams or double stitched ripstop edges, reinforced spar pockets, tight vinyl or rubber fittings, non-slip bridle attachment points, robust spar caps and strong elastic sail tensioners should all be examined on purchase of what amounts to the most expensive division of the whole commercial kite spectrum.

In summary,what we term the 'third generation' of swept wings, are those with the myriad gimmicks and fancy names among the hundreds of stunters that emerged so quickly when common availability of carbon tubes made the rigid frame possible.

Simultaneous with the emergence of the swept wing shape and its cambered sails stretched either side of a central spine, there were other innovations which arrived during the same development period.

Top: Train of Speedwing scoops, a simple though very entertaining kite. Above: Cat's Cradle on a typical scoop kite links leading edge and the cross-spar with two triple bridles.

Scoops

Quite apart from the desire for steady and precise steering qualities as offered by the conical camber of the Rogallo sail forms, there always have been kite flyers who were less interested in 'Figurekiting' and more inclined to the spectacular fast flying trains of smaller kites with linked bridles.

These generally took the form of Deltas with a spine and single cross-spar. Two points on the spine were used as bridle attachments and a third at the cross-spar connection to the leading edge. Some variations had extended trailing edges and consequently longer spines. They performed well - many have sustained their popularity for years mainly because they give excellent performance for low cost.

Then the 'Scoop', or single cone kite emerged in 1988. The spine was omitted, and a 'cats-cradle' bridle cross-linked the trihedron arrangement of the bridle lines so that each end of the cross-spar and the leading edge nose joint carried two of the lines. Patented in Germany and the USA by Edmund Heid of Berlin, the concept was not entirely new to British kite flyers. Refined as the 'Speedwing' in production with a variation on the patent, it has become very popular worldwide. Simplicity is possibly its greatest feature, and with the use of a doubled line link from the nose to the bridle on the production Speedwing to give adjustment for wind strength, the advantages are emphasised all the more.It flies best when coupled in pairs or triples.

The language of the patent, ambiguous as it always is, emphasised how elimination of the spine could remove the risk of distortion and subsequent uneven distribution of pressure. It was referring only to the single cross-spar stunter where two bridle points are used fore and aft on the spine.

The real advantage of the single cone lies in its lighter weight for its area, and the deeper cone in the sail which generates significant lift and is capable of very fast rotation when the centre of pressure is displaced either side of centre. For some it is too fast, particularly in stronger winds, hence the trend to 'stack' in pairs or triples to slow them. The pull of a triple demands 80kg (200lb) breaking strain line.

There is one penalty - noise! The full length of unsupported trailing edge on the comparatively deep cone will vibrate at considerable amplitude. As the kites are designed to be flown with verve and pulled through sharp manoeuvres, they buzz like a horde of hornets. All great fun for the flyer but not for those inclined to complain about noise pollution.

A larger version, called the Super-Speedwing was introduced for 1992 with extra panels, a deeper cone and a trailing edge hem, all of which combined to reduce the noise element to claims of 'near silent'.These findings were subsequently applied to the standard size kite.

David Clarke of Windy Kites with his 'Silent K(n)ite' creation using trailing edge stiffening.

Silence is golden?

Coincident with the attention to reduce noise on the German Speedwing, David Clarke of Windy Kites in the UK had been toying with the concept of a kite with a rigid trailing edge. Choosing to follow the Rogallo principle to the letter, he evolved a near perfect conical camber design with conventional structure of leading edges,cross-spar and spine in carbon tube. The difference came in the trailing edge where 3mm glass fibre rod was retained in a hemmed sleeve.

Shaped as a parallelogram with equal length sides inclined at 60 degrees the sail was pulled into two deep cones by controlling the length of the cross-spar. Being in compression, the trailing edge hoops retained a natural shape. To distribute compression loads on the spars and spine, the bridle trihedrons went direct to each extremity.What could be more simple?

Flying this kite among the otherwise conventional swept wings on a breezy day was initially eerie. It was absolutely silent.In strong wind it was as nimble as one might wish and had exceptional stability at the extreme sides of the

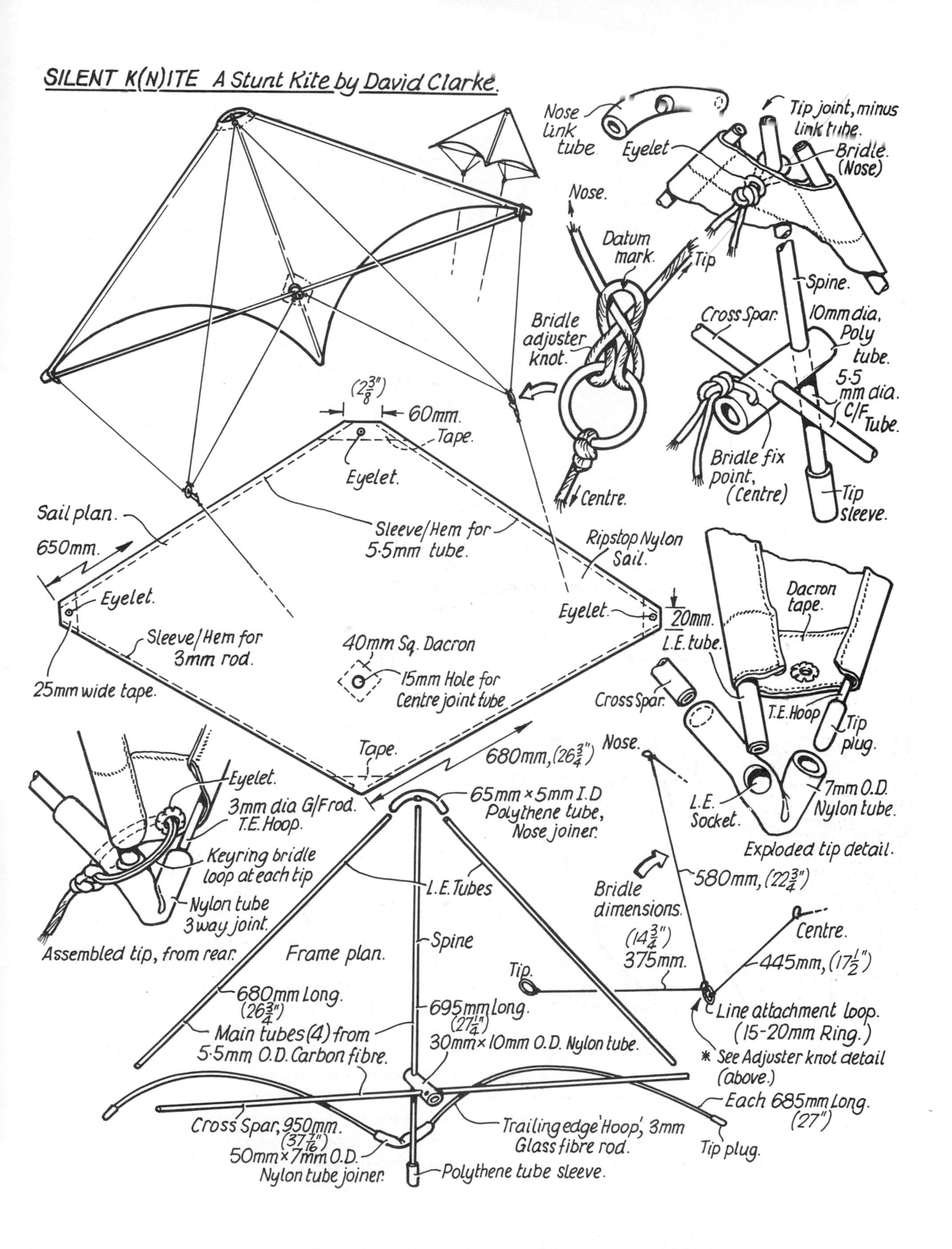
SILENT K(N)ITE A Stunt Kite by David Clarke.
Nose link tube
Tip joint, minus link tube.
Eyelet
Bridle. (Nose)
Nose.
Datum mark.
Tip
Bridle adjuster knot.
Centre.
Spine.
Cross Spar.
10mm dia. Poly tube.
5·5 mm dia. C/F Tube.
Bridle fix point, (Centre)
Tip sleeve.
(2 3/8")
60mm.
Tape.
Eyelet.
Sail plan.
650mm.
Sleeve/Hem for 5·5mm tube.
Ripstop Nylon Sail.
Eyelet.
Eyelet.
20mm.
L.E. tube.
Dacron tape.
Sleeve/Hem for 3mm rod.
40mm Sq. Dacron
15mm Hole for Centre joint tube
25mm wide tape.
Cross Spar.
T.E. Hoop
Tip plug.
Tape.
680mm, (26 3/4")
Nose.
L.E. Socket.
7mm O.D. Nylon tube.
Exploded tip detail.
Eyelet.
3mm dia G/F rod. T.E. Hoop.
65mm × 5mm I.D Polythene tube, Nose joiner.
Keyring bridle loop at each tip
Nylon tube 3 way joint.
Assembled tip, from rear.
L.E. Tubes
Bridle dimensions.
580mm, (22 3/4")
Frame plan.
Spine
(14 3/4") 375mm.
Centre.
445mm, (17 1/2")
Tip.
680mm Long. (26 3/4")
695mm Long. (27 1/4")
Line attachment loop. (15-20mm Ring.)
Main tubes (4) from 5·5mm O.D. Carbon fibre.
30mm × 10mm O.D. Nylon tube.
* See Adjuster knot detail (above.)
Each 685mm Long. (27")
Cross Spar, 950mm. (37 7/16")
Trailing edge 'Hoop', 3mm Glass fibre rod.
Tip plug.
50mm × 7mm O.D. Nylon tube joiner.
Polythene tube sleeve.

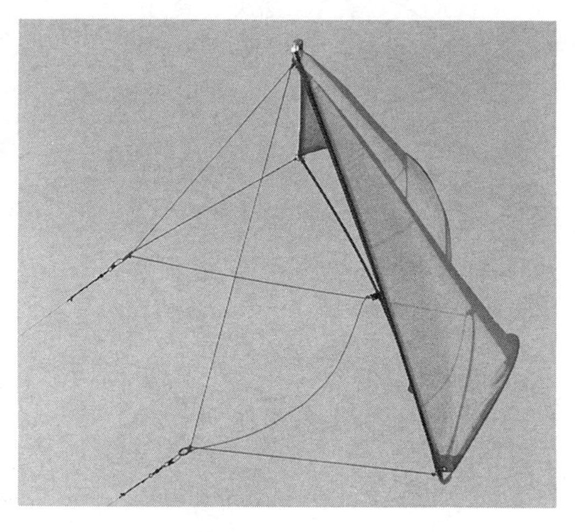

Silent K(n)ite being held at the edge of the wind window, displays its Rogallo form.

wind window. Calmer conditions limited the stunts one could perform; but here at last was a totally Silent K(n)ite!

Other designers were also on the track of trailing edge resonance. Extra sail stretchers either in the form of rigid stand-offs made of carbon fibre rod, or flexible glass fibre rods forced by compression into a curve which push the sail upwards and backwards, have become regular features.These serve to dampen the flutter, and to hold the sail form during the critical launch stage in low wind speeds.Trailing edge stiffening with mylar film or mesh overlay is used on some designs.

Flying wings

Similarity with flying wing aircraft had become apparent as the rigid frame evolved. In aircraft ranging from the Lippisch deltas to Horten sailplanes and the more recent hang gliders, the wing profile or aerofoil is twisted from a positive angle at the centreline to negative at the wingtips. As we have already explained, such a change in angle of attack is formed naturally by the conical sailform on a delta; but it needs to be held in place on sharply tapered types.

Among the third generation of swept wings, many had an included leading edge or nose angle greater than the basic 90 degrees - more usually 105 degrees. The shallower angle meant that the aspect ratio of chord to span was higher. Additionally the spine lengths were reduced and sweepback on the trailing edge made the tip panels narrow and very pointed.

This was intended to make the turn radius smaller and it worked as long as there was a strong wind. Otherwise, a tip would stall, lose lift, and the kite flipped over and fell. So to ensure that the tip panels were reflexed, and the trailing edge was taut, extra stand-offs or stretchers (Flexifoil called theirs 'Flexors' on the Scorpion) came into vogue. Apart from improving efficiency and reducing noise nuisance, the twist on a swept wing harks back to Don Dunford's original statement concerning relative incidence fore and aft making his kite so stable and controllable. The tips of a swept wing are well aft of the centre of lift, and *have* to be negative in relation to the centrebody forward of the bridle points.

The scoop - type single cone kite is an exception to this, but Messrs Hoffrei of Herford in Germany introduced a novel 'X' shaped cross-spar on the Vision II which enabled sail tension and a taut spinal form to be maintained with a single stand-off. This also permitted a closer coupling of the leading edge bridle attachments, while at the same time the 'cat's cradle' cross-linking of

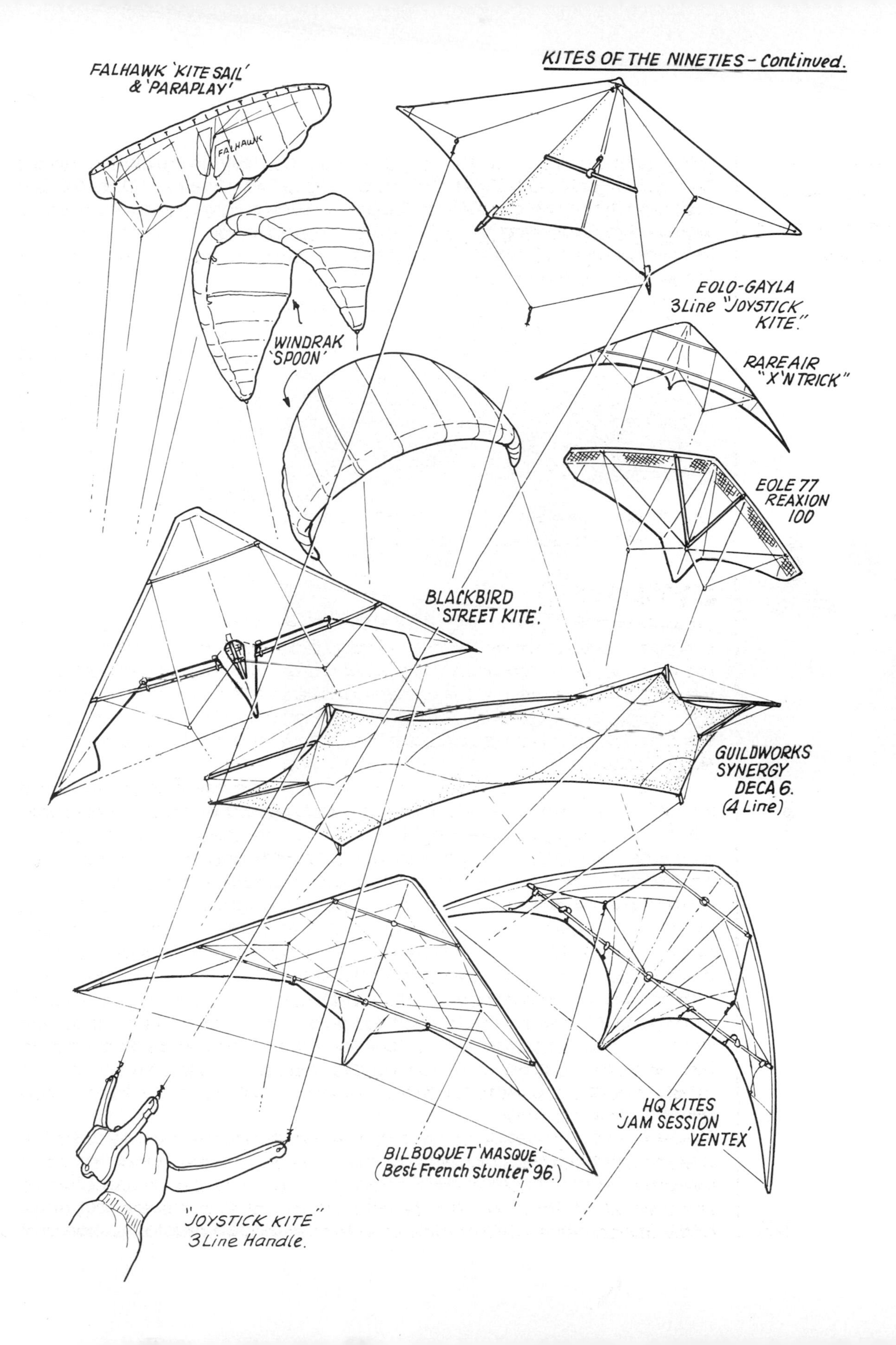
KITES OF THE NINETIES - Continued.
FALHAWK 'KITE SAIL' & 'PARAPLAY'
FALHAWK
WINDRAK 'SPOON'
EOLO-GAYLA
3Line "JOYSTICK KITE."
RAREAIR "X'N TRICK"
EOLE 77 REAXION 100
BLACKBIRD 'STREET KITE'.
GUILDWORKS SYNERGY DECA 6. (4 Line)
HQ KITES 'JAM SESSION VENTEX'
BILBOQUET 'MASQUE' (Best French stunter '96.)
"JOYSTICK KITE" 3Line Handle.

the inner bridle lines served under flight tensions to restrain the leading edges from spreading apart. The result was a 150g (5.25oz) kite of 110cm (43.3in) span with 32dm2 (3.4 sq ft)of sail area. With such a light loading the Vision II would have a sparkling performance in a broad range of wind speeds; but its inverted vee form is a bit of a paradox after all we have stated in favour of the twist in a swept wing!

Foils

Created in Germany and made in the USA, this fun-fly soft kite comes in Snoopy and Red Baron variations.

Another contradiction of the twisted wing theory is the air-filled steerable soft kite, as originated by the Flexifoil mentioned earlier. Having a double surface which is held to an aerofoil section by the risers sewn into the body, the foils are quite different to the otherwise flat plate single surface Deltas.

In the case of the Flexifoil, the wing section or aerofoil is one with a distinct reflex. In other words, the undersurface curves upwards as it approaches the trailing edge.In fact this is so prominent in the Flexifoil that the section could be said to be upside down!

The foil is straight, it is a rectangular shape without sweep, and the reflex is adopted from 'Flying Plank' aircraft technique to provide stability. The Flexifoil also has its only spar in a low set leading edge, above which there is first a narrow open mesh air intake slot, and then a steeply cambered upper surface. This curvature is contained in the first quarter of the wing so that the centre of lift is also well forward while the stabilising reflex on the underside provides a down force at the trailing edge.

As a result of this arrangement, the Flexifoil does not require a bridle - the lines are connected directly to the spar ends, and therein lies another unusual feature.As the inflated Flexifoil is launched and airloads take effect with movement, the spar which is tapered in its end components flexes into a curve to create a 'channel wing' like an inverted 'U' with the control lines at each of the cusped ends. In this form, the Flexifoil acquires its own optimum angle of attack to the 'apparent' airflow (more on this later under 'The Flight Envelope') and achieves speeds far greater than actual wind strengths.This is a resultant of the thrust component derived from a forward centre of pressure together with the efficiency of the developed aerofoil section. Just as with the Rogallo wing, or the single cone scoops, any lateral shift of the Flexifoil will steer the kite; but there is a difference in that the single line attachment point allows the Flexifoil to adopt its own twist, and the angle of attack is reduced on the inside of the turn. By sweeping back and forth through figures of eight, a Flexifoil will build up speed even when wind strength is such that without movement or being flown as though on a single line, it will just hover with the spar nearly straight.

Flexifoil sizes vary from the 122cm (48in) Hot Shot with an aspect ratio of 2.7 to the 393cm (154.5in) Hyper 12 of 5.7 aspect ratio. The higher the AR (aspect ratio) or span-over-chord, then in theory, the greater the efficiency. In their mid range, Flexifoil have four kites with the same 61cm (24in) chord. The Stacker and Splitz are similar at 2.9 AR with 178cm (70in) span and the Pro-Team extends to 249cm (98in) for 4.1 AR while the Super-10 goes to 286cm (112.5in) and 4.7 AR.

Although the same in chord (width), the Stacker and Super-10 thus have a difference of 108cm (42.5in) in span or 62% in area; but in terms of efficiency in lift the increase is nearer to 400%. No small wonder the Super-10 is emphasised as *not suitable for children!*

Collectively known as 'Power Kites', the Flexifoils have a wide repertoire, from Individual and team ballet performances to sprint flying through a speed trap, man-lifting power-kiting, beach skiing, buggy towing, cable riding and establishing a world sailing speed record of 25.03 knots (46.6km/h or 29mph) with fifteen Super-10s towing a catamaran.

Sparless foils

The air-filled Parafoil has long been established as a stable single line kite so its development as a steerable stunter for two (or more as we shall see) lines came initially as a surprise knowing only too well how difficult it is sometimes to get a Parafoil started.

The 'Sparless Stunter' is a 6 cell foil with an AR of less than one. It is longer than it is wide, and it is modelled directly on the Parafoils. Another US product is the 'Skynasaur Stuntfoil' which comes in various sizes and incorporates an airfilled leading edge chamber to aid lateral rigidity. This has an AR of two and upwards,and is not at all like the Flexifoil in concept. Nearer to the shape of the Flexi, though sparless and using a multiple line bridle system, is the Wolkensturmer Paraflex with a span of 240cm (94½in)and sixteen cells which give it exceptional lifting power. It maintains the rectangular format, with the square tips carrying keels for the outer bridles. One wonders how long this shape will persist in the face of rapid changes that have revolutionised the Paragliding and Parapenting sports. In very few seasons, the practice of jumping off mountains with a ram-air parachute in tow, and prolonged descents with steerable harness has accelerated design at a rapid rate.

The man-carrying steerable parachute has adopted elliptical shapes with downturned tips. Collectively known as 'canopies',they have utilised brilliant colours in ripstop nylon to add to the attraction of their shapes, and just as with our kites, have created flash names to compete in a crowded marketplace. Taboo, Magic Kiss, Amour, Elan, Apex, Typhoon, Voodoo, Cobra, Firebird and The Twist are typical. Designed for glide ratio rather than aerobatics, the paragliders nevertheless can teach the foil-kite designer something, just as we strongly suspect they in turn, had already learned from the Flexifoil.

The ubiquitous Peter Lynn was first to apply the tapered planform to a foil for steerable aerobatics. In deference to its similarity to a piece of an orange it was dubbed the 'Peel' and like the man-carrying paragliders of similar shape, it was immediately adopted by the kite fraternity worldwide.

Peter Lynn, the New Zealander who features at many international festivals, in action on his buggy towed by a 'Peel' kite at Weymouth.

With bridle lines cross-linked and adjusted in length to hold a curved form and also to distribute the lift loads, the Peel comes in two sizes both of which are comparatively large for stunters. The 3.6 metre (142in) has a surface area of 2.5 sq m(27 sq ft) and its larger variant with 5 metres span (16½ ft.) has 5 sq m area or 54 sq ft! No wonder it tows the Peter Lynn Buggy (with himself steering on board) all over wherever land surfing and similar buggy racing is permitted.

Let's go flying!

You've been to the kite shop and bought your first steerable. One of the first discoveries will most likely be the paucity of instructions! Despite the high cost of these kites, due mainly to the involved sail, carbon spar and a fair chunk of distribution costs through the marketing chain, the fact remains that few of them are fully supported by adequate instruction on how to assemble or fly the device. It is as though steerable kites are regarded like bicycles, to be controlled by intuitive and natural reaction, therefore no need for instruction. Let nature take its course!

It could well be said that once the 'feel' of control is acquired then all you need is practice, and just like riding the bicycle, all the arm movements become a spontaneous impulse.But life isn't that easy, and in our experience, the 'prangs' have to be suffered before the 'knack' of dexterous control is accomplished.

Maybe you've found the cost of a sophisticated stunt or sport kite too expensive and want to make your own. In that case we firmly say make it simple and don't aim for the moon at first try. Something like the 80-degree tri-angled shape with straightforward plain panels, glass fibre tube spars with basic joints, no stand-offs or stretchers and bridle points that will permit a wide range of trimming is an ideal introduction.

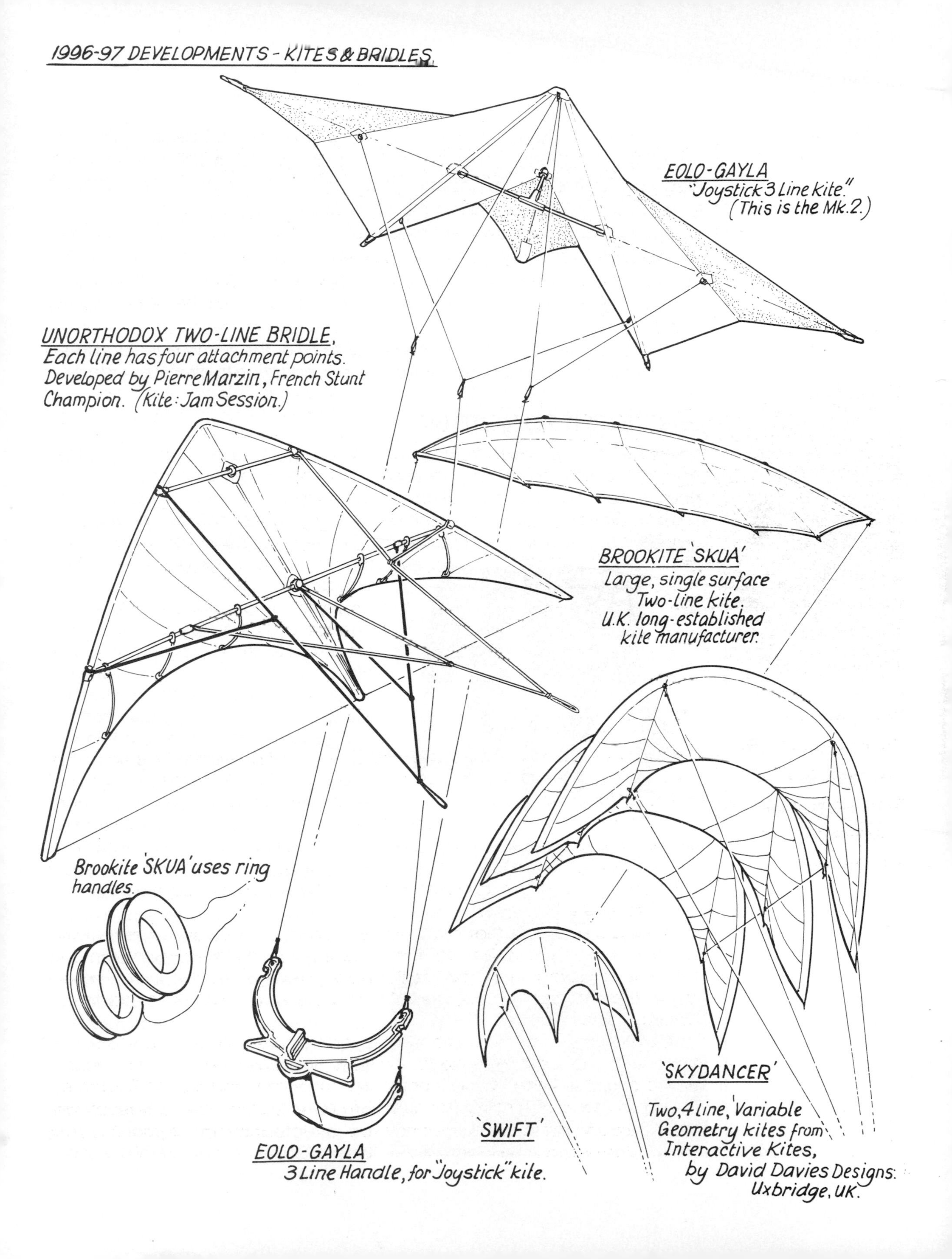
1996-97 DEVELOPMENTS - KITES & BRIDLES.
EOLO-GAYLA
"Joystick 3 Line kite."
(This is the Mk.2.)
UNORTHODOX TWO-LINE BRIDLE.
Each line has four attachment points.
Developed by Pierre Marzin, French Stunt
Champion. (Kite: Jam Session.)
BROOKITE 'SKUA'
Large, single surface
Two-line kite.
U.K. long-established
kite manufacturer.
Brookite 'SKUA' uses ring
handles.
'SKYDANCER'
Two, 4 line, 'Variable
Geometry kites from
Interactive Kites,
by David Davies Designs.
Uxbridge, UK.
'SWIFT'
EOLO-GAYLA
3 Line Handle, for "Joystick" kite.

Or maybe you would prefer to go straight into the 100 degree kite as in Tim Benson's Scorcher opposite. This is a popular design which has proven ideal as a starter design, capable of all the intricate manoeuvres sufficient to satisfy the novice and even the most demanding competition flyer. Of course, if carbon tubes are used instead of glass fibre, and the sail is extended into angular camber by fitting a pair of vertical stand-offs from the rear cross-spar to the sail trailing edge, then performance will be enhanced. This is a good design which accepts experiment but still offers all the novice seeks, as it comes.

Whatever the situation, with the ready-to-fly or the home-built sport kite, there still remains a set sequence of assembly, common to almost every type.

Getting it together

When the kite is removed from its bag, it is most likely to be folded short. The rearmost sections of the leading edges might be detached from the joining ferrules. The first clue to this is to find that the bag itself is folded over and much longer than its contents. Spread the sail and lay it flat as possible with the underside uppermost. The bridle lines should be straightened and laid so that the connector rings fall evenly either side of the spine.

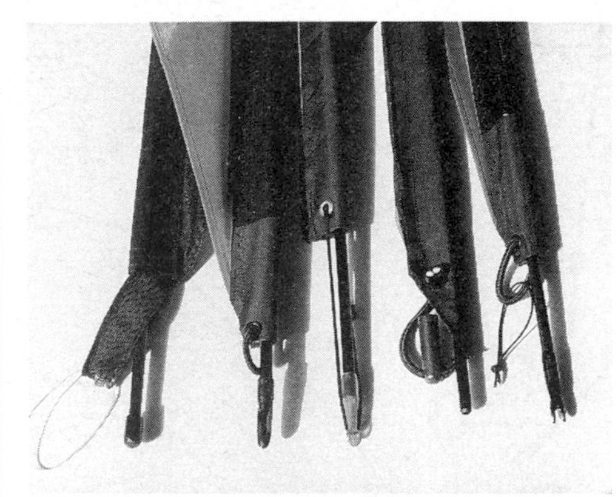

Tail ends of leading edges, L to R, the Liteflite elastic pocket, Scorcher rubber bungee cord and nock, Jester elastic cord and nock, Eolo-Gayla Wind-walker bungee and end cap, Scorpion bungee with pull cord and nock. Each used to stretch the sail.

Centre point – keep bridle lines clear of the cross-spar.

Assembly should be in the sequence of 1) leading edges 2) front cross-spar 3) rear cross-spars 4) stand-offs and stretchers. At each stage, ensure that the bridle lines are free of the spars, but secure at their proper anchor points. The tangled bridle is a common mistake. Starting with the leading edges, push the loose rear section forwards in its sleeve with one hand while holding the tubular end of the joining ferrule in the other. This enables you to guide the rear section home without catching the inner surface of the Dacron sleeve and helps to make sure that the sections are fully joined. The rearmost ends will project beyond the sleeve and may have arrow knocks fitted. In such a case, the tube must be rotated before the elastic sail tensioner is stretched into position in the knock slot. It has to be aligned properly, otherwise the tensioned sleeve can be affected by the torsion of the elastic. Some kites avoid this by using

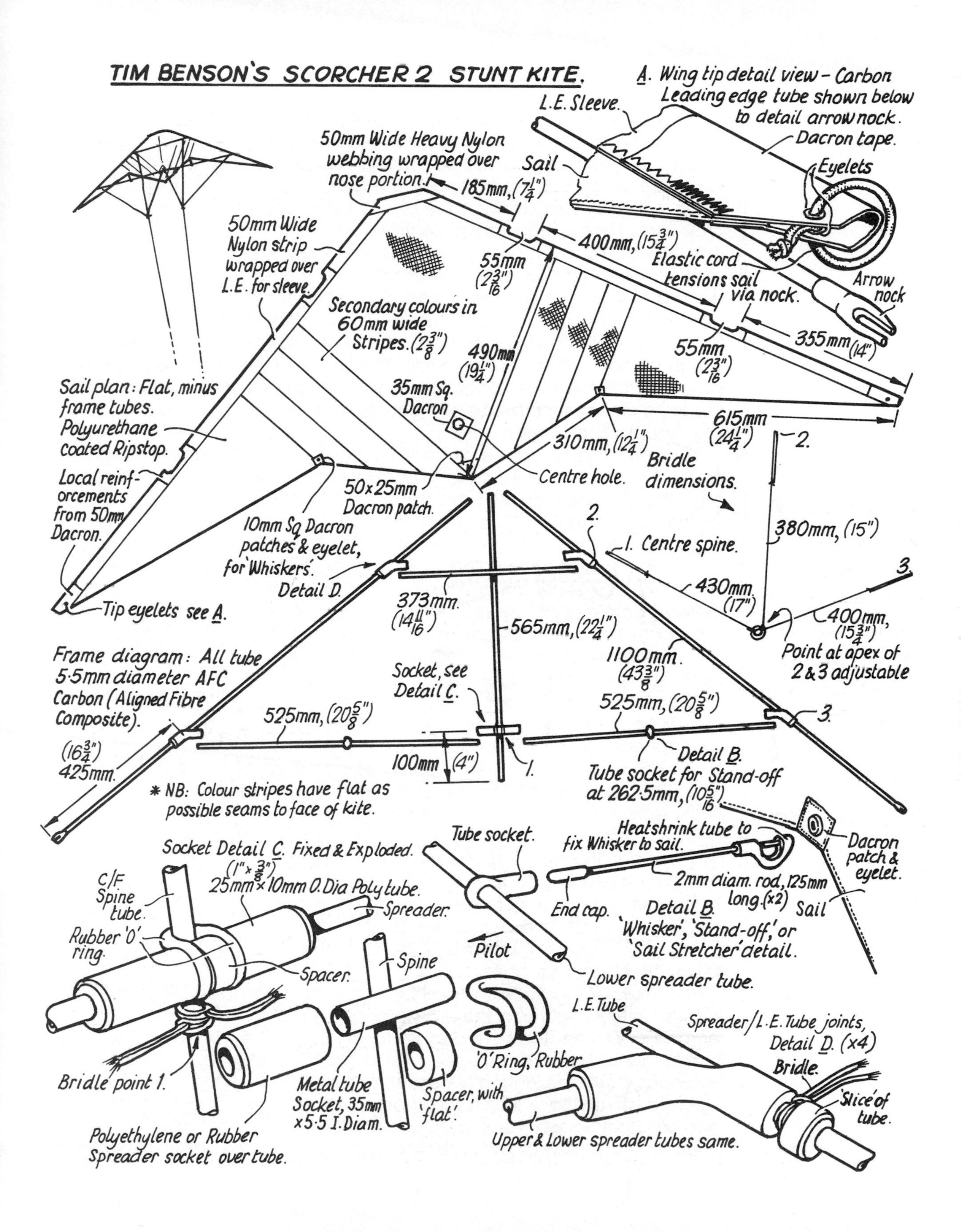
TIM BENSON'S SCORCHER 2 STUNT KITE.
A. Wing tip detail view - Carbon Leading edge tube shown below to detail arrow nock.
L.E. Sleeve.
Dacron tape.
Sail
Eyelets
Elastic cord tensions sail via nock.
Arrow nock
50mm Wide Heavy Nylon webbing wrapped over nose portion.
185mm, (7¼")
400mm, (15¾")
55mm (2 3/16")
55mm (2 3/16")
355mm (14")
50mm Wide Nylon strip wrapped over L.E. for sleeve.
Secondary colours in 60mm wide Stripes. (2⅜")
490mm (19¼")
35mm Sq. Dacron
615mm (24¼")
310mm, (12¼")
Centre hole.
Bridle dimensions.
2.
Sail plan: Flat, minus frame tubes. Polyurethane coated Ripstop.
Local reinforcements from 50mm Dacron.
50x25mm Dacron patch.
10mm Sq Dacron patches & eyelet, for 'Whiskers'.
Detail D.
Tip eyelets see A.
2.
1. Centre spine.
380mm, (15")
430mm. (17")
3.
400mm, (15¾")
Point at apex of 2 & 3 adjustable
373mm. (14 11/16")
565mm, (22¼")
1100mm. (43⅜")
Frame diagram: All tube 5·5mm diameter AFC Carbon (Aligned Fibre Composite).
Socket, see Detail C.
525mm, (20⅝")
525mm, (20⅝")
3.
(16¾") 425mm.
100mm (4")
1.
Detail B.
Tube socket for Stand-off at 262·5mm, (10 5/16")
* NB: Colour stripes have flat as possible seams to face of kite.
Socket Detail C. Fixed & Exploded.
C/F Spine tube.
(1" x ⅜") 25mm x 10mm O. Dia Poly tube.
Spreader.
Rubber 'O' ring.
Spacer.
Spine
Bridle point 1.
Metal tube Socket, 35mm x 5·5 I. Diam.
Spacer, with 'flat'.
Polyethylene or Rubber Spreader socket over tube.
Tube socket.
Pilot
Lower spreader tube.
'O' Ring, Rubber
Heatshrink tube to fix Whisker to sail.
Dacron patch & eyelet.
2mm diam. rod, 125mm long. (x2)
End cap.
Sail
Detail B. 'Whisker', 'Stand-off', or 'Sail Stretcher' detail.
L.E. Tube
Spreader/L.E. Tube joints, Detail D. (x4)
Bridle.
'Slice' of tube.
Upper & Lower spreader tubes same.

wide elastic pockets, or plastic end caps on the tensioners. Nocks are vulnerable when landing; but can themselves be capped with a length of vinyl tube. When pushing the leading edges into place, it is inevitable that some movement will displace the plastic tube, rubber or vinyl joints for the cross-spars. Locate these fittings central in each of the sail cutouts,and check that they are horizontal once the sail is tensioned.

The front or upper cross-spar is normally one-piece. It may carry fittings for stand-offs, in which case the cross-spar must be rotated so that the fittings point in the correct direction,then the spar can be inserted into the vinyls. Now fit the rear or lower cross-spars. It is usual for them to come in two equal length pieces to fit a central ferrule. This is part of a tee piece projecting vertically from the spine, located according to the sail shape at the end of the spine or within the profile of the extended 'bustle' tail. Early short spine kites had horizontal tee pieces.

An alternative to the tee piece is the clever use in Tim Benson designs of an 'O' ring to hold the ferrule or socket directly against the spine, then polyethylene rubber or vinyl tubes form combined stops and securing sockets for the cross-spars (see 'Scorcher 2' drawing on previous page).

Whichever the centre fixing, there are two vital precautions to observe *before* the cross-spars are fitted, First, check the positions of the stand-off fittings. It is almost certain that they will not be central on the spar halves so there will be a *right* way around. Align the spar half with the sail (not forgetting to bring the tip closer to the spine to simulated the sail form in flight) and check the stand-off junction with the sail. When there are more than one stand-off per side, there is little doubt as to which way round the spars are fitted.

The second precaution is to pull each of the bridle rings up and away from the sail before positioning the cross-spar first into the centre ferrule with the spar *beneath all the bridle lines*. It is most important that the lines which we will call 'B' and 'C' as in the drawings, are *free and not in any way tangled with either the leading edge vinyl, spine fitting or the cross-spar*. Otherwise the bridle setting and consequent flight performance will be sadly deformed!

Having made sure that the rear cross-spars are correctly located, the bridle is clear and the stand-off sockets in their right places, push the spar ends well into the leading edge vinyls.

In some instances where sail stretchers are in flexible glass fibre (as distinct from the rigid carbon fibre rod used for stand-offs) it may be necessary to fit the stretchers *before* the spar is pushed into the leading edge vinyl in order to get the correct sail deflection.

In all cases, do not alter the position of any fittings on the cross-spars - they should be a very tight fit for the very purpose of staying where set! All that now remains is to fit the stand-offs and stretchers (also called whiskers or whiskas in some kites) which enable the kite to adopt its conical shape, and as described earlier, to induce 'washout' or reflex angle at the tips.

Stand-offs on the forward cross spar provide positive if angular camber to form an aerofoil, but if used to force a batten into such a profile, the effect of further stand-offs from the rear cross-spars on the same batten is to form a gentle reflex, not unlike the wing section of a soaring bird. Mark Cottrell's 'Radial' was among the first to have this feature.

With all the pieces in place, we suggest that the bridle should now claim full attention as its symmetry and adjustment will be critical for best performance.

Bridle check

The most common bridle on narrow angle swept-wing sport kites has three lines, which we will refer to as a Trihedron. There are variations, the most distinctive being the addition of a fourth line to the leading edge, midway between the two cross-spars. This is to relieve the compression loads which are considerable. When glass fibre tubular spars were standard practice, leading edge distortion would form an otherwise straight line into a swan-neck shape! Similarly, the spine would distort under air loads, hence the rather short spines of earlier designs. Carbon fibre tubes have cured much of this;but the fourth bridle line is used on larger sport kites even though it requires accurate adjustment to balance with the other lines.

Top pic: The WRONG way to link the bridle to the leading edge, not connecting with the cross-spar as in the lower pic where the bridle wraps both leading edge and cross-spar,

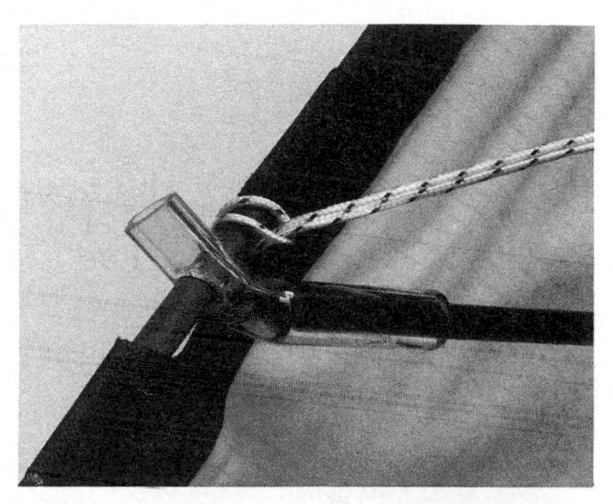

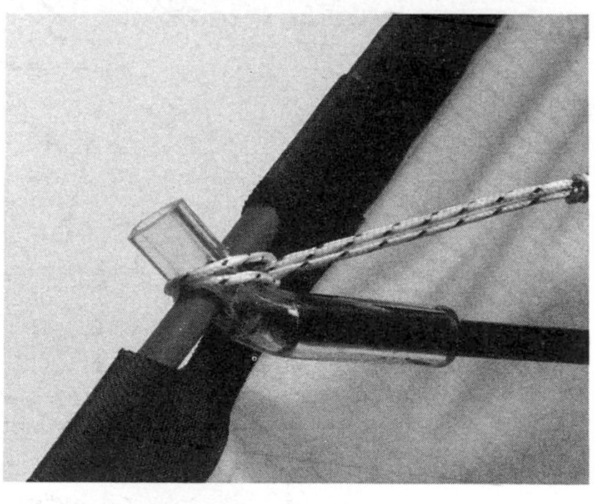

Manufacturers set the bridles with a secure knotted loop at each of the spar attachments. Check to see that they *are* secure and looped the same way on each half of the kite. Some have a loop around the spar, between stops, others have loops which are intended to go around the spar *and* the extension of the vinyl sockets. The latter relax annoyingly when not in use and tend to slide out of place. So check thoroughly - there is no point in our going further to describe how small adjustments on the connecting ring can affect performance when the bridle loops to the leading edge are allowed to slide out of position!

Leaving bridle line 'C' slack, pull on the connector ring (sometimes a snap and swivel is fitted instead) so that lines 'A' and 'B' are taut and the ring is held down at the level of the rear cross-spar. The two rings, either side of the spine, should be at the same distance below the cross-spar, and symmetrical either side of the kite centreline (spine).

If they are not, then the manufacturer set-up might be wrong. However this is most unlikely. Before making any alterations, let line 'B' be slack and pull 'A' and 'C' for a second check on symmetry. This will quickly identify which line is short, or possibly too long, and the loop connection to the leading edge on that line should be examined. It may have slipped. If so, the cure will be obvious; but there is still one other possibility. The 'Lark's Head' may have slipped on the connector ring.

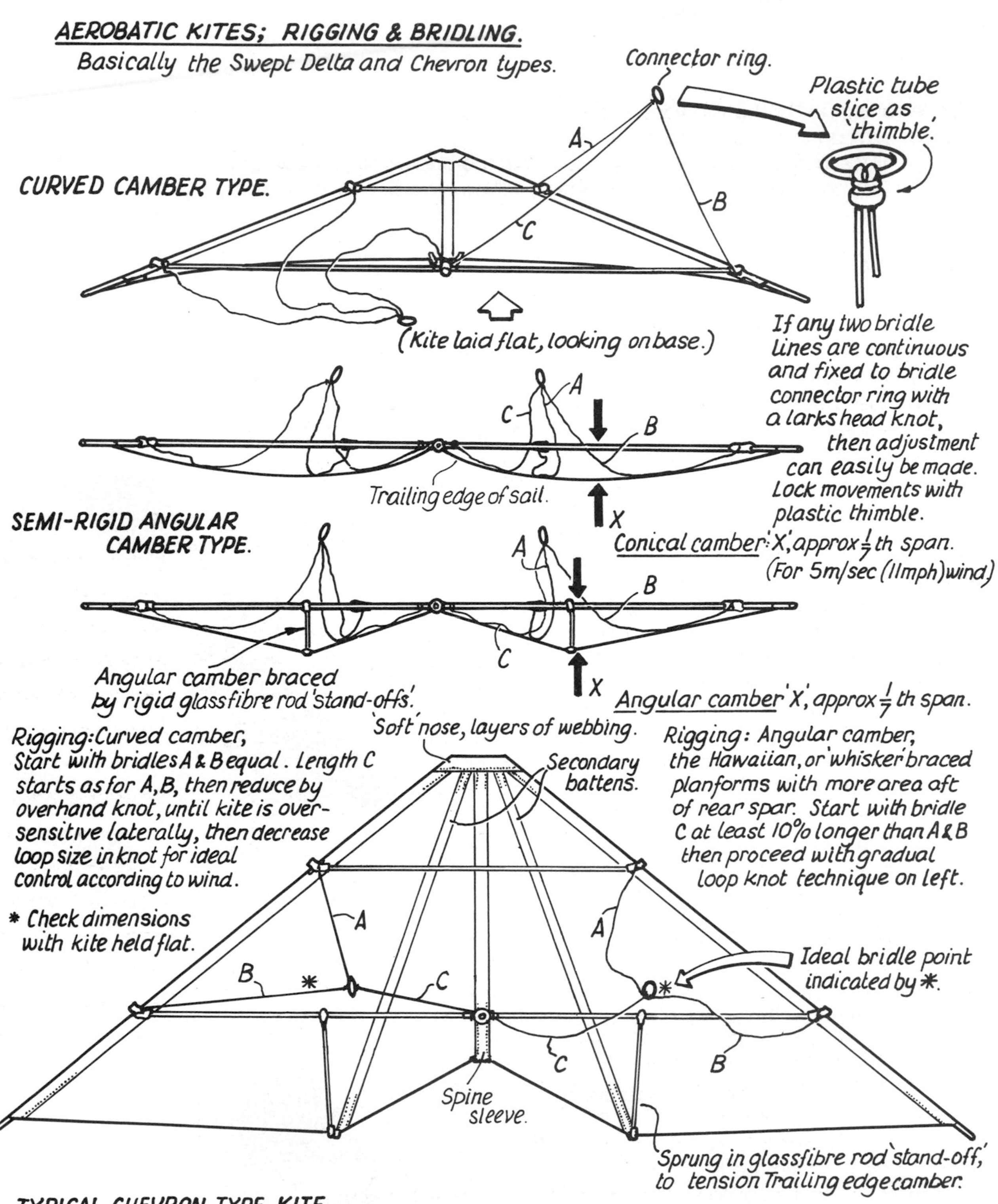
AEROBATIC KITES; RIGGING & BRIDLING.
Basically the Swept Delta and Chevron types.
Connector ring.
Plastic tube slice as 'thimble'.
CURVED CAMBER TYPE.
A
B
C
(Kite laid flat, looking on base.)
If any two bridle lines are continuous and fixed to bridle connector ring with a larkshead knot, then adjustment can easily be made. Lock movements with plastic thimble.
C
A
B
Trailing edge of sail.
X
SEMI-RIGID ANGULAR CAMBER TYPE.
Conical camber 'X', approx 1/7 th span. (For 5m/sec (11mph) wind)
A
B
C
X
Angular camber braced by rigid glassfibre rod 'stand-offs'.
Angular camber 'X', approx 1/7 th span.
'Soft' nose, layers of webbing.
Secondary battens.
Rigging: Curved camber, Start with bridles A & B equal. Length C starts as for A, B, then reduce by overhand knot, until kite is over-sensitive laterally, then decrease loop size in knot for ideal control according to wind.
Rigging: Angular camber, the Hawaiian, or 'whisker' braced planforms with more area aft of rear spar. Start with bridle C at least 10% longer than A & B then proceed with gradual loop knot technique on left.
* Check dimensions with kite held flat.
A
B
*
C
A
Ideal bridle point indicated by *.
C
B
Spine sleeve.
Sprung in glassfibre rod 'stand-off', to tension Trailing edge camber.
TYPICAL CHEVRON TYPE KITE.
Sail bridle check: A=B; C=A+10% Or more.
Check symmetry of both sides before flight.
Trimming: Generally, Reduce A = Slower reaction, Reduce B = Stall on Fast pull, Reduce C = Increasing sensitivity.

Lark's Head adjustment

The connector ring is in almost every case, attached to two of the bridle lines by a slip knot (Lark's Head) so that the ring can be moved towards either end of the continuous line.

The majority of kites are supplied with lines 'A' and 'B' continuous and line 'C' of fixed length. Among exceptions are Mark Cottrell's 'Liteflite' series and Terry Crumpler's Windwalker/Gayla 'Wasps' which have lines 'A' and 'C' continuous and line 'B' of fixed length. There is a distinct difference as we will detail; but for the moment we have to return to the basic check for symmetry.

The slip knot should be marked with a spot of colour on the line. If the rings do not align symmetrically, look for a mark on either bridle. One should be on or very near the Lark's Head itself so that will be the correct trihedron. The other must be adjusted by slipping the ring after slackening the knot. If *both* have identical mark positions but are still asymmetric then the fault can only be analyzed by measuring line lengths to determine which is wrong.

A simple test of any sport kite bridle is to take the connector rings in each hand and lift the assembled kite off a level surface. The spine will be suspended *horizontally* or at most slightly nose-down, unlike the 15-degree 'angle of dangle' described for single line kites in Chapters 1 and 3. This is a handy tip for setting a home-built kite bridle. There is always the old technique of adjusting bridle length with an overhand hitch to shorten the line; but for sport kites, the knot itself takes up a lot of line. One solution is to have long bridles, so reducing the effect and in a way de-sensitising the bridle changes.

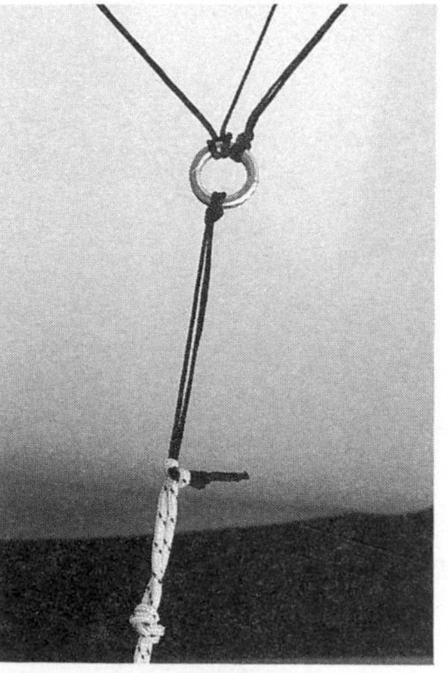

Left: Using two connecting rings on a short loop of bridle from the upper cross-spar, it is possible to adjust the bridle more accurately. The line to be held in place is pinched between the rings.

Bottom: Derived from linkage used by Peter Powell for kite trains, many stunters are beginning to eliminate metal links and use a Lark's Head knot to join flying line to a cord loop on the connecting ring.

This is counter productive, and until someone invents a simple method of adjusting the trihedron in small increments of say, 1/4in or 6.5mm around, the Lark's Head will have to suffice. There are systems of fitting a bead on one line into any of a series of keyholes on another but to date only for light loads on small kites.

Why adjust the bridle?

Why adjust anyway? Some solo flyers claim never to change bridle settings. The reason is more likely to be that they only fly in ideal wind conditions. To be able to perform a full schedule of aerobatics in a range of winds from 8 to 40 km/h (5-25mph), some bridle adjustment is essential for 'tuning'.

We mentioned that the fixed length line was commonly 'C'. Study the triangular pattern of a vertical projection and consider line C as a fixed radius. (See drawing of Japanese ARK kite on page 145). Movement of the connecting ring 'upwards' will shorten 'A' and lengthen 'B'. At the same time the ring will move *inwards.*

Now consider the alternative, with line 'B' fixed, and the ring moved 'upwards' to shorten 'A' and lengthen 'C'. The ring will then move *outwards*, away from the spine.

The opposite will happen in each case when the ring is moved 'down' the continuous line, i.e. *outwards* when line 'C' is fixed and *inwards* when line 'B' is fixed.

For lower wind speeds, the ring has to be moved 'down' the line. This will increase the angle of the bridle and load the sail. In strong winds, the reverse is the case. The angle of the kite to the relative or 'apparent' wind is best reduced, so the ring is moved 'up' and this reduces the load on the sail, enabling better control. In normal wind strengths,this would slow up reaction.

Pitch and yaw

When the connecting ring is moved 'upwards', we are reducing the pitch. In simple terms, making the angle of the kite shallower when in flight. When the ring is moved 'outwards', we are able to increase the yaw, or sideways angle of the kite.

The result of this is that when line 'C' is fixed, the rate of turn brought about by the lateral position of the connecting ring is reduced when the ring is moved up, and increased when the ring is moved down. When line 'B' is fixed, the opposite happens and the rate of turn is faster when the ring is moved up, and vice versa. Mark Cottrell's Liteflite Radial makes use of this feature in a bridle where line 'A' is continuous from side to side. This permits the ring to move down on the line under greater tension, and up on the opposite side which would be on the outside of a turn. This increases the rate of turn and reduces the risk of stalling.

The effect of relative line lengths in the bridle will also de-

SKY DANCE was the top Sport Kite Team of 1996, and with three members, Mark & Jeanette Lummas and Stephen Hoath, won the VIIth World Cup in Japan, the World Grand Prix in Malaysia, the European Championship in France and the British National Champs, all in 1996. They fly the Top of the Line North Shore range. With equally impressive wins and placings from previous years they are the most successful 3-person team ever. Opposite: A spin-off kite (see sketch on p. 129) at the ready with bridle lines clear of the spars and checked for symmetry.

pend on the actual length of the lines and the positions of the cross-spars. One thing is certain and that is the better tolerance of having line 'B' as the fixed line in a standard trihedron rig. We prefer to generalise and to treat the three lines independently. Shortening line 'A' reduces pitch and offers slower reaction. (There is a good case for making all adjustments on this line only.) Making line 'B' shorter will enable faster turns until you reach the stage where the tip which is inboard of the turn will stall. Shortening line 'C' will make the kite sensitive in direction because the moment arm is reduced and your own arm movements will have greater effect.

All of these generalisations depend of course on a constant wind speed sufficient to fly the kite within its designed wind window.

The launch

We discussed preparation of the lines in Chapter Three, so all we want to emphasise now is that you remember to take a stake (preferably fitted with a brightly coloured flag so that you do not lose it, or tread on it) and that the line ends have swivel connections of adequate strength. It's of no use to have 70kg (150lb) lines and links of lower strength values. Ideally, you should have an assistant, if only to reduce the number of 100metre (330ft) round trips up and down a fully paid- out set of lines, which always seems to happen when you are on your own! The most important pre-flight check is to ensure that you pick up the handles so that right and left align with the correct bridles. If not, then that first flight is going to be a short one.

An assistant can hold the kite vertically by the leading edges, and as a ground level breeze fills the sail, releases rather than throws the kite , with the lines kept taut. Flexifoils and similar foil types have to fill out their shapes by taking in air. This happens rapidly in a steady wind but can be helped by shaking the kite. This separates the surfaces and opens the intake mesh. Some experienced foilers hold on to the spar end of a Flexifoil, shaking it while nearly vertical then when air-filled, it is brought down to the horizontal for the assisted launch.

Solo launches are easy to perform once you have a little experience. Rigid frame steerables can stand vertically on the frame ends, and be held in that attitude by line tension if the handles are pegged to the ground by a stake. All you have to do then is to take up the handles maintaining line tension, and either take two steps backwards smartly or pull the arms backwards and downwards in a smooth tugging action to set the kite aloft.

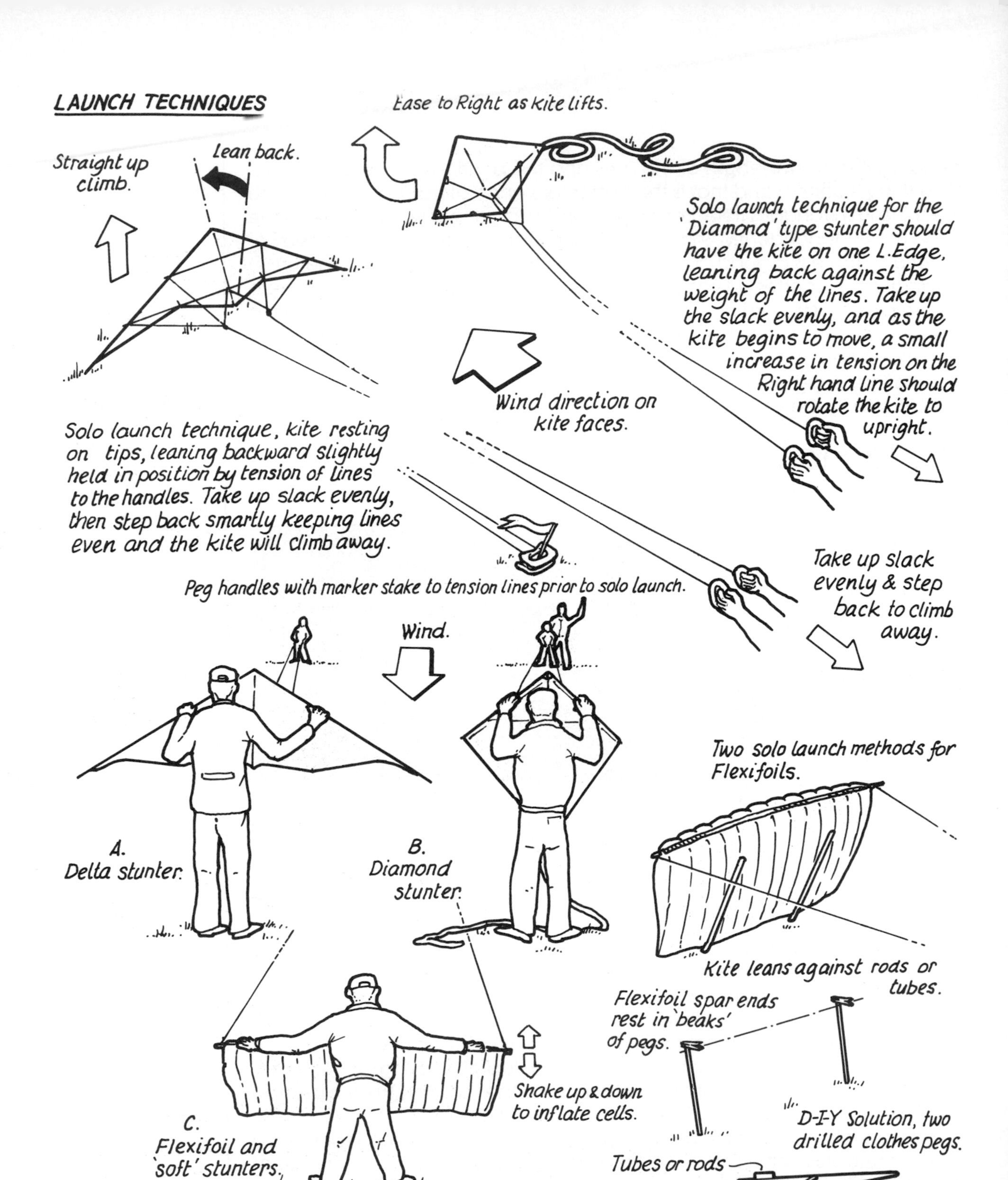

A, B & C. Show launching methods with a helper.
(Helper always behind or downwind of kite)

BASIC STUNT FLYING TECHNIQUES.

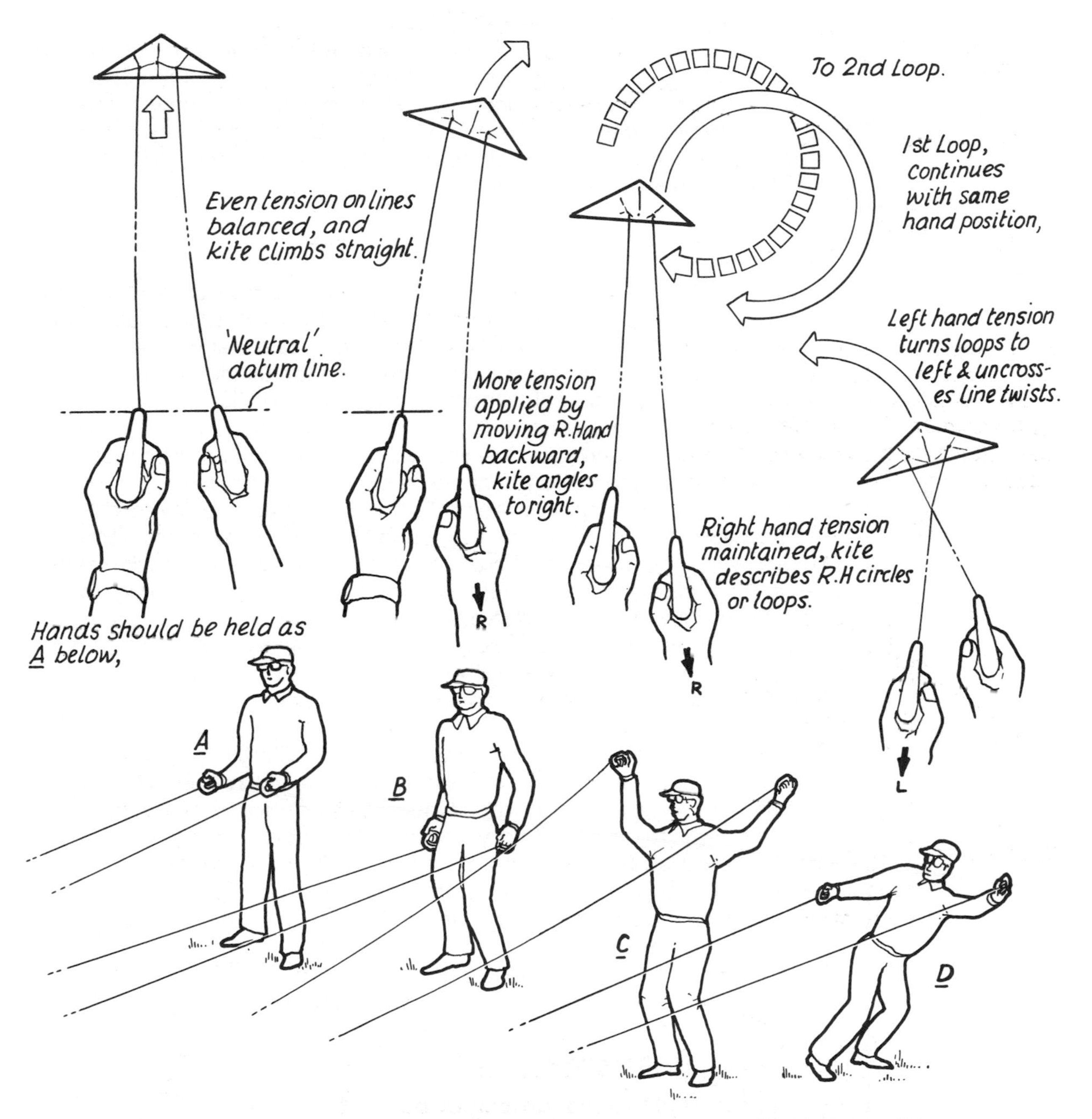

Flying stances: A Flier is relaxed, hands and arms at a comfortable position.

B With a slackening wind the flier has dropped his hands (maintaining Neutral tension) to increase the line tension to help keep control. If he pumps both arms back & forth the kite will climb upwards in steps.

C This flier has troubles; Not enough wind, Not enough line out, Very uncomfortable position, and little chance of being able to balance his line tensions.

D No! Never run to maintain tension; Certainly never backwards!!

The Golden Rule is to ensure that you have Mother Nature on your side, or rather, on your back. No wind - no fly. This applies equally to the choice of flying site. If there are obstructions upwind to turbulate the airflow, then even the perfectly bridled stunter will have difficulty in breaking through the rough air at low level. Steady wind in an open area is what you should be seeking long before the lines are laid out.

Take off

Our advice for the first launch is - hands together, arms outstretched (*but never above your head as is so often attempted by first time flyers*) and as the kite rises, sustain an even line tension. If the kite turns either way left or right, then increase the pull on the opposite side, right for turning to the right and left for a left turn. Be content with small movements at first.

Maybe the kite rose reluctantly,then rotated rapidly and fell. Try again before making any changes, the kite will have stalled in most cases;but it *might* be necessary to adjust the bridle by moving the ring up the continuous line. Check the colour mark on the bridle line and by slipping the Lark's Head knot and pulling on the lower line ('B') bring the mark *no more than* 6mm (1/4in) below the ring. Try again, the decreased angle of the kite ought to cure the launch problem but as future experience will show you, the original trim set by the manufacturer will have been the ideal compromise for rate of turn and sensitivity. Ensure that you know the location of the bridle mark and are familiar with its movement around the rim of the connecting ring at any stage of adjustment. It's a good idea to use fine electrical tie-wraps to lock the Lark's Head in position.

Initially you will adopt a tense stance , concentrating the mind on which action brings about what result. The first loop, or sideways circle will be erratic and recovery somewhat haphazard; but just as with riding a bicycle, all the reactions begin to come naturally and in no time at all, even as early as the third flight, you will be diving and climbing at will.

Most of the expert stunt flyers operate with hands at waist level pulling back from extended arms to hips, and this range of push and pull enables them to fly through an incredible range of manoeuvres. Totally relaxed, but never still on their feet, the experts make full use of the wind window from self launch to controlled landing.

The flight envelope

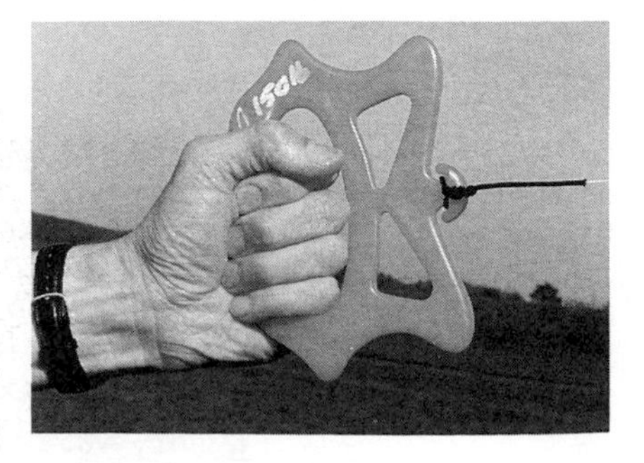

Before you feel brave enough to steer the kite around fanciful shapes, it is best to understand what is meant by the 'edge of the wind' because in effect, this is the limit line within which the kite will fly.

You have first to appreciate that although wind will be flowing in a straight direction from behind the flyer, the kite is limited to a radius of action, horizontally and vertically. It will be flying in half a hemisphere, or less. It is the 'less' bit that concerns us. As the kite climbs up vertically from a launch, it starts with a steep angle to the wind, and at the peak of its climb is at a shallow angle. Meanwhile the lift and drag forces are at work, and the kite has achieved the point of no further progress. It will hover, and dependent on wind strength, the lines will be at 85 degrees or so to the ground, the kite stationary nearly overhead if it is a good performer. Take the kite down to ground level and at either side, it will again reach a point where the aerodynamic forces are balanced, and it hovers nose into wind, the spine horizontal and cross-spars vertical. It will be short of being directly at right angles to the flyer by up to 30 degrees due to the contribution of gravity working at 90 degrees to the lift force; but never mind about that, the demonstration illustrates how the flight envelope embraces a wide band of air. Within that band the kite will move from steep to shallow angles against the oncoming wind. Since the kite has a lift component to enable it to accelerate at certain positions around the hemisphere it also has what is known as an 'apparent wind' and it is this which sustains the motion within the flight envelope. Pat has sketched the band of usable airspace for light or strong wind conditions.

Above: Padded straps are comfortable to grip this way – note arms extended, NOT raised by George Mattison. Opposite: The Peter Powell heavy duty handle. End of the Spectra line is sleeved, in this case with fine shoe lace.

The design of the kite plays its part in determining the size of the flight envelope just as much as does wind strength. Kites with good lift coefficients have a broader envelope, and are often more difficult to control into a landing. Others with higher sail loadings and lower lift coefficients are generally faster in acceleration and yet when line tension is relaxed, will for example, make controlled landings at will. It is very much a case of selecting your kite for the conditions. That is why the keenest kite flyers carry a variety of stunt kites to the meetings.

Basic movements

Now that we have a better understanding of the flight envelope it is time to explore! Take the kite straight up, pull to make a 180-degree turn and you've turned the kite nose-down into vertical dive. Halfway to the ground pull again for horizontal flight. Hold it! Let the kite zoom around the circuit and find that 'edge'. Now try to keep the kite in hover. If at first you don't succeed, then try and try again.

These basic movements plus the complete 360-degree circle or loop will

STUNT FLYING TECHNIQUES : Continued.

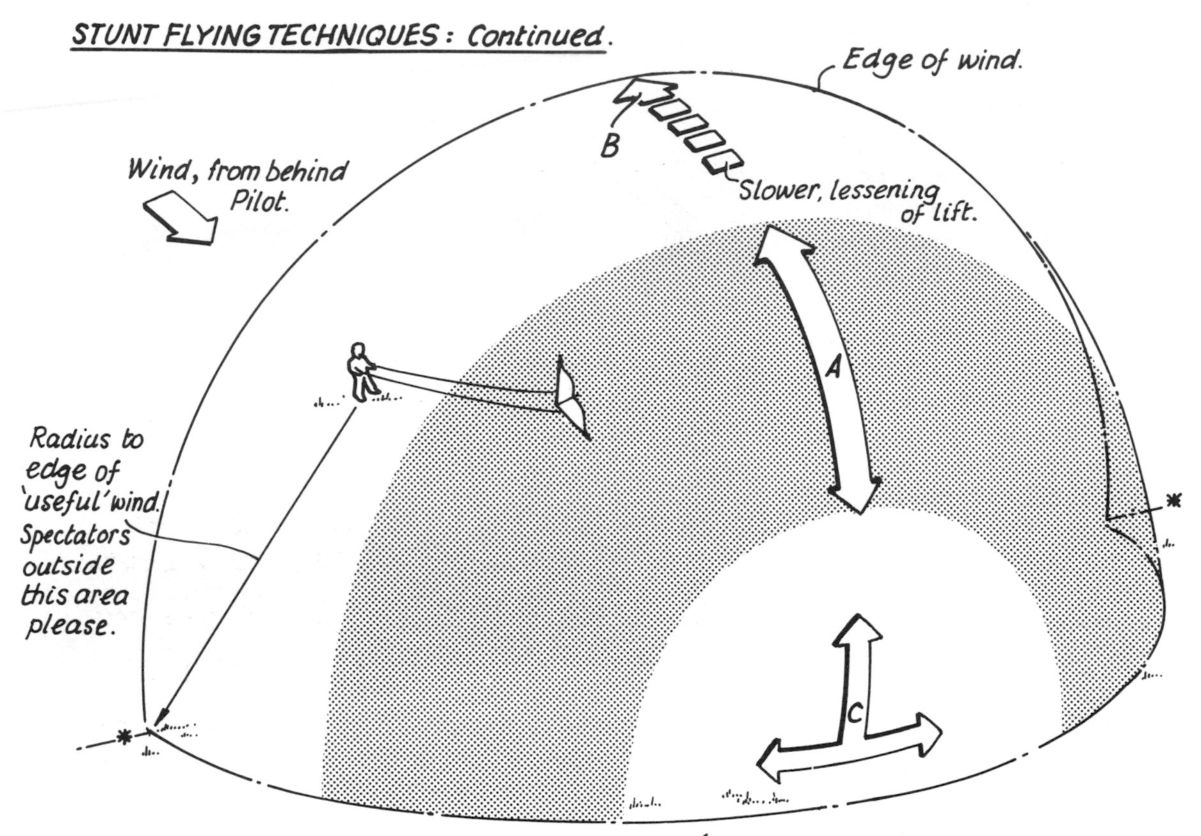

All Stunt flying takes place within a hemisphere (Imaginary & Invisible)
In light winds the useful part or zone in which flight is possible with full control is shown above.
In stronger winds the useful zone is as shown below.

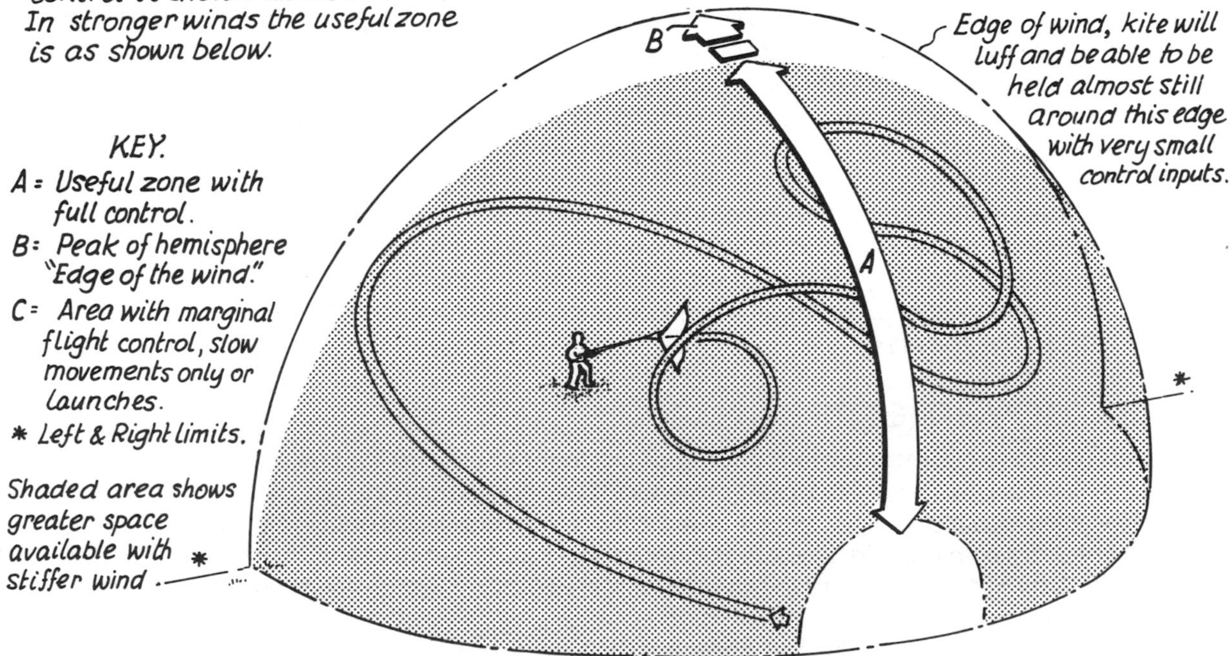

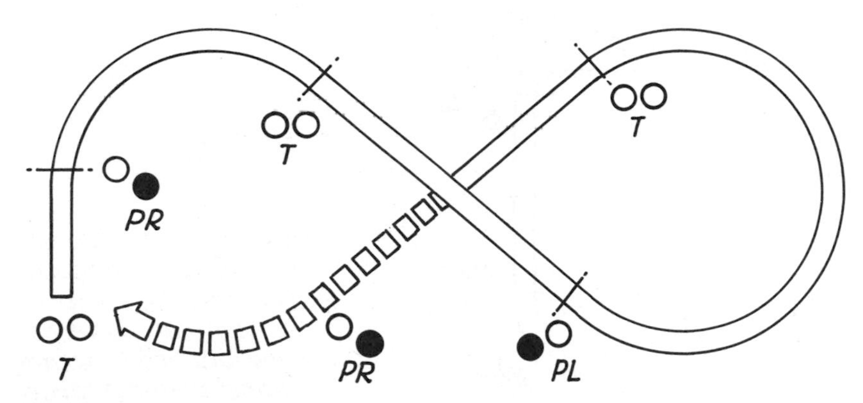

Take Off.

Hand position KEY.

T: ○○ Hands together. (Neutral)

PL: ●○ Pull Left hand.

PR: ○● Pull Right hand.

¦ Transition line between figures.

The above diagram illustrates the basic 'loop' manoeuvre– extending into a horizontal figure of eight. Careful practice of these loops & eights, to smooth out jerky transitions and trying for even size figures will help your style, as nearly all stunt patterns are based on these or combinations of them.

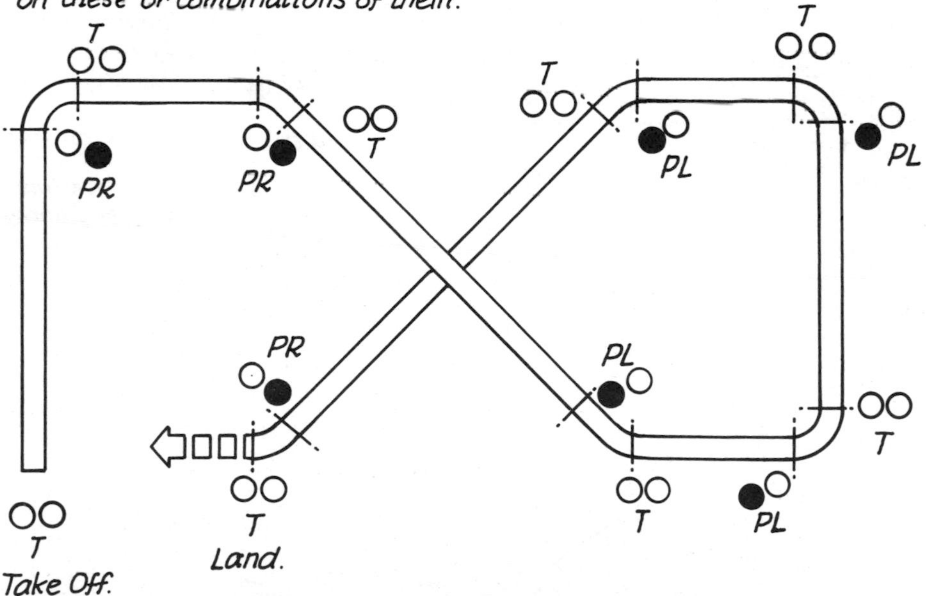

Take Off.

This diagram follows the basic figure of eight pattern above but the corners have been sharpened & 'squared' by abrupt hand movements, holding small neutral breaks between turns. To polish this manoeuvre, practise separate square loops, linking the odd one together.

STUNT FLYING TECHNIQUES Continued.

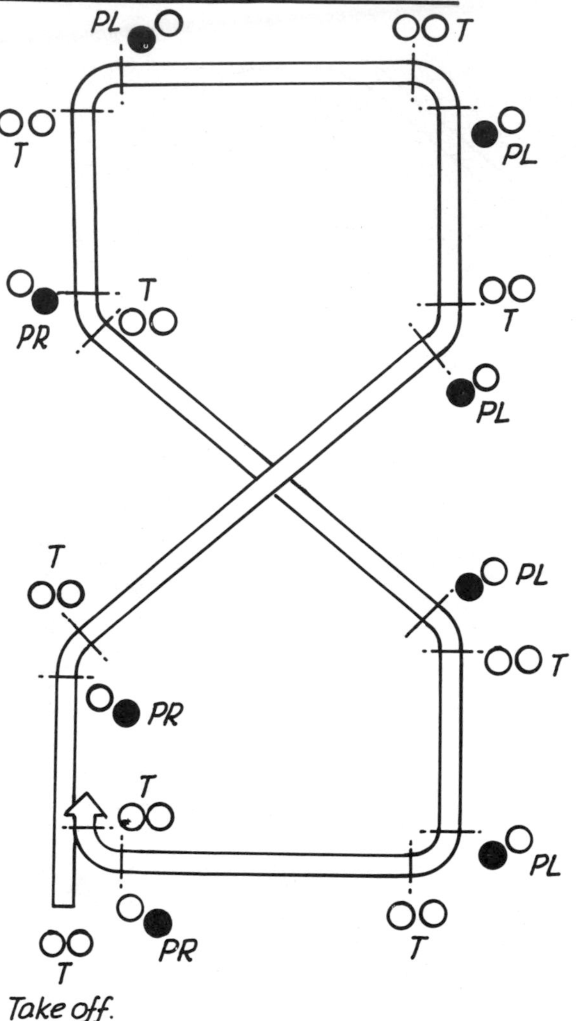

Having perfected your technique on horizontal figures, then progress to vertical patterns, you will need a steady wind giving a large hemisphere of control and enough altitude to complete the figures.

Hand position KEY.

T: ○○ Hands together. (Neutral)

PL: ●○ Pull Left hand.

PR: ○● Pull Right hand.

| Transition line between figures.

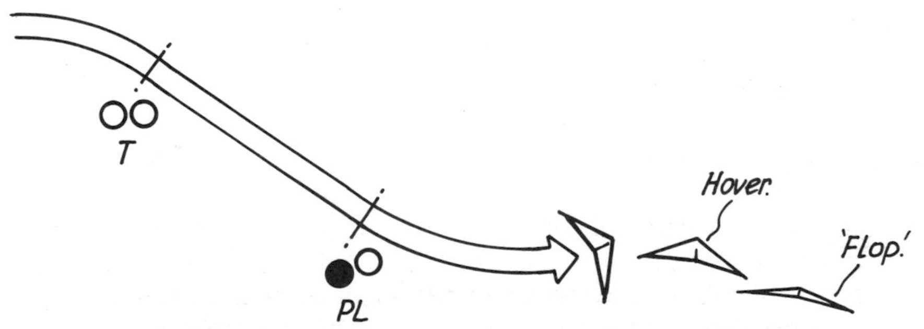

This shows the landing procedure, if you bring the kite down near the ground, a little lift can be eased on to the kite, and by walking toward the kite, the lack of tension enables the kite to 'float'. By juggling a little with tension & control the kite can be hovered on to the ground almost at a chosen spot.

lead you into the lazy eight. This is a most satisfying stunt when you can make each half of the eight equal in diameter. Start from a vertical climb directly downwind. Pull right and hold until the kite is heading at 45 degrees downwards, then check with hands level and pull left to start the second half of the figure. Three-quarters of the way around, check with hands level again, and then pull right to complete the eight. As you can see from the drawing, this and the 'square' eight make simple yet testing manoeuvres. Now go and try to execute them as precisely as drawn!

It follows that the vertical eight will be next in the practice schedule followed by an intentional landing as distinct from those earlier arrivals. This is made all the easier after first experiencing the hover at the 'edge'. Take the kite around the circuit, and hold a hover then walk towards the kite, allowing the lines to slacken and if it is a relatively standard steerable kite, it will land.

There are some very light kites which float for ever, never land but eventually flop. These are the high performance light wind kites with large sail area. In contrast, the smaller and faster stunters with higher sail loading can be pirouetted over a designated spot and dropped onto their tails with little more than a quick lunge forwards by the flyer.

Line breaks

We've already designated Spectra or Dyneema line for your stunter and for the good reason that this material is not only lightest for its strength, but that it also has the property of sliding on itself even with as many as fifteen twists between flyer and kite. (The world record is 328 wraps before loss of control!)

There is a disadvantage. If these lines come into contact with others of 'foreign' material such as a polyester, then they melt through in a trice and your kite is away, floating down on the breeze leaving you with two rather expensive short ends. For this there is only one solution. Avoid others, find your own space and keep clear.

The other possibility is a single line failure, or more likely, a single line clip failure. Here we have a dangerous situation of a kite flailing on one line, possibly towards onlookers and certainly not under control. If the way is clear, run towards the kite and relieve the tension on the single line. This way, it is possible to make a sort of controlled flop and avoid any damage to kite or third party (see page 164).

No wind flying

We mentioned the 'unlandable' lightweight kites and have explained how the lift component can enable a kite to fly faster than the wind speed. A combination of the two features can result in the remarkable ability to fly in near zero wind conditions.

Using shorter (24m or 80ft) and lightest lines, with a light 240cm (96in) kite it is possible to fly right around 360 degrees in horizontal flight and throw in a loop or two for good measure. This is made possible by whipping up speed with a continuous pull on the lines bv the flyer. At the same time he or she is backing up in an eccentric centre circle so that the flight pattern resembles that of the snail design.

The stunt makes those frustrating windless days tolerable. Lee Sedgwick's party piece involves self-launching a quad-line 'Revolution' javelin fashion, snatching the lines into tension and flying a routine largely through sheer manpower, then concluding in the reverse of the opening by catching the kite from a stall! He calls it '3-D'.

Lee also pioneered ground hook flying, as seen at the Wildwood New Jersey Championships in 1987. First he flew on a full radius but with the lines passing through a ground hook. As the flight progressed he walked towards the kite so that he was underneath it, then flew to the tune of *Lady in Red*, stroking the kite in the sand and with his hand as the song reached the `cheek to cheek' section.

Ron Reich demo's the ground loop system at Dieppe, passing his Top-of-the-Line kite behind him!

Changing over handles is a solution to the reversal of control caused by the ground hook system. Like all other unusual accomplishments, it only nears perfection after hours of determined practice; but once the technique is acquired, the potential for a stunning showpiece is obvious.

STACK

What began as a London UK-based co-ordinating body for organisation of stunt kite competitions has grown into a worldwide association known as Stunt, Team And Competitive Kiting (or STACK) with representatives in over l4 nations. It runs a League, and is involved in establishing criteria for judging standards for a series of compulsory figures. Their diagrams have become the official guide for events everywhere, both individual and for team flying. All serious sport kite flyers are advised to join and so keep up to date with their official manual.

The manual is the nearest one could get to an International *Code Sportif* or Sporting Code which enables all competitively minded flyers of steerable or stunt kites to operate on identical standards. It includes basic information for the newcomer either as a solo performer or as a team flyer. Diagrams cover a vast range of figures and general rules cover the requirements for various competition classes. These are divided into Junior and Senior sections

FLYING THROUGH A GROUND HOOK.

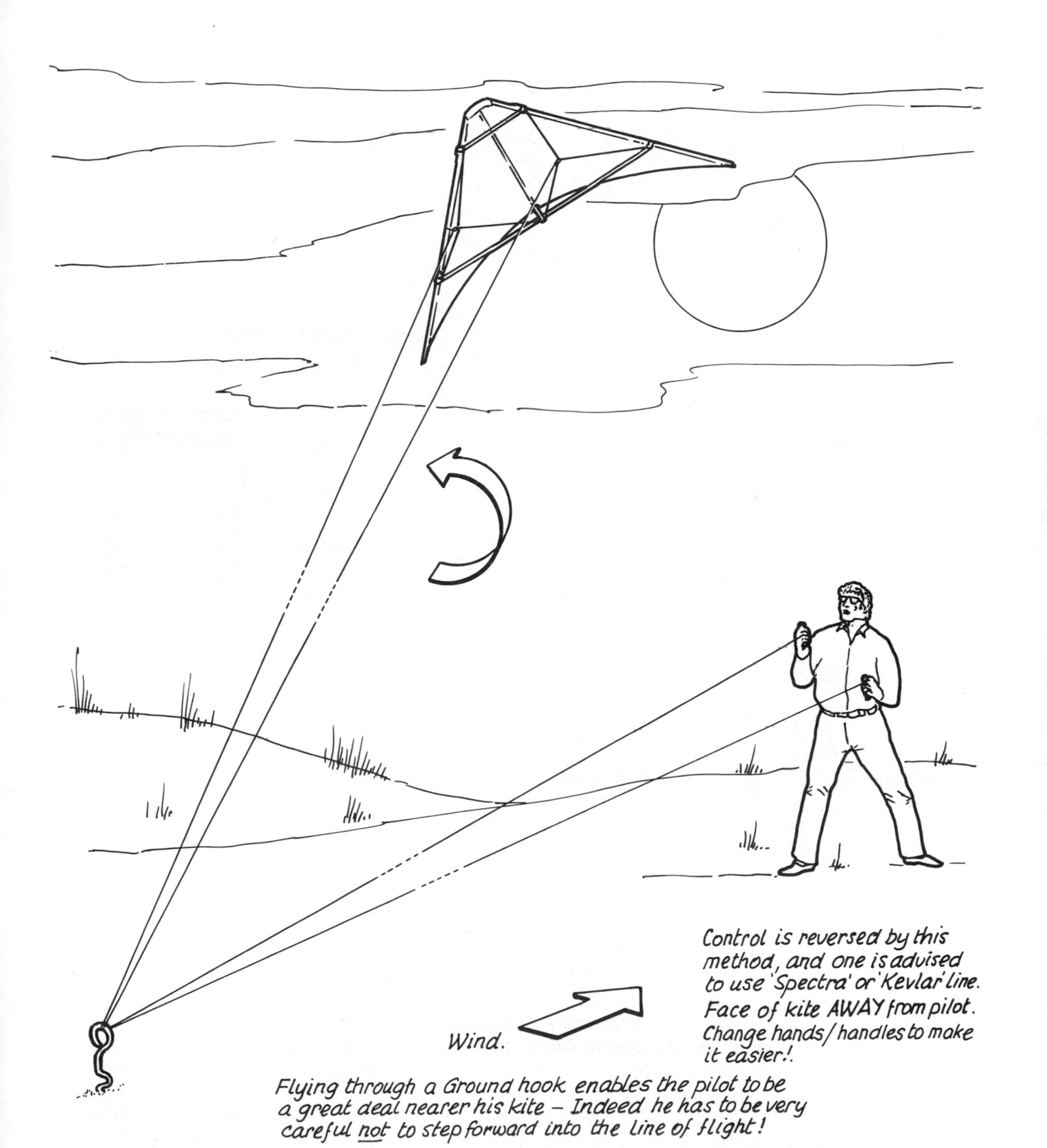

STUNT FLYING TECHNIQUES : Continued.

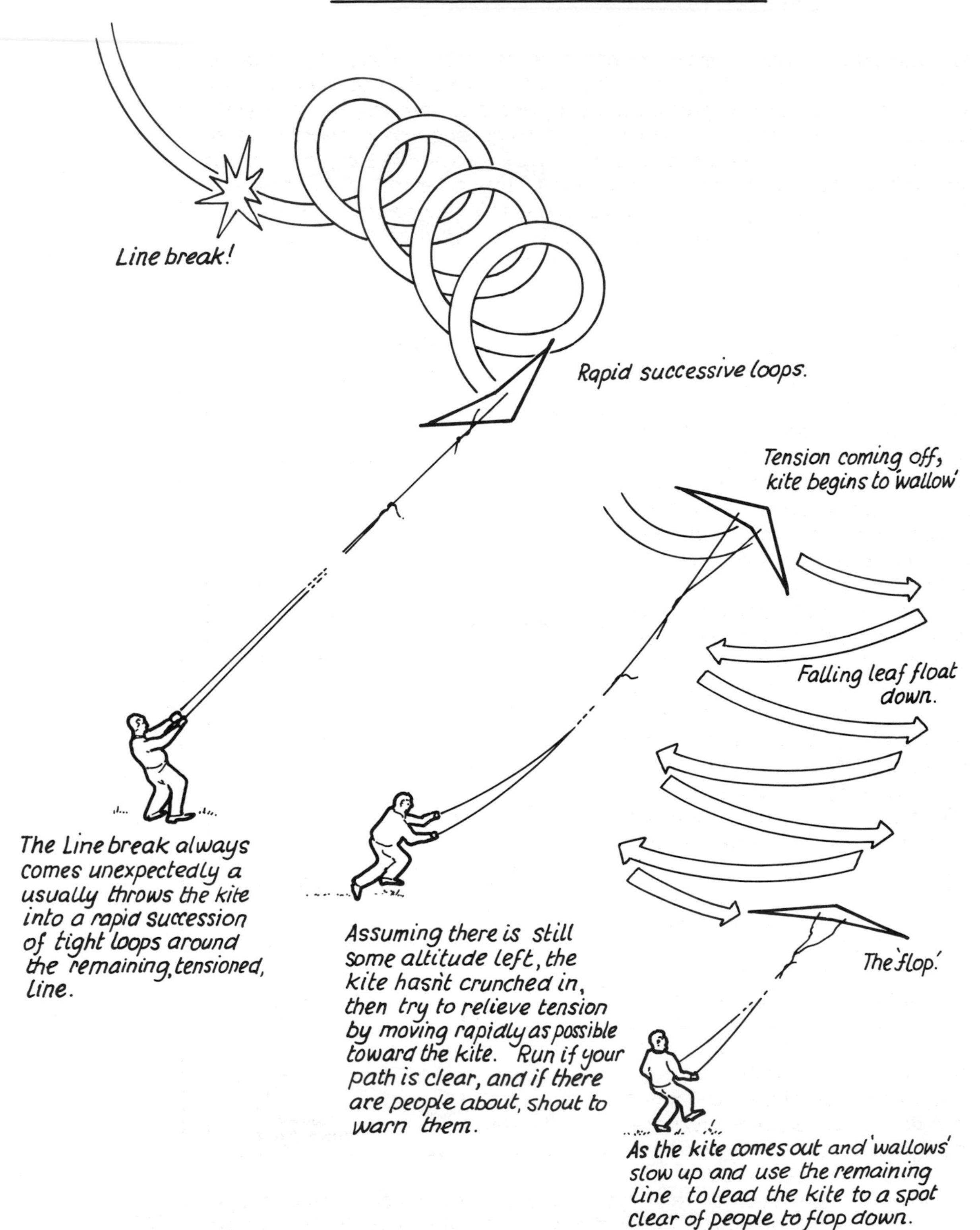

for Individual or Team precision flying plus the same divisions for Ballet to musical accompaniment. It is the latter which has greatest popular appeal, rather akin to the following for ice dancing where choreography and style in tempo with the rhythm offer great opportunities for graceful interpretation.

Even if not in any way competitive, STACK membership is advantageous for the individual stunt flyer, not only for the informative manual, but also for its newsletter with a regular calendar of activities worldwide and tips on team flying. This implies two, three or four kites being flown in formation - a daunting challenge for anyone, especially those who feel they have their hands full when trying to control a solo kite!

Team flying

Team flying offers a thrilling extension of excitement to the participants, but only after hours of practice. Co-ordination with the other flyers through a schedule of figures can only be achieved with full understanding, and a strong team leader who acts as a caller to signal all the changes of movement.

Practice with 'sticks' (miniature kites on the end of a l metre (39in) dowel) forms a good start for understanding both the figures and the jargon which has developed to describe the shapes. Unfortunately this is the least standardised aspect of team flying. There is as yet no official glossary and to the outsider, the terms bear little relationship to the shapes, much more to the flip lingo of rock music. Nothing wrong in that - it is a reflection on the lively approach of a youthful new generation of kite enthusiasts.

Quite apart from the vernacular which ranges from 'circle' for a loop or 'infinity' for a figure of eight, to 'the rug' for expanding loops around a constant centre - or two of these for a 'Gautier brassiere' there are varied modes of command from the team leaders.

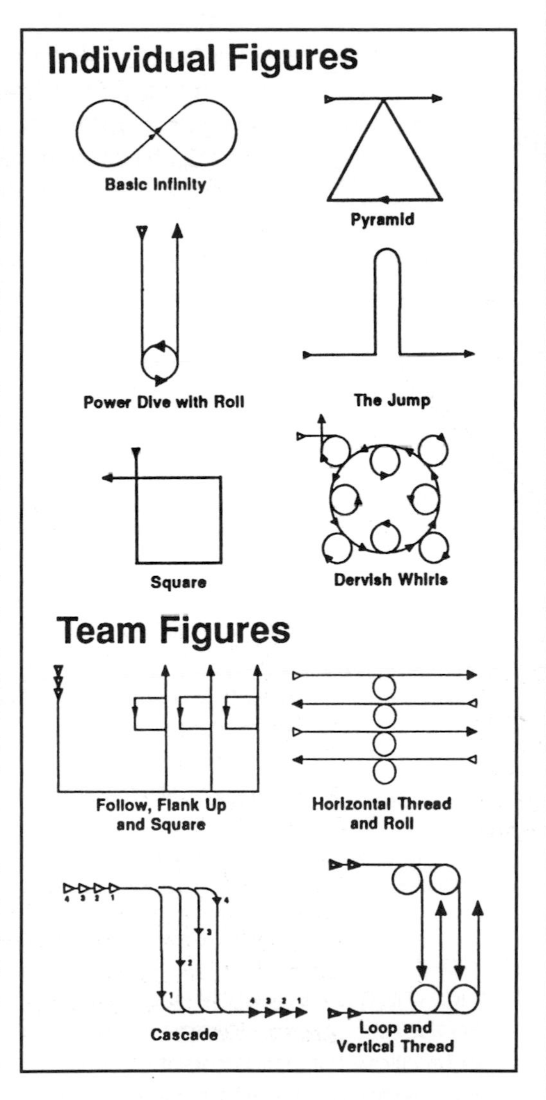

Synchronised movement is critical, so the command has to be brief and so timed by a delayed 'one-and-*pull*' sort of order. This gives the kite flyers a

Perfect spacing, identical speeds, absolute harmony in the Top-of-the-Line team display at Bristol.

precise notice of timing for making any change of direction.

Warming-up figures like the infinity or horizontal figure of eight, and straight passes in trail are the introductory manoeuvres which involve least complication and yet teach the flyers the arts of timing and speed control. 'Laddering' or a series of ascending (or descending) passes in trail to form a stepping sequence is another basic stunt which tests the degree of compatibility of each flyer and the kite.

Speed control is exercised by body movement. Experienced teams are constantly moving back and forth, speeding up by pulling hard on the lines or conversely running forward to slow the kites. They cover a lot of ground so the designated flight area has to include very generous margins of up to an extra 50% over the line lengths and, most important, this area has to be clear of all spectators and especially other kites.

Team kites fall into several categories, chosen for precision flying or ballet and to suit the wind conditions. So for smooth open stunts, the Flexifoil has been used, and for team precision, the quad-line Revolutions. In between and by far the most popular are the range of 245-275cm (8-9ft) swept-wing Delta shapes, and in some cases the 3/4 scaled down variants.

The Delta kites are selected for their consistency rather than extreme rates of turn or agility as would be the case for a sport kite flown solo. Obviously, all the team kites should be of identical design for aesthetic and performance reasons. They should also be of a type that has simple and quickly altered bridle trimming. This ought to harmonise the general flight speed of the 3 or 4 kites in the team, and their rate of turn.

What is not readily appreciated by the newcomer is the stagger of the line lengths. A variance of 1-2m (3-6ft) is typical between each of the flyers. This reduces the risk of airflow interference from kite to kite which at close quarters can have surprising effects! A similar and perhaps better appreciated situation can happen in yacht racing where it is possible to 'take the wind' out of another's sail. This is in fact a deliberate tactic which is frequently exercised.

So there is an advantage in using handles of the Peter Powell heavy duty type to allow for adjustment on the flying field. Many of the expert demonstration teams use wrist straps with pre-set line lengths. They are said to relieve cramp in the forearm which can arise from long periods of practice flying; but here another caution should be observed. Any sign of tingling fingers or triggering motion in the thumb or fingers must be taken as a warning.

Carpal Tunnel Syndrome is a latter day symptom that has affected keyboard operators and kite flyers alike. Basically it means that the nervous system which passes through the wrist in the carpal tunnel has been stressed and the best cure is rest , or in the case of trigger finger, an injection or even an operation. This situation is best avoided by arranging the straps over the back of the hand and *not* around the wrist. Pass the straps between the forefinger and the thumb, then by closing the hand, the loop is gripped firm, and one retains full sensitivity as though the line were to be connected directly to the arm. If the straps are padded with foam plastic, especially in the area that comes across the back of the hand, then so much the better. It goes without saying that the strap loops must be large enough to enable them to be cast off the hands quickly in any emergency. All of this applies to kites flown on two lines, for solo sport, individual precision aerobatics or team flying.

When flying on four lines where wrist action is part of the control system, the handles will be entirely different in their shape and operation.

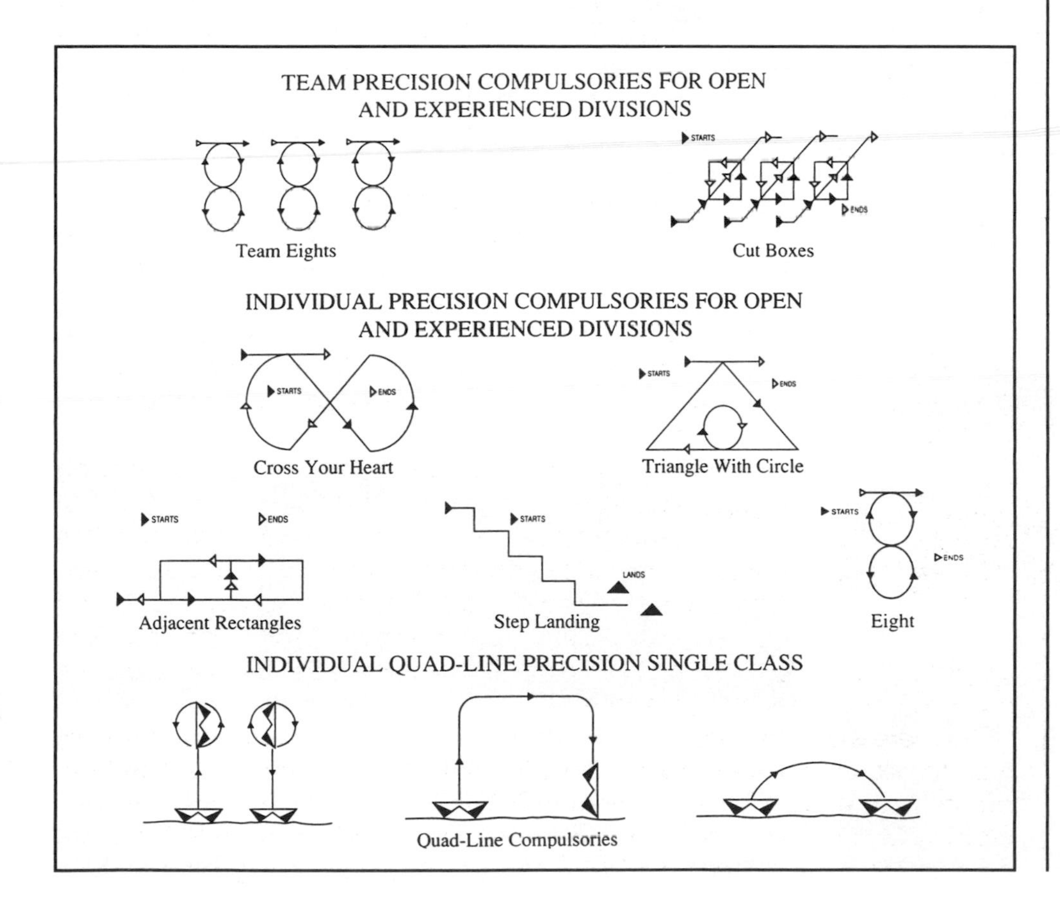

Quad line kites

There really isn't anything new in flying with four lines to control a kite with deflections in pitch as well as laterally. For many years a method of flying a Malay or Diamond kite using a cruciform control handle to turn the kite upside-down at will, was an amusing diversion.

From 1974 to 1976, such a stunt was often seen at the British Kite Flying Association rallies, held at Old Warden in Bedfordshire. Phil Morley could hold his striped Malay stationary though inverted, with the streamer tail trailing down from the uppermost part of the kite. Phil developed the system by adding two more kites in train.

Another 4-line enthusiast of the same period was Trevor Hubble who chose to use six sided kites but, in his case, they were flown using two separate handles, each with a pair of lines. This offered a wider range of control; but still the concept failed to influence other enthusiasts and although remarkable for the precision of their flight, these kites gradually disappeared from view with the emergence of the more spectacular narrow delta shapes.

Holding a cruciform 'puppetry' style control bar, Phil Morley (below) could hold the simple Malay kite in vertical or inverted positions.

Neos Omega

The scene was to change when Joe Hadzicki and his brother in San Diego created what was first called the Neos Omega, now known as the Revolution.

This linked two sail elements on a common leading edge. One and a third square metre (14 sq ft) of surface area could articulate about the leading edge spar so that it could even twist like a propeller. With the pitch of either side held constant, it could be held stationary, advance in climb or reverse direction to fly backwards. Between the sail and the leading edge spar, a plastic coated open mesh venting screen enabled airflow to escape in tight turns and particularly in reversed flight. Without such a vented area, the characteristics of the Revolution are lost.

Left: Ready for take-off, a Revolution shows its unique bridle system. Note the leading edges. Below: Mark Cottrell shows the gentle touch on Revolution control handles, held at their upper ends.

The success story of the Hadzicki invention is due as much to the special control handles as to the kite itself. Using a hockey stick curve and held at the upper ends so that rotation of the wrist can make the lower end of the handle move from extremes of full forward to full back, the reverse flight relationship of the lines is thus amplified.

For the two-line steerable kite flyer, initial control is totally foreign since the push-pull reaction no longer produces the desired effect! There is no substitute for training by video in this instance. First to show how the handles are held, next to illustrate how instead of extending the arms, the elbows are locked back into one's sides, and then to show how the wrists are pivoted so that the pitch of the kite is varied from side to side.

Many kite flyers have gone direct to the quad-line Revolution from single lines and had less trouble assimilating the method. It is very difficult to discard the natural habit of pulling from side to side as with a 'normal' stunter; but once the artform is acquired, the Revolution, or its smaller successor the Revolution II, becomes one of the most satisfying kites one can buy.

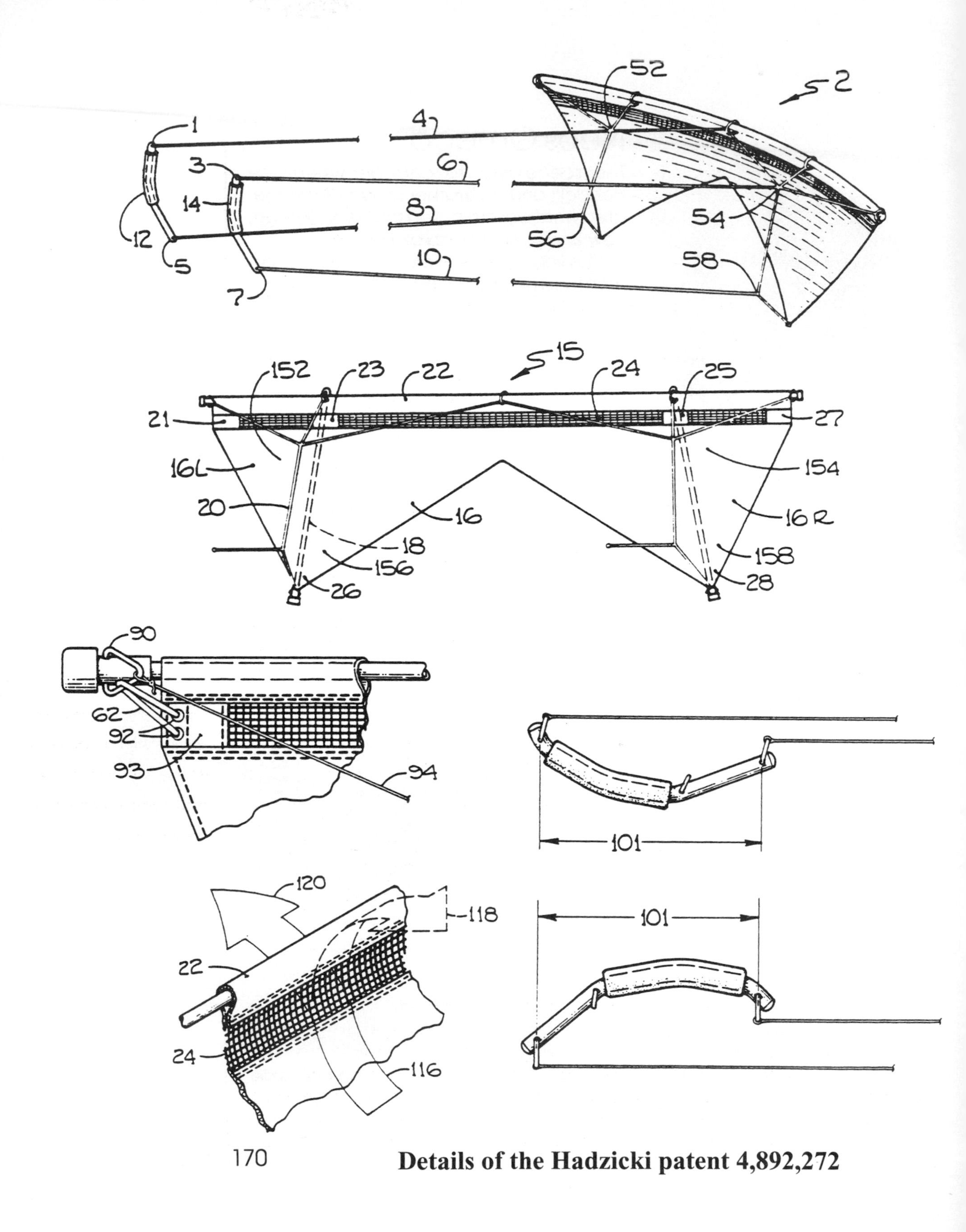

Details of the Hadzicki patent 4,892,272

Soft quads

In the wake of the Hadzicki invention and its protection by US Patent granted in January 1990, there have been many commercial options with clever variations on the four-line principle of changing the camber of the kite surface. Among them have been double surface airfilled 'soft' kites.

The Quadrifoil 25 is a twelve-cell parafoil type 240 x 100cm (7ft x40in) with additional shroud lines to the rear or trailing edge as well as the traditional shrouds to the cells at the forward position. So the four lines are connected to each of the bridles, and can be manipulated to twist the Quadrifoil laterally and alter the angle of attack to the airflow. It will fly sideways, rotate about itself, and with care can go into reverse! A novel feature is that it can be flown right out of its carrying bag which has a tethering strap to stake it to the ground while the soft fabric shape is pulled out on the taut lines.

Quad conversions

The standard triple leg bridle on a 2-line kite lends itself to ingenious conversions that will take another set of lines. Al Hargus, a dedicated kite flyer who was a founder of the famous Chicago Skyliners, kite festival organiser, member of the Chicago Fire stunt team and among many achievements, flew a stack of 84 (!) Flexifoils, also played about with quad-line conversions to the extent of self-publishing a booklet about them. Al's first objective was to slow forward motion and reduce the line tension especially with a stack of kites. His experiments led to three simple methods for changing a stock narrow angle Delta.

His *Quick Quad* conversion is to measure both 'up' and 'down' from the standard bridle point (connector ring) positions at about 75mm or 3in. Mark these points with a touch of paint and then make them the new connector points for the four lines. You now have an arrangement that is symmetrical about the original 2-line system, and to which it can easily be re-converted. Al has a Spin-off kite with glass fibre rods sewn into its trailing edges that will fly well in reverse using this method.

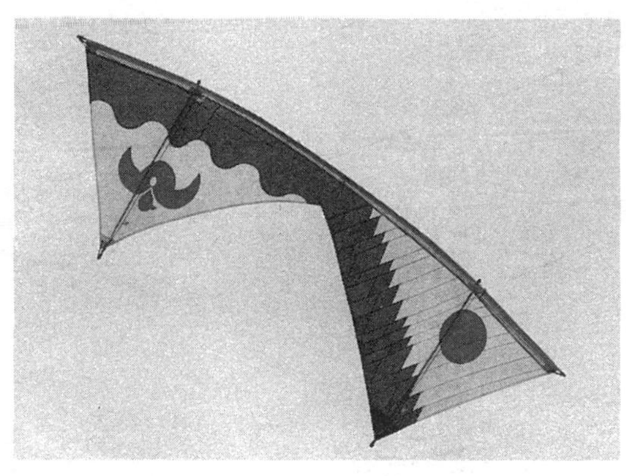

Decorations on Revolution sails have become distinctive. This one allows for viewing when flying in the reverse - or backwards.

His second system is to add a second bottom bridle between the lower cross-spar joints at the spine and leading edges. This bridle will form two legs and should be about 30cm or 12in longer than the cross-spar halves. The connector point should be marked so that it is in-line with the standard bridle point above. Then shift the standard connector 'up' 75mm or 3in and the conversion to Al's *One Line Quad* method is complete. He has a Fire Dart that flies best with this system.

EXPERIMENTS & VARIATIONS IN QUAD LINE by Al Hargus.

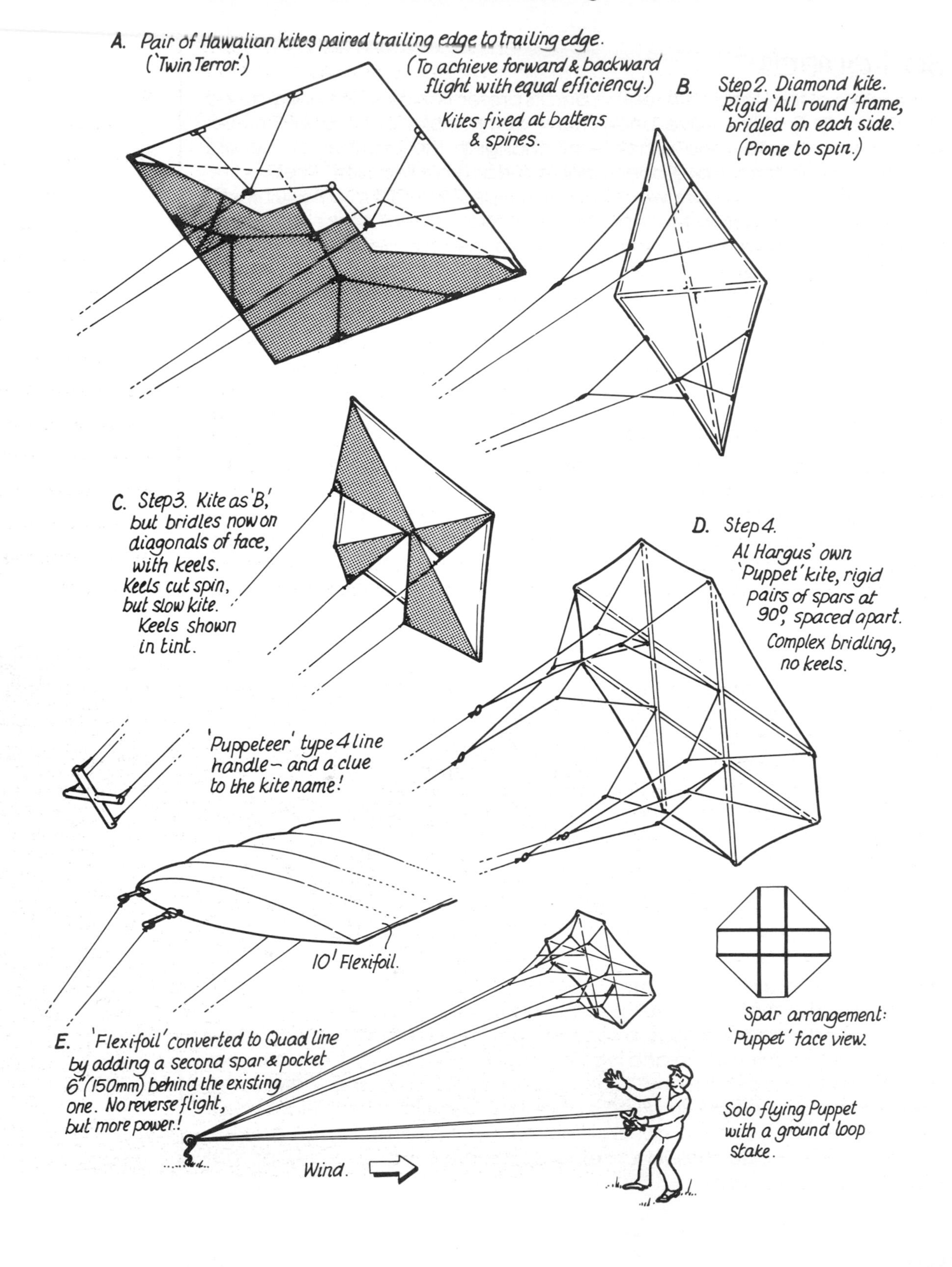

A *Double Quad-line* method followed to achieve reverse flight more easily, and has been successful with a 183cm (6ft) Peter Powell among others. This involves making precise duplicates of the existing bridles for 2-line and attaching them at the same points. The original connectors are moved 'up', and the new connectors moved 'down' by 75mm or 3in.

This third method offers best overall control; but still requires patience in learning how to operate. Holding the handles at their tops so that the lower lines are pulled or relaxed is the first key to success. Above all, do not expect to emulate the specially designed Revolution because that cross-spar and rigid frame will prevent any articulation. Fun flying is the objective and Al Hargus takes us right into this aspect with his other inventions.

With Lee Sedgwick, he hatched the idea of mating two Hawaiian kites rear-to-rear on the basis that a rigid leading edge would ease flight in any direction. Linking the battens and leading edges produced a 'Twin-Terror' that pulled too much in good wind but proved the concept. A 'Solid Diamond' followed with a lighter structure as Step 2. This moved in all directions and rotated at the drop of a hat, so Step 3 came next with keels to slow the rotation. They also slowed any sideways movement, so Al took a break, and after a year came up with his 'Puppet' kite which he calls a 'combination of sleds and deltas' (Step 4). This one does not spin, but moves in any direction at will. It flies slowly and because on lines of up to 30m or 100ft it was difficult to observe all that was happening in tests, Al flew using a ground stake tether for a close look as he controlled with a cruciform puppeteer handle.

David Davies has a range of 'Inter Active' variable geometry quadline kites with an aero-elastic quality that enables them to vary form with wind and airspeed. There are no bridles. The four lines are direct to chordwise braces from the flexible leading edge. One of six shapes, this is 'Swift' with an appliqué 'Puff the Dragon' decoration. An outstanding design of the late 1990s.

One other experiment came out of a conversation with Ray Merry, co-inventor of the Flexifoil. This involved sleeving a second spar 15cm or 6in behind the main spar and attaching a second set of lines. Ray had forecast variation of the pull on the lines as the angle of the foil was adjusted and when Al tried it he found the lift appeared to be doubled!

As with the back-to-back Hawaiians, the weight increase needed good wind strength, and reverse flight for the Flexifoil was not possible, though the designer's prediction was fully confirmed.

Batman

Oliviero Olivieri's Batman design is symmetrical, unlike the Revolution, and cleverly adopts contemporary graphics.

Oliviero Olivieri in Rome had wanted to design a character kite ever since he was involved in running repairs to Malcolm Goodman's world travelling 'Superman' inflatable kite. His four children were caught up in the Batman craze with black and yellow logos around the house so the basis of a quad-line Batman was entered into Oliviero's IBM on Autocad software. However, as founder of the *Associazione Italiana Aquilonisti* he was much involved in their fine newsletter and in any case, the challenge was as much one of curvaceous sewing as simple design. So that Batman rested a while.

Meanwhile in Belgium, Wim Gaudean had similar ideas, and his finished design appeared in *Nouveau Cervoliste Belge*. This inspired Oliviero to wake up the files and resume his project which though similar in its spar arrangement, is quite different in detail. As Pat has noted on the two sketch pages, the sewing challenge to produce those additional hems for the inside curves of the unsupported sail has called for some inventiveness. Oliviero solved the problem by cutting ripstop into 25mm (lin) strips with the grain at 45 degrees, folding and then stretching the folded edge to form spirals. The semi-circular outer edges are another matter. They have to be more robust, and sleeved with Dacron to take the GF rod outline. Here the ripstop joiners or doublers have to be cut on the curve, joined to make up the length then sewn to either edge of the Dacron. Then the 'bicycle tyre' shape is sewn in turn to the sail edge with two parallel rows of stitching. Of course there is the option to go for a fully decorative sail, or to make it all black with yellow detail. This is a quad-line kite that gets away from straight lines yet has full flexibility about the central spar, and it does fly on lines 50m (l65ft) long! Those specially made lugs which Oliviero cuts from l2mm (1/2in) nylon may cause problems but substitution with plastic tube, split rings and stock fittings is, we're sure, within the grasp of most kite-makers.

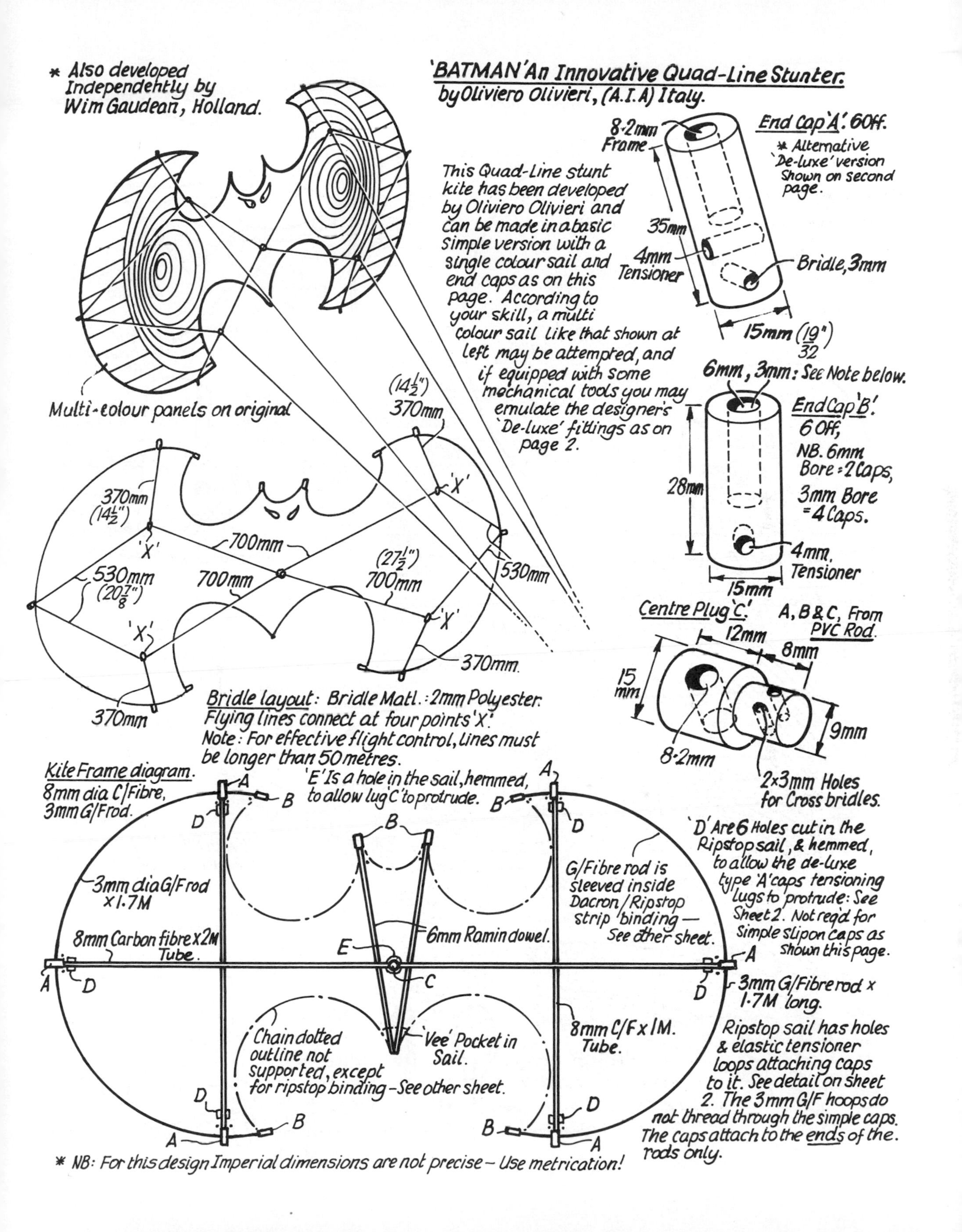

* Also developed Independehtly by Wim Gaudean, Holland.
'BATMAN' An Innovative Quad-Line Stunter.
by Oliviero Olivieri, (A.I.A) Italy.
This Quad-Line stunt kite has been developed by Oliviero Olivieri and can be made in a basic simple version with a single colour sail and end caps as on this page. According to your skill, a multi colour sail like that shown at left may be attempted, and if equipped with some mechanical tools you may emulate the designer's 'De-luxe' fittings as on page 2.
8·2mm Frame
End Cap 'A'. 6 Off.
* Alternative 'De-luxe' version Shown on second page.
35mm
4mm Tensioner
Bridle, 3mm
15mm (19/32")
6mm, 3mm: See Note below.
End Cap 'B'. 6 Off,
NB. 6mm Bore = 2 Caps, 3mm Bore = 4 Caps.
28mm
4mm. Tensioner
15mm
Centre Plug 'C'.
A, B & C, From PVC Rod.
12mm
8mm
15 mm
9mm
8·2mm
2x3mm Holes for Cross bridles.
Multi-colour panels on original
(14½")
370mm
370mm (14½")
'X'
700mm
530mm (20⅞")
700mm
(27½")
700mm
530mm
370mm.
370mm
Bridle layout: Bridle Matl.: 2mm Polyester. Flying lines connect at four points 'X'.
Note: For effective flight control, Lines must be longer than 50 metres.
Kite Frame diagram. 8mm dia C/Fibre, 3mm G/Frod.
'E' Is a hole in the sail, hemmed, to allow lug 'C' to protrude.
A
B
D
3mm dia G/F rod x 1.7M
8mm Carbon fibre x 2M Tube.
6mm Ramin dowel.
E
C
G/Fibre rod is sleeved inside Dacron/Ripstop strip binding — See other sheet.
'D' Are 6 Holes cut in the Ripstop sail, & hemmed, to allow the de-luxe type 'A' caps tensioning lugs to protrude: See Sheet 2. Not req'd for simple slip on caps as shown this page.
3mm G/Fibre rod x 1·7M long.
Chain dotted outline not supported, except for ripstop binding - See other sheet.
'Vee' Pocket in Sail.
8mm C/F x 1M. Tube.
Ripstop sail has holes & elastic tensioner loops attaching caps to it. See detail on sheet 2. The 3mm G/F hoops do not thread through the simple caps. The caps attach to the ends of the rods only.
* NB: For this design Imperial dimensions are not precise – Use metrication!

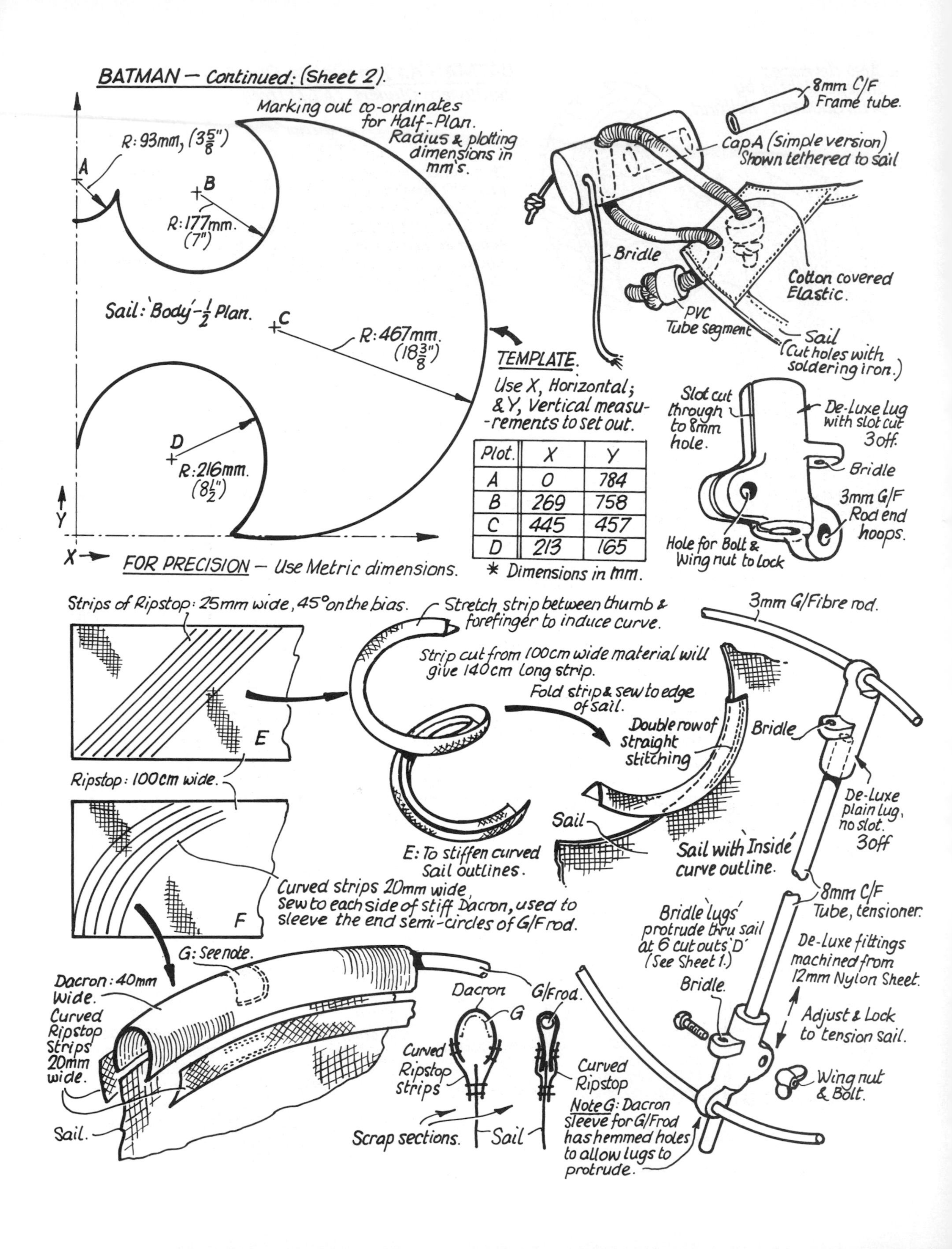

Plot.	X	Y
A	0	784
B	269	758
C	445	457
D	213	165

Machismo

Kite-power can be impressive. Inflatable wings and Parafoils are capable of lifting the flyer for considerable distances and to heights of 20m or 65ft in strong winds. Over sand dunes or winter snow, the sport of power-kiting has spread around the world, and not always with happy results. There have been fatalities and serious injuries, most of them due to loss of kite control.

The craze ranges from straight jumps after speed has built up on snow or sand skis, to being hauled in a harness along a cable stretched between two anchor poles. Another option is the towed cart which becomes more of a risk to onlookers than the operator, or the tethered flyer mode where the 'pilot' becomes a floating ballast between the anchor pylon and a train of foils which he manoeuvres. Each of these exploits has had its serious incidents.

Regrettable as the fatalities and personal injuries are, It is even more distressing that official reaction to these adventures is to close down *all* kite flying at some sites in the USA and Europe. So it is very much up to the enthusiasts themselves to impose regulations to preserve safety and in turn, the flying areas.

The dichotomy which already divides the single line kite flyers who build and fly for peace of mind, from the two and four-line steerable operators who need ground space for their energy sapping excitement is tolerated amicably. But there are signs that kite *power* induces danger.

TAKE CARE!

Left: Jumping in Cornwall, not for your average kite flyer!

Below: High level lift under a powerful kite is not to be recommended for the nervous.

CHAPTER NINE

Nineteen selected variations

If you are an avid kite flyer and collect books on kite-making, you will know the vast range of shapes and varied modes of construction which have already been described in detail elsewhere.

We all have special favourites and, from correspondence over the years since our first book was published, we have also learned which designs have the greatest attraction for the home constructor. The eleven types in this Chapter provide a sampling of designs that meet the primary requirement. In other words, *they work*!

All are for single line flying, as distinct from the steerable designs in Chapter Eight. There are five designs which have derived from long-standing traditional kites, and six which are representative of the many changes which have arisen over the past 15 years, as we have earlier described. They will all fly in a broad range of wind speeds, and are sufficiently self-stable to be anchored to a corkscrew ground peg. Given a steady wind, they will fly as long as you wish without undue attention.

We'll start with an all-time favourite from London's Kensington Palace Gardens and the famous Round Pond which has been a weekend kite flying site for almost a century.

Pearson's Roller

The *Roloplan* stems from Germany and is at least 80 years old as a concept of an almost square main lifting sail, stabilised by a tandem-mounted, second surface with a ventral keel. Originally, the two surfaces were linked by apron strings or multiple tapes, but use of cross-spars at both of the edges either side of the 'letterbox' slot and bracing ties at the spar ends both simplifies and modernises this classic design. Transformed by Alexander ('Alick') Pearson, the great character who was always persuaded to sell his 'last' Roller to a generous tourist, then later to 'find' a reserve one in his kite bag, this is one of our special favourites.

It could be called the suitcase kite because, if a tube joint is used in the spine, the longest length of dowel is around 750mm (29 ½ in), just right to pack diagonally in most family suitcases ready for the annual holiday. Other cross-spars are four-off 600mm (23 5/8 in) and two-off 550mm (21 5/8 in). The sails can be folded while still connected by the bracing ties and a central tape. While Alick Pearson's originals were mostly in cotton cambric, and much prized still by those lucky enough to have acquired one, ripstop nylon is ideal; but use of a tape reinforcement all around each of the sails is essential.

Commercial Rollers as made by Jilly Pelham of Vertical Visuals and the 'Monday Lunch' generally use a distinctive colour for the edge binding which adds to the vintage character.

Two forms of bridle have been used. The rear keel is essential for either, and the difference is that a forward ventral fin or keel offers a convenient eyelet position for the front line. Some Rollers have adopted a spring tensioner in the rear line, as used in Germany. You'll learn why if your Roller is launched in a stiff wind. They take off like a rocket and climb quickly on strong line pull. An 'elasticated' rear bridle line helps to take the strain. Front line for this size of Roller should start at 1020mm (40in) and rear at 1525 (60in).

There's only one awkward aspect of making a Roller. It has to have a generous dihedral angle of 15 degrees on either side of centre. This is fixed by bending alloy tube after it has been flattened in the centre. Ideally, you should use a bench vice into which a steel bolt or similar of the same size as the spine dowel is sandwiched horizontally between the jaws and the tube which is vertical. If the tube has been annealed by heating over a gas flame concentrated on the area to be flattened, it will be softened and will work easily. Smear the area with soap before heating and, when it turns black, the time is right to squeeze those vice jaws. Another long bolt inside the tube will act as a lever to bend the tube while still held vertically in the vice, before it has a chance to cool off. Quench in water to normalise.

Far left: Alex Pearson and one of the many Rollers he made for London visitors.

Above: The colourful sunburst design on Dave Thompson's example makes the Roller more attractive.

Needless to say, you must find a tube that is of the same internal diameter as the *outside* diameter of the dowel (8mm or 5/16in is fine for the given size of Roller). If you want to use carbon fibre tube, then brass connectors are usually available from the same source and these have to be a perfect sliding fit, so are very convenient. However, these connectors are of hard brass,

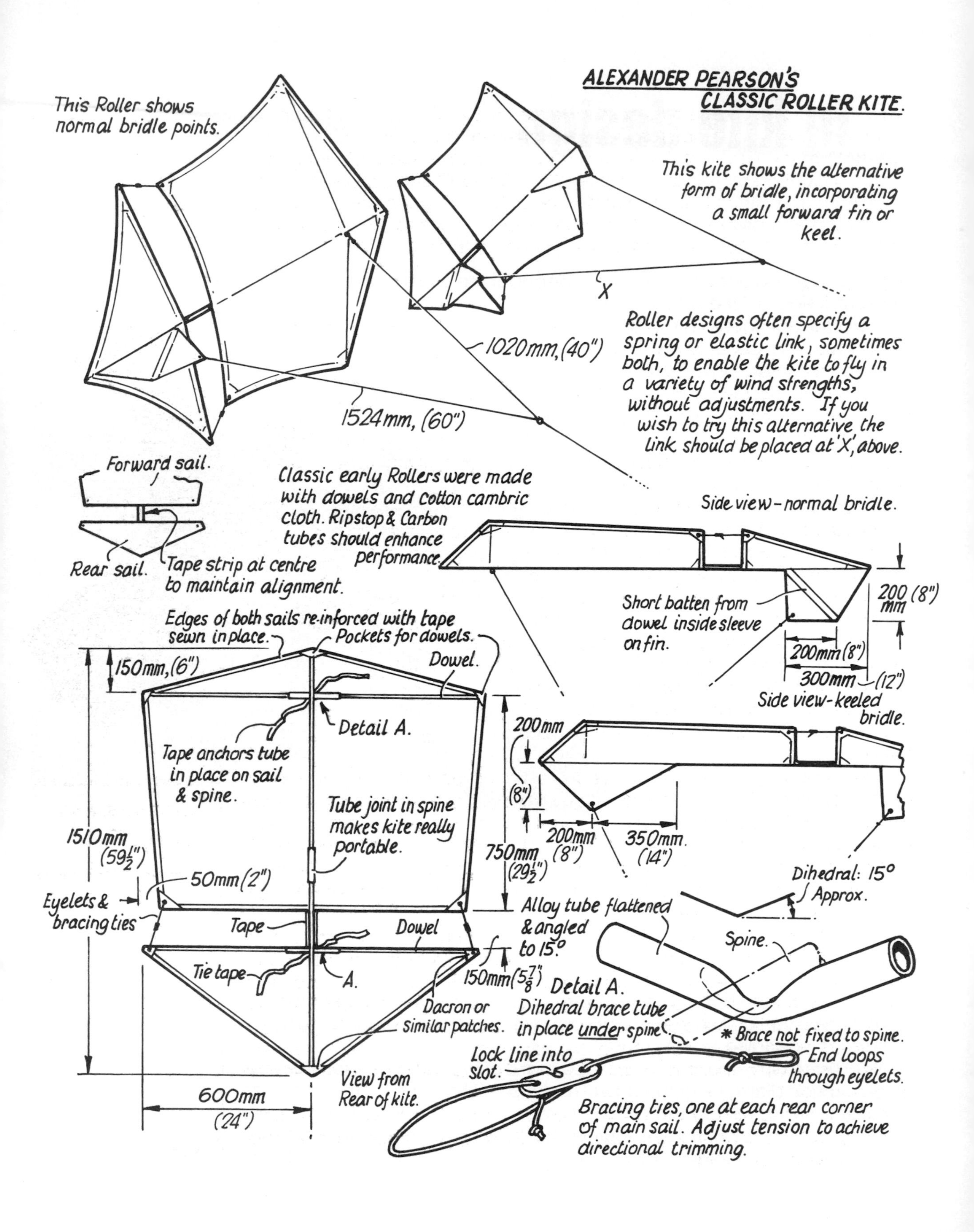

ALEXANDER PEARSON'S
CLASSIC ROLLER KITE.
This Roller shows normal bridle points.
This kite shows the alternative form of bridle, incorporating a small forward fin or keel.
X
1020mm, (40")
1524mm, (60")
Roller designs often specify a spring or elastic link, sometimes both, to enable the kite to fly in a variety of wind strengths, without adjustments. If you wish to try this alternative the link should be placed at 'X', above.
Forward sail.
Rear sail.
Tape strip at centre to maintain alignment.
Classic early Rollers were made with dowels and cotton cambric cloth. Ripstop & Carbon tubes should enhance performance.
Side view - normal bridle.
Short batten from dowel inside sleeve on fin.
200 (8") mm
200mm (8")
300mm (12")
Side view - keeled bridle.
Edges of both sails re-inforced with tape sewn in place.
Pockets for dowels.
150mm, (6")
Dowel.
Detail A.
Tape anchors tube in place on sail & spine.
Tube joint in spine makes kite really portable.
1510mm (59½")
750mm (29½")
50mm (2")
Eyelets & bracing ties
Tape
Dowel
Tie tape
A.
150mm (5⅞")
Dacron or similar patches.
600mm (24")
View from Rear of kite.
200mm
(8")
200mm (8")
350mm. (14")
Dihedral: 15° Approx.
Alloy tube flattened & angled to 15°
Spine.
Detail A.
Dihedral brace tube in place under spine.
* Brace not fixed to spine.
Lock line into slot.
End loops through eyelets.
Bracing ties, one at each rear corner of main sail. Adjust tension to achieve directional trimming.

which has to be annealed as described, or it may split as it is flattened.

There is always a temptation to fly to tremendous heights with a Pearson Roller. When joyrides were flown over London from Croydon and Heston, kites were frequently reported at 600 metres (2000ft) over Clapham Common and Hyde Park. They were tolerated until a deHavilland Dragon Rapide collided with one, fortunately without further accident. Inevitably many of those offenders were Rollers - so, be warned and resist temptation.

Balinese bird

The Eastern nations have always favoured insects and bird shapes for kite design, and we've found that many newcomers in the UK also have a fixation that their first kite should be like a bird.

One of the Round Pond specialists, along with Alick Pearson, produced magnificent bird kites. John Robson could turn out a Buzzard with patterned feathers and twitching tail that drew many admirers. But it was not for everyone because of its intricate detail.

So we have another traditional kite, though modified from the original Balinese, to use a ripstop nylon cover and a foam plastic head. These concessions will make the project all the easier to make for the Westerner; but there is no escaping the tapering of the spine and leading edge spars if you want to reproduce the fluttering flight. Split cane can be shaved with a single-edged razor blade or even the raw edge of a piece of glass. It might be replaced by thin glass fibre rod, as has been done with latest Indian or Malaysian fighting kite frames; but, If you want a Balinese bird, then why not keep to some of the original features?

Using similar structure to the Balinese Bird, this example had multi-coloured panels of ripstop.

Start by enlarging the bird profile for a cutting pattern. The squares will offer a 1200mm (47 ¼ in) birdspan. If you prefer something smaller, then we would suggest making the squares 35mm (33in), but certainly not less. Use a heavy grade nylon, if possible, as the greater part of the wing around its trailing edge remains unsupported and should not be completely limp. In flight, this part of the 'sail' will fluctuate and give the flapping effect, though it also has to contribute lift.

Once the sail plan is cut, the frame can be assembled and trimmed to fit. There is nothing that is absolutely critical, and dimensions on Pat's drawings are nominal. This is the kind of kite where a little imagination can be applied, especially in the final decoration and the head. Keep the head as light as possible. Expanded polystyrene is the best material. It can usually be found discarded as packing blocks round the waste from shops such as TV stores. A razor blade will carve it, and acrylic paints from an art shop will offer vivid colouring.

Note that the central spine must be tapered in width to permit some flexibility; but has to be strong as, along with the cross-spar behind the head, it carries the 3-point bridle.

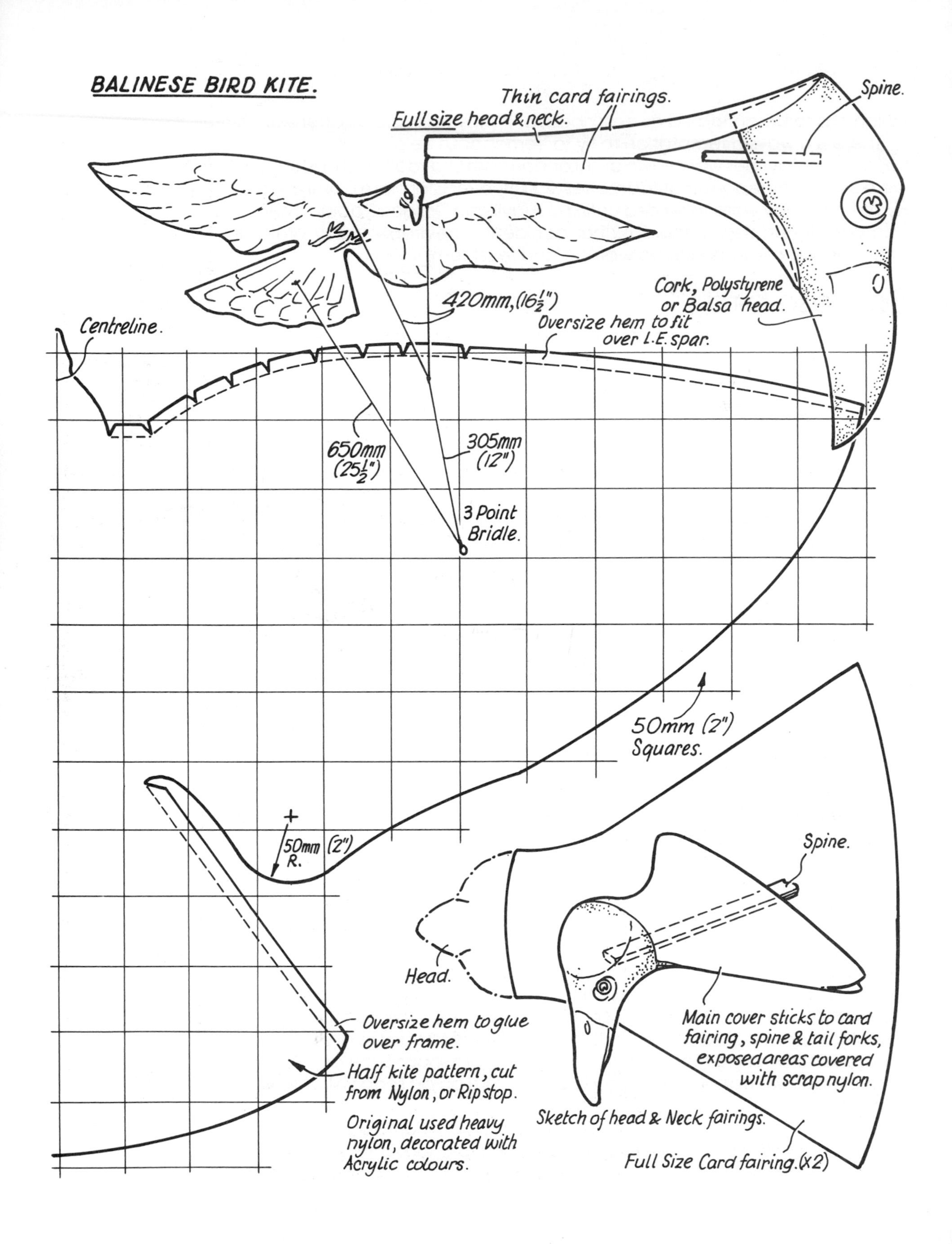
BALINESE BIRD KITE.
Thin card fairings.
Full size head & neck.
Spine.
Cork, Polystyrene or Balsa head.
420mm, (16½")
Oversize hem to fit over L.E. spar.
Centreline.
650mm (25½")
305mm (12")
3 Point Bridle.
50mm (2") Squares.
50mm (2") R.
Spine.
Head.
Oversize hem to glue over frame.
Half kite pattern, cut from Nylon, or Ripstop.
Original used heavy nylon, decorated with Acrylic colours.
Main cover sticks to card fairing, spine & tail forks, exposed areas covered with scrap nylon.
Sketch of head & Neck fairings.
Full Size Card fairing. (x2)

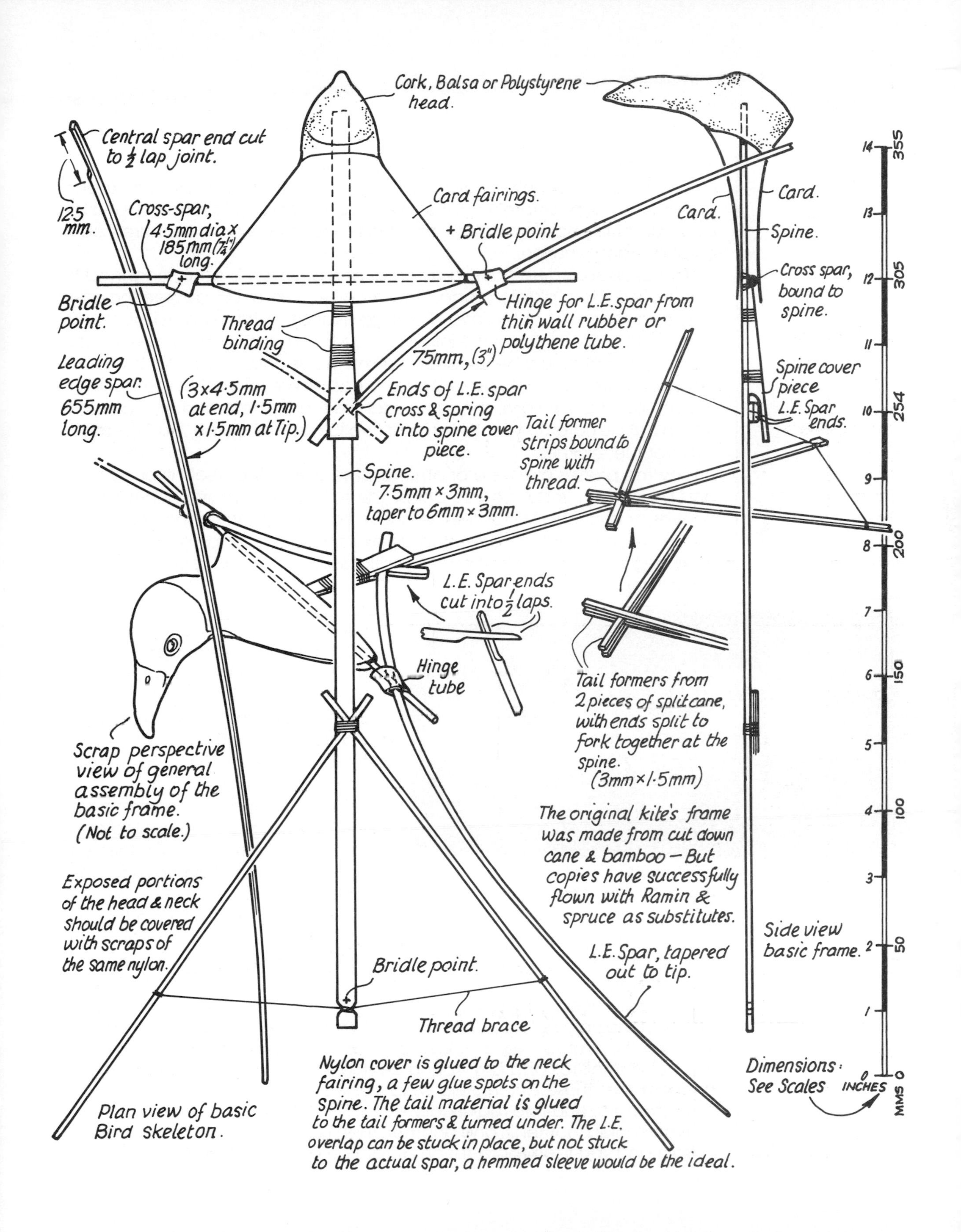
Cork, Balsa or Polystyrene head.
Central spar end cut to ½ lap joint.
12.5 mm.
Cross-spar, 4.5mm dia x 185mm (7¼") long.
Card fairings.
+ Bridle point
Bridle point.
Thread binding
Hinge for L.E. spar from thin wall rubber or polythene tube.
75mm, (3")
Leading edge spar. 655mm long.
(3x4.5mm at end, 1.5mm x 1.5mm at Tip.)
Ends of L.E. spar cross & spring into spine cover piece.
Spine. 7.5mm x 3mm, taper to 6mm x 3mm.
Tail former strips bound to spine with thread.
L.E. Spar ends cut into ½ laps.
Hinge tube
Scrap perspective view of general assembly of the basic frame. (Not to scale.)
Tail formers from 2 pieces of split cane, with ends split to fork together at the spine. (3mm x 1.5mm)
The original kite's frame was made from cut down cane & bamboo – But copies have successfully flown with Ramin & spruce as substitutes.
Exposed portions of the head & neck should be covered with scraps of the same nylon.
Bridle point.
L.E. Spar, tapered out to tip.
Thread brace
Nylon cover is glued to the neck fairing, a few glue spots on the spine. The tail material is glued to the tail formers & turned under. The L.E. overlap can be stuck in place, but not stuck to the actual spar, a hemmed sleeve would be the ideal.
Plan view of basic Bird skeleton.
Card.
Card.
Spine.
Cross spar, bound to spine.
Spine cover piece
L.E. Spar ends.
Side view basic frame.
Dimensions: See Scales
INCHES
MMS
14
13
12
11
10
9
8
7
6
5
4
3
2
1
0
355
305
254
200
150
100
50
0

Seagull

Made commercially, the extremely popular Tern or Seagull kite is a product of the 1970s. In a remote way, it is a very high aspect ratio form of Delta kite by virtue of its single cross-spar attached to leading edges that are free to pivot at each junction.

The cross-spar must have a dihedral angle, which is easily arranged by fitting a standard plastic brace. These come with a 145-degree included angle, so you get an automatic 17.5 degrees under each wing. All the dowels can be 4mm (5/32in) and the joiners made from plastic tubing as sketched. White material should be used for realism, in this case the standard ripstop nylon, or lighter if possible as this is a fair-weather kite for soft breezes. Enlarge the profile which will give the same span as the Balinese bird of 1200mm or 47 ¼ in but be careful - you have to cut two full profiles, as the 'body' is made into double thickness.

Lay the profiles over one another and sew from the neck, around head and beak, then under-belly to below the tail. Make two stitch lines to form sleeves at front and tail for a spine and, after strengthening the breast area with a piece of wire or cane, add an extra twill tape binding to secure the towing zone. A selection of eyelet positions will offer changes of single point line attachment.

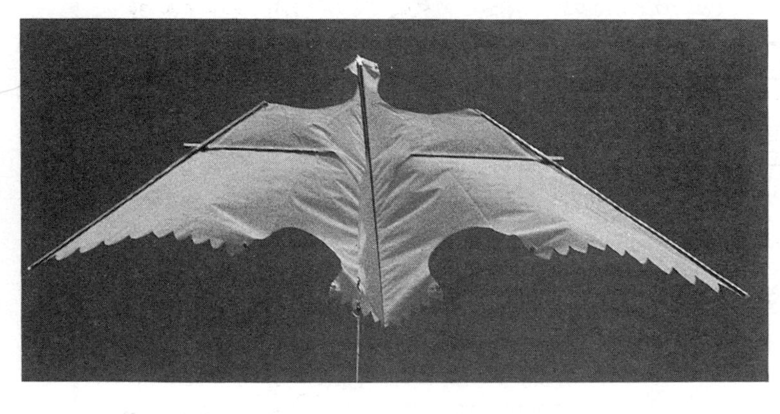

Variation on the Seagull, using the same articulating leading edge.

At the tips, a hem offers a sleeve for the dowel which can be sewn in place permanently as breakages here are rare. Don't forget to fit the plastic tube brace or joiner first! The central spine must also have the dihedral brace fitted before it is slotted into the body sleeves, and then we are ready to trim the cross-spar halves to length. There must be just enough tension in the 'gull' part of the wings to prevent the cross-spars from slipping out of either end fixing in flight. They also have to be of equal length. Make two spares while you are about it.

You'll now have the basics of a Seagull. What it cries out for is a touch of colour. Spraypaints, or acrylics can make a world of difference to the appearance - with a dash of black and a yellow beak, Jonathan Livingston comes to life!

Not quite as easy to launch as most kites, the Seagull is best started with plenty of line paid out first. Use the eyelets to vary the towing position, shifting forwards if it pitches over, backwards if it is reticent to climb. Being a bird, it has but a small tail and any trailing device is unthinkable. Instead, you will discover that those articulating wing panels take care of any variation of wind strength, and flap realistically.

Now make another,... and another, then tie them centre brace to eyelet in train and you'll really create an attraction on the beach!

SEAGULL KITE.

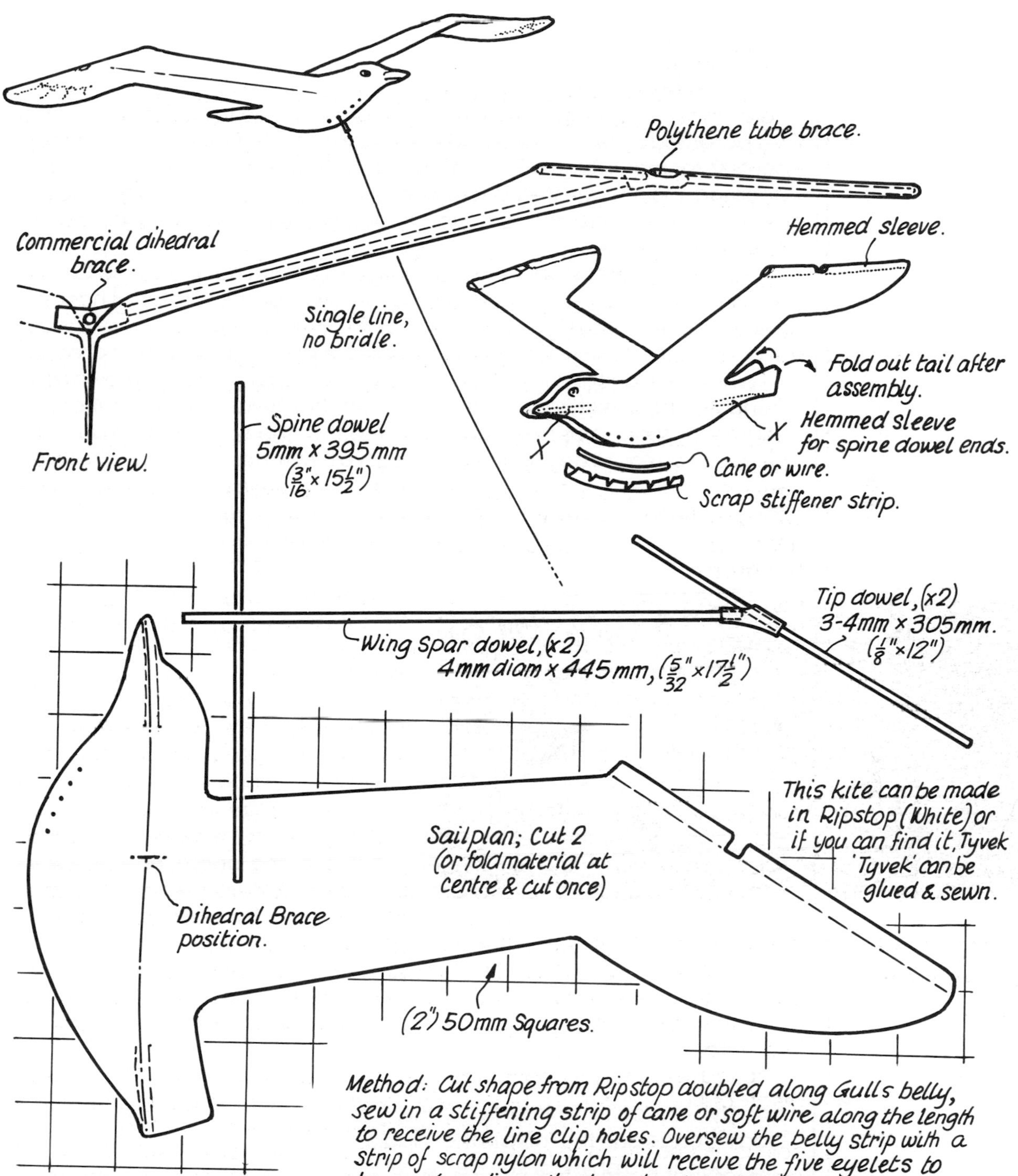

Polythene tube brace.

Commercial dihedral brace.

Hemmed sleeve.

Single line, no bridle.

Fold out tail after assembly.

Hemmed sleeve for spine dowel ends.

X

X

Front view.

Spine dowel 5mm x 395mm ($\frac{3}{16}$" x $15\frac{1}{2}$")

Cane or wire.

Scrap stiffener strip.

Tip dowel, (x2) 3-4mm x 305mm. ($\frac{1}{8}$" x 12")

Wing spar dowel, (x2) 4mm diam x 445 mm, ($\frac{5}{32}$" x $17\frac{1}{2}$")

This kite can be made in Ripstop (White) or if you can find it, Tyvek 'Tyvek' can be glued & sewn.

Sailplan; Cut 2 (or fold material at centre & cut once)

Dihedral Brace position.

(2") 50mm Squares.

Method: Cut shape from Ripstop doubled along Gulls belly, sew in a stiffening strip of cane or soft wire along the length to receive the line clip holes. Oversew the belly strip with a strip of scrap nylon which will receive the five eyelets to be used as line attachment points. The beak & tail areas indicated at 'X' have a sleeve stitched in to accept spine dowel. Wing tips have an overlap to make a sleeve for the tip dowel & cut-out for the dihedral brace, made from polythene tube.

Paint on eyes, beak & Wingtip patches. Consult a Bird Book!!

The Conyne

Here is another quite simple design that has stood the test of time ever since first patented by its creator Silas J.Conyne in the USA in 1902. It won a major contest for man-lifting in France during 1905, when the specialist, Maillot, adopted the shape after discarding his own octagonal design. Soon it became commonplace among amateurs, being featured in *The Boy Mechanic* of 1908.

Britain's longest established kite-makers, Brookite, have had the Conyne type in their range since it was first flown. Known as their 'Cutter', it has been made in innumerable thousands and, as the 'Master' with two boxes instead of one. It is the oldest, most worn and reliable kite in our personal collection.

When the French Army adopted the shape, it was generally known as the French Military Box Kite, and that title, too, has held for ages. Often, the 'diamond' shape has been converted so that the flares either side of the triangular box are extended and curved like bird wings. Will Yolen, that great international ambassador for kite flying, flew this type wherever he went. Europe, the Middle East, India and South America as well as his native Sheeps Meadow in Central Park, Manhattan, were his playgrounds. The Yolen crusade for kites impinged on royalty, politicians and the famous and infamous alike. Will wrote three books on kites and his best *The Complete Book of Kites and Kite Flying,* Simon and Schuster, NY, 1976, confirmed his romance with the Conyne type which he flew around the world, preferring to call it simply a triangular Box kite with wings.

By now, you will have realised that this is a number one recommendation if you want something that looks big, lifts well, is easily made and transported. The real difference between the Conyne and the commonly accepted Box kite is that the triangular cells do not have any internal bracing to form rigid panels. They rely on the tension from the lower longeron, or spar, to which the bridle is fitted, and the lift of the flares to stretch the sides. So the main sail shape is held by the two longerons 'A' over the length, and across the width, the cross-spar 'C' stretches the flares. The latter is the only dowel that needs to be detachable so that the Conyne can be rolled up.

All the sail edges will benefit from a taped hem; but this depends on the material. Brookite 'Cutter' kites were made with cotton cambric for decades, the porosity enabling them to fly in very strong winds, and making them excellent carrier kites. Ripstop is both lighter and proofed, and has the greater requirement for an edging.

While it has always been the practice to make the rear boxes longer, stability was found to improve if the forward box was set back from the leading edge and the 'hole in the middle' extended, more than is usual. We are indebted to physicist Ray Biehler for this discovery which also locates the two bridle points neatly at the leading edge of each box on the lower longeron 'B'.

Make one - you'll never regret it.

IMPROVED CONYNE KITE.

A revised version of the familiar favourite Conyne or Pilot kite, the obvious difference is the shorter front cell, dimensions too have been optimised. These revisions are attributed to Ray Biehler, and reduce drag & improve stability and performance.

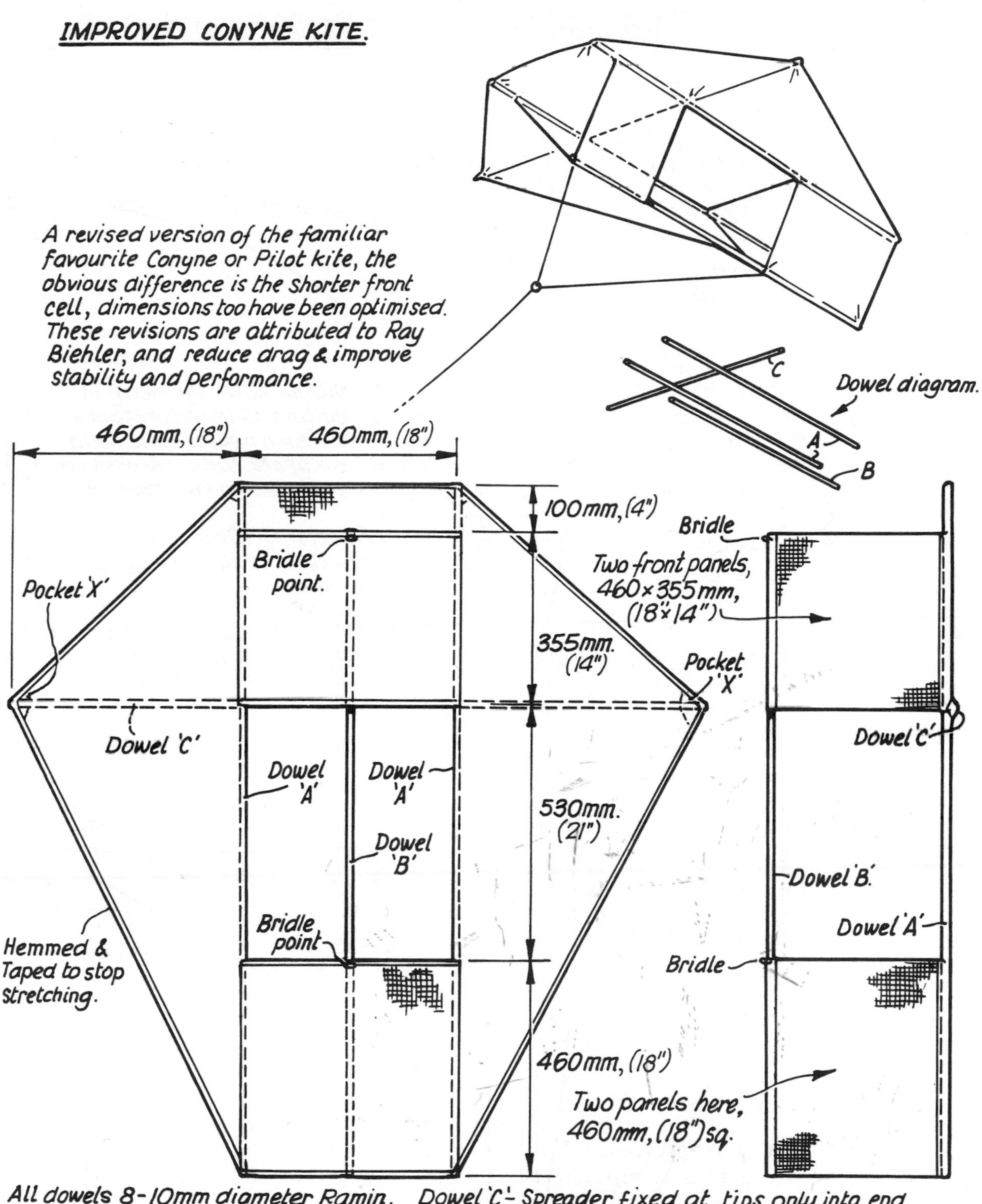

All dowels 8-10mm diameter Ramin. Glassfibre or Carbon tube could be used.

Sail: Ripstop, hemmed & taped at edges.

Dowel 'C' - Spreader fixed at tips only into end pockets in sail at X.

Dowels 'A' - Slid into sleeve hems. Dowel 'B' - 100mm shorter than 'A', Sleeved at ends into cell hems.

MARTIN POWELL'S HIGH ASPECT RATIO DELTA – BÉLIER (RAM) KITE.

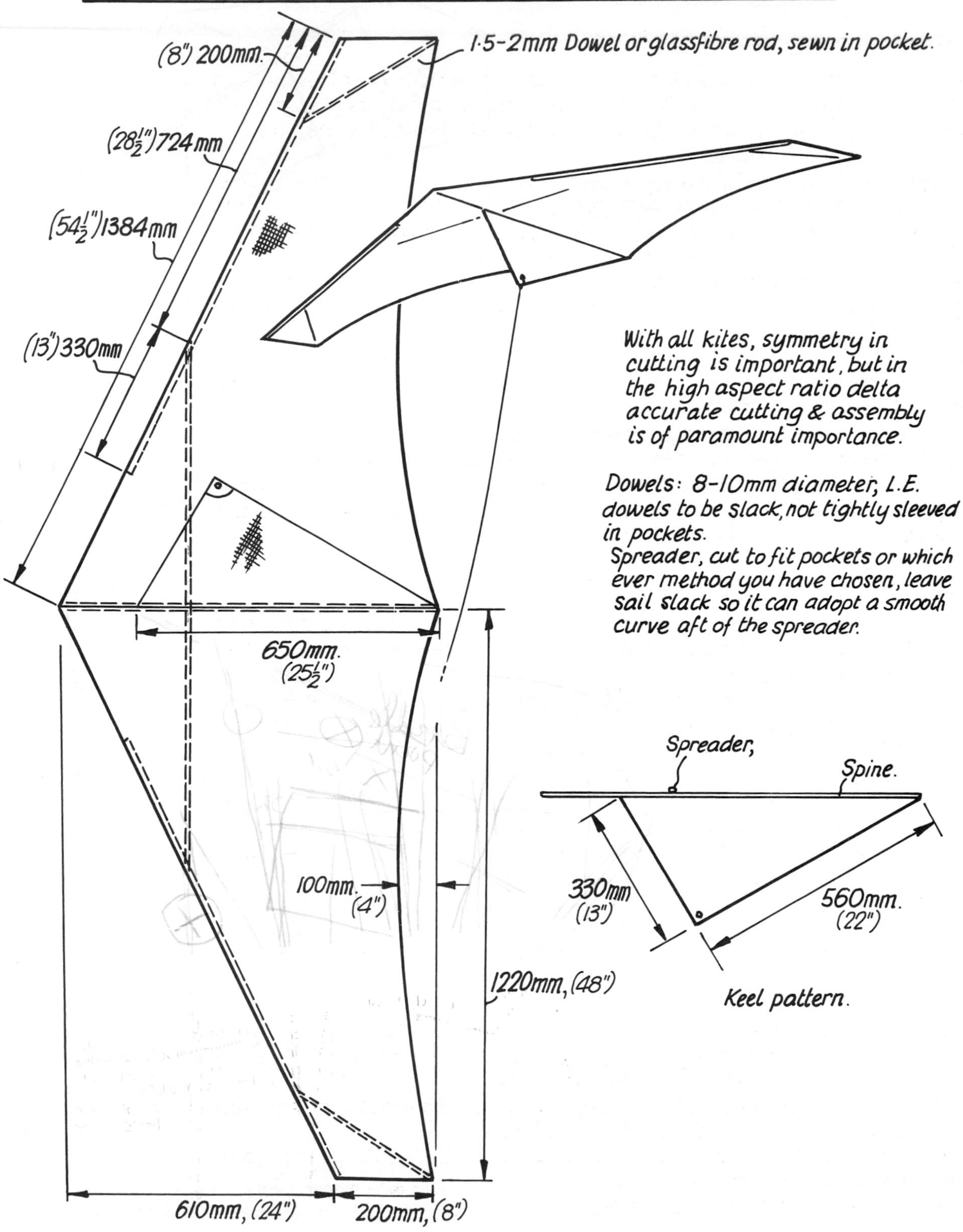

With all kites, symmetry in cutting is important, but in the high aspect ratio delta accurate cutting & assembly is of paramount importance.

Dowels: 8-10mm diameter, L.E. dowels to be slack, not tightly sleeved in pockets.
Spreader, cut to fit pockets or which ever method you have chosen, leave sail slack so it can adopt a smooth curve aft of the spreader.

Martin Powell's Belier

This comes out of competition, hang glider influence, and has quite a development history. Martin Powell of the British Northern Kite Group spent four years trying to tame the high aspect ratio, 130-degree Delta before he perfected this one in late 1983.

In the process of evolution, he tried tunnel keels, wing battens, spineless and extended leading edge variations, when he found that the simple conventional delta keel and light tip battens plus a change to the trailing edge curve gave the stable performance he was seeking. He refined the prototype as we see here and named it the Belier, or French for Aries the Ram, which was a pun on its predecessors' efforts to butt other kites out of the sky.

At 2440mm or 8ft wingspan, Martin's Belier is lightly loaded and will soar with a thermal on balmy days, yet hold its own in a strong wind. Those flexible leading edges which pivot on the spreader or cross-spar (as described in the Seagull) enable it to ride the gusts.

Note the suggested grain direction for the ripstop nylon, parallel to the leading edges. This enables the sail to adopt its own aerofoil in flight. Keep the tip battens light by using thin GF or carbon rod, and check the lateral balance before flight testing. It's important to select the leading edge dowels to ensure they are of equal weight.

Navy Pattern Dove Kite

This one is a lesser-known design from the early days. It was unearthed from the archives by Paul Chapman, another member of the inventive club known as the Northern Kite Group. Paul relaxes from his full scale aerospace commitments with his collection of traditional kites. His work has enabled him to uncover original drawings of kites that were created for military purposes, and the Navy Pattern Dove Kite, No D1587, is typical.

Originally tested in the Royal Aircraft Establishment's 24ft wind tunnel for use as a naval anti-aircraft barrage during the 1939–45 World War, it was designed to lay flat, presumably when on board ships. This makes it more attractive than, say a Cody or Lecornu, where rigid poles are arranged diagonally to hold their centre box forms. Because it is rather confusing in its plan elevation, Pat provides two perspectives which illustrate how it is not dissimilar to the Conyne. Instead of a single central triangular box, the Navy Dove has extended cells at front and rear. There are no rigid vertical braces and, when placed flat on the ground, the upper sails collapse down onto the lower sails.

In the air, it is rather like a pair of biplane wings, separated by the three upper and two lower longerons. It has an unusual characteristic of a vibration or rattle on the rear cell flaps. This was the reason for those tests in the wind tunnel, and the same rattle also showed up on Paul's replica, though ten years of flying it have not revealed any ill effects.

It is not exactly small! Length of the longerons is 1650mm (65in) and the two upper cross-spars are each 1920mm (75 ½ in) but, being another 'roll-up' type, transporting it is not usually a problem.

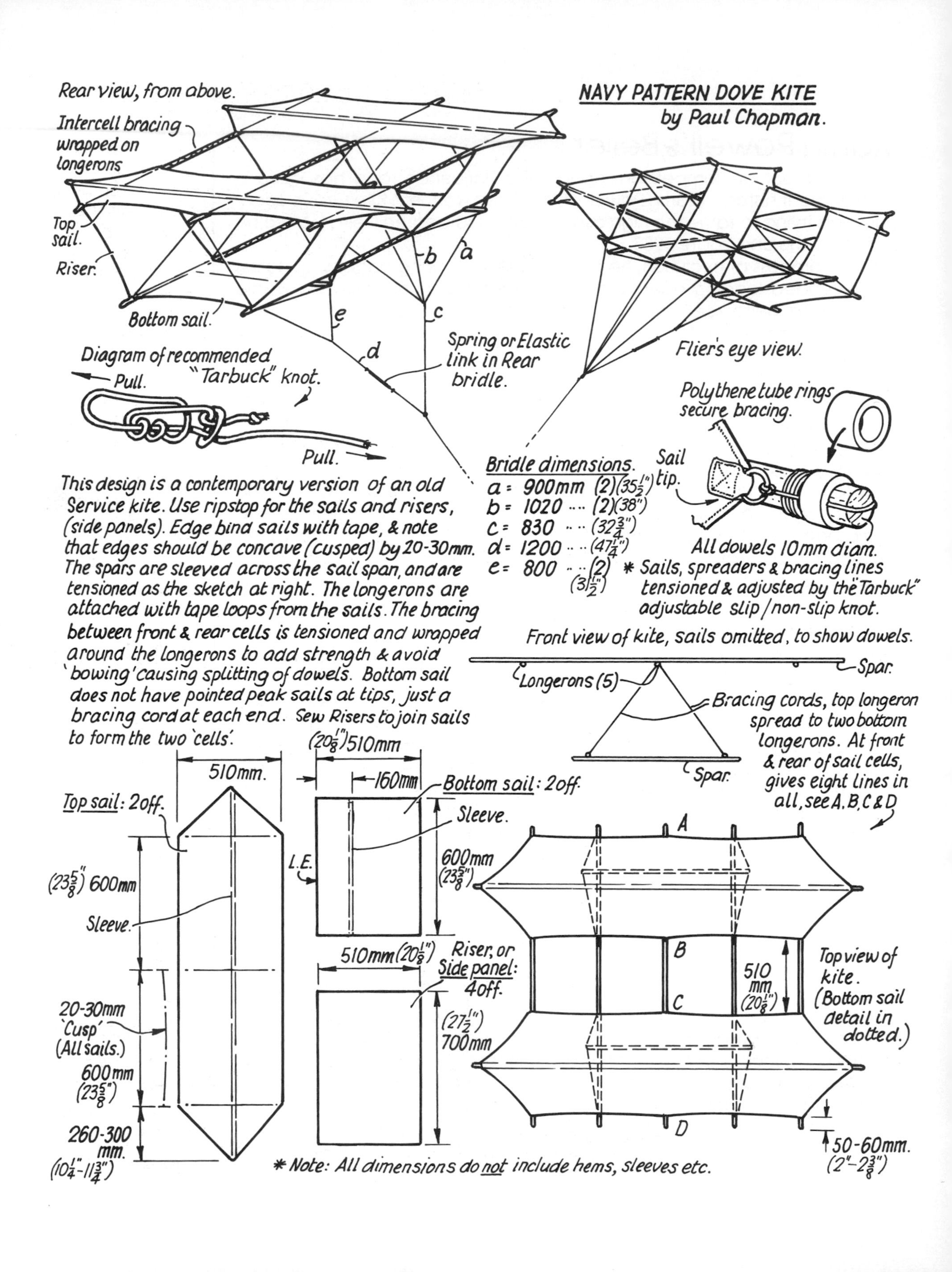

NAVY PATTERN DOVE KITE
by Paul Chapman.
Rear view, from above.
Intercell bracing wrapped on longerons
Top sail.
Riser.
Bottom sail.
a
b
c
d
e
Spring or Elastic link in Rear bridle.
Flier's eye view.
Diagram of recommended "Tarbuck" knot.
Pull.
Pull.
Polythene tube rings secure bracing.
Sail tip.
All dowels 10mm diam.
Bridle dimensions.
a = 900mm (2)(35½")
b = 1020 ···· (2)(38")
c = 830 ···· (32¾")
d = 1200 ···· (47¼")
e = 800 ···· (2)(31½")
* Sails, spreaders & bracing lines tensioned & adjusted by the "Tarbuck" adjustable slip/non-slip knot.
This design is a contemporary version of an old Service kite. Use ripstop for the sails and risers, (side panels). Edge bind sails with tape, & note that edges should be concave (cusped) by 20-30mm. The spars are sleeved across the sail span, and are tensioned as the sketch at right. The longerons are attached with tape loops from the sails. The bracing between front & rear cells is tensioned and wrapped around the longerons to add strength & avoid 'bowing' causing splitting of dowels. Bottom sail does not have pointed peak sails at tips, just a bracing cord at each end. Sew Risers to join sails to form the two 'cells'.
Front view of kite, sails omitted, to show dowels.
Spar.
Longerons (5)
Bracing cords, top longeron spread to two bottom longerons. At front & rear of sail cells, gives eight lines in all, see A, B, C & D
Spar.
510mm.
Top sail: 2off.
(23⅝") 600mm
Sleeve.
20-30mm 'Cusp' (All sails.)
600mm (23⅝")
260-300 mm. (10¼"-11¾")
(20⅛")510mm
160mm
Bottom sail: 2off.
Sleeve.
L.E.
600mm (23⅝")
510mm (20⅛")
Riser, or Side panel: 4off.
(27½") 700mm
A
B
C
D
510 mm (20⅛")
Top view of kite. (Bottom sail detail in dotted.)
50-60mm. (2"-2⅜")
* Note: All dimensions do not include hems, sleeves etc.

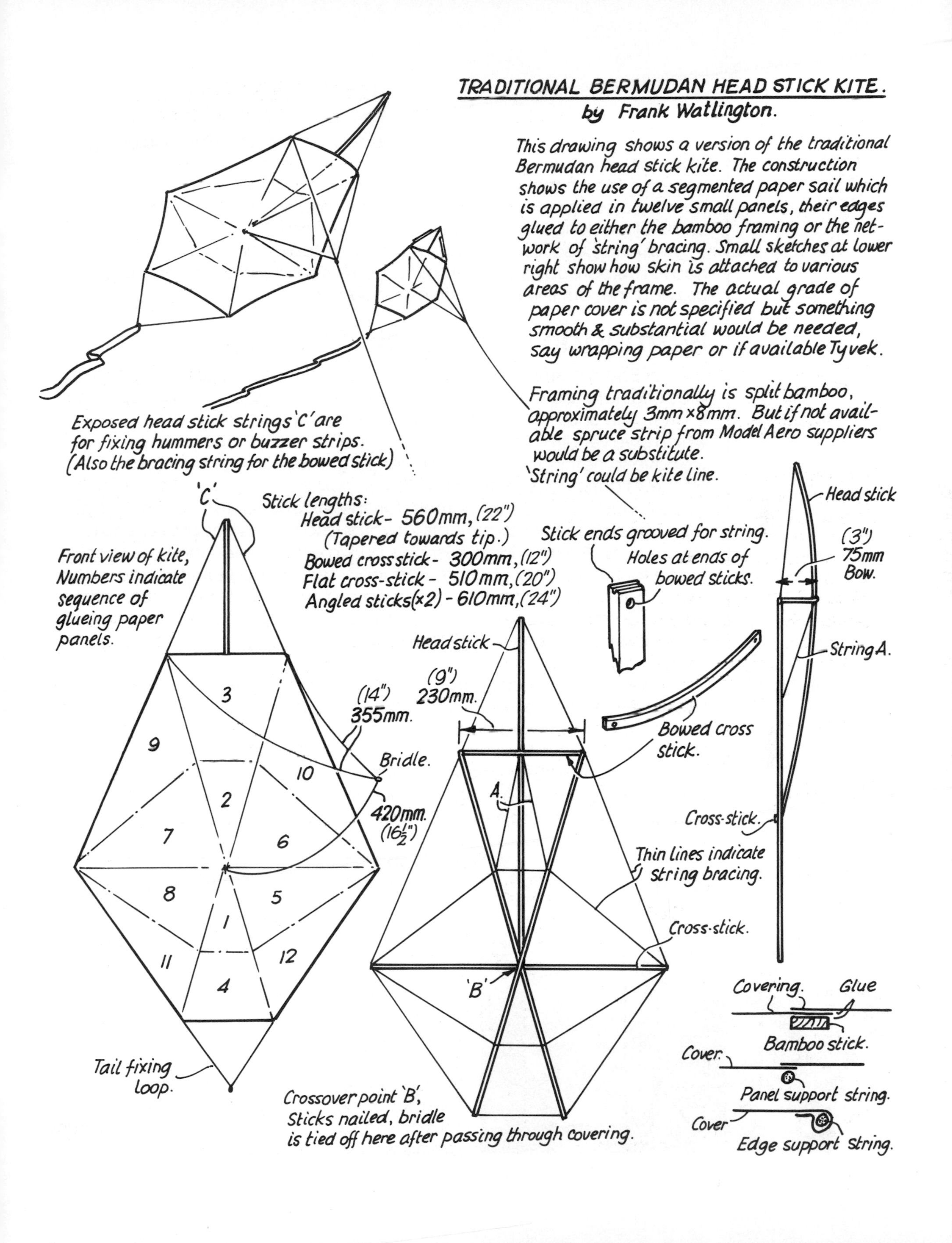
TRADITIONAL BERMUDAN HEAD STICK KITE.
by Frank Watlington.
This drawing shows a version of the traditional Bermudan head stick kite. The construction shows the use of a segmented paper sail which is applied in twelve small panels, their edges glued to either the bamboo framing or the network of 'string' bracing. Small sketches at lower right show how skin is attached to various areas of the frame. The actual grade of paper cover is not specified but something smooth & substantial would be needed, say wrapping paper or if available Tyvek.
Framing traditionally is split bamboo, approximately 3mm x 8mm. But if not available spruce strip from Model Aero suppliers would be a substitute.
'String' could be kite line.
Exposed head stick strings 'C' are for fixing hummers or buzzer strips. (Also the bracing string for the bowed stick)
'C'
Stick lengths:
Head stick - 560mm, (22")
(Tapered towards tip.)
Bowed cross stick - 300mm, (12")
Flat cross-stick - 510mm, (20")
Angled sticks (x2) - 610mm, (24")
Front view of kite, Numbers indicate sequence of glueing paper panels.
3
9
10
2
7
6
8
5
1
11
12
4
(14")
355mm.
Bridle.
420mm.
(16½")
Tail fixing loop.
Head stick
(9")
230mm.
A.
'B'
Crossover point 'B', Sticks nailed, bridle is tied off here after passing through covering.
Stick ends grooved for string.
Holes at ends of bowed sticks.
Bowed cross stick.
Thin lines indicate string bracing.
Cross-stick.
Head stick
(3")
75mm
Bow.
String A.
Cross-stick.
Covering.
Glue
Bamboo stick.
Cover.
Panel support string.
Cover
Edge support string.

Paul's construction notes are included on the drawing, including the 'Tarbuck' knot which he uses to adjust the eight diagonal bracing lines between the central longeron at the top and the two lower longerons. Note how the cross-spars are all sleeved into the sails, and that the spars go on top of the upper longerons and below the lower pair.

With its large sail area, this one has considerable lifting power (the originals had to lift steel cables) so it will call for investment in some high breaking strain braided line; but it will be worth every penny.

Bermudan Head Stick

The Far East is normally regarded as the zone where traditional celebratory kites originate; but they do not hold exclusive rights, as any serious student knows. The South American nations have their own particular kite shapes which owe very little to those more familiar Chinese or Japanese types. Here, we have something different again, and it comes from the island of Bermuda, off the Atlantic coast of the USA.

The 'Head Stick' is so named for the extended spar on an irregular hexagonal shape which projects forwards and carries paper buzzers on two rigging lines. Flown in large numbers during each Easter, they are specially made for the occasion, but are frequently seen throughout the year on the many beaches, trailing tails and weaving while they hum. Sometimes, Bermudan kites have been left to fly through the night. One was claimed to have flown for days on end, though it was of a more symmetrical hexagonal design.

It might not be appreciated at first glance that, when the head stick is fitted (this kite can be made 'flat' without one), there is a stiff wire or bamboo bow across the forward ends of the main spars. This, in turn, becomes the fulcrum about which the head stick is curved. It is not a large kite - the width is only 510mm or 20in - and it calls for little more than bamboo and tissue paper. Kiteline can be tied tightly around the full outline first, at the same time pulling the bow into the head stick, then an inner line bracing is tied in support. This allows the paper covering in various colours to be applied to both the sticks and the string. See the detail in Pat's sketch.

The paper is applied in numerical sequence, wrapping the last panels around the outline, but first securing the ends of the towing bridle to the central cross-stick point and the ends of the main spars. A tail is advised, either as a straight streamer or with paper bows on a length of kiteline, and you are away. Bright sunshine from an azure sky and a background of rap may help to transport you to Bermudan fantasy; if not, then tune the hummers and make your own music, with experiments in material and widths of strip wrapped to the head stick lines.

Bermudan 'Buzzer' kite flown at London kite festivals.

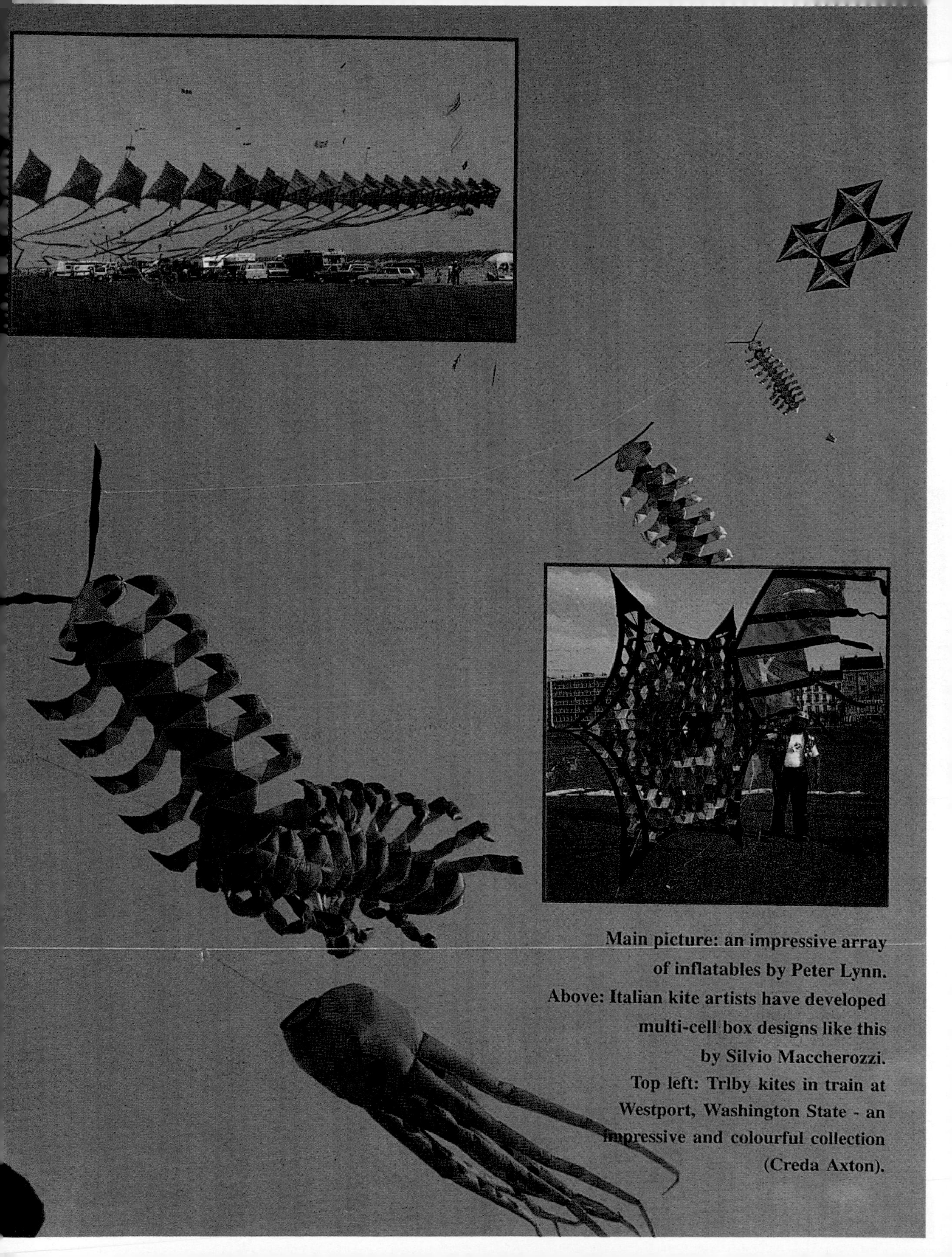

Main picture: an impressive array of inflatables by Peter Lynn.
Above: Italian kite artists have developed multi-cell box designs like this by Silvio Maccherozzi.
Top left: Trlby kites in train at Westport, Washington State - an impressive and colourful collection (Creda Axton).

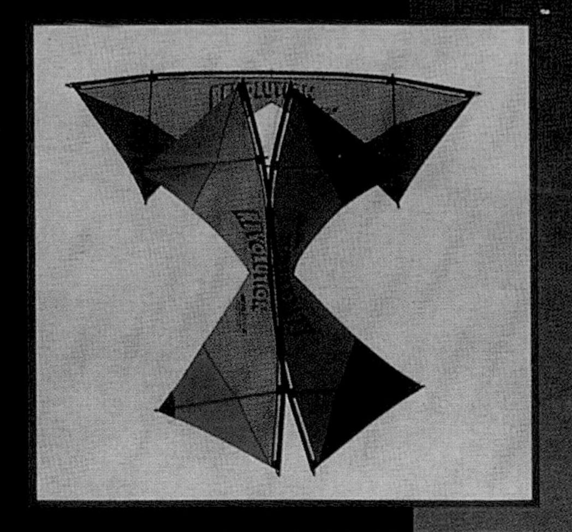

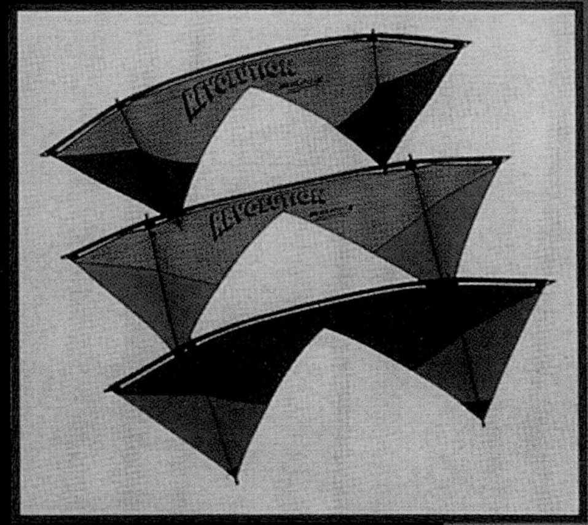

Main picture shows the sun setting on an unforgettable aerial display.
Above & below: Linking the Quad-line Revolution kites into one figure is a speciality of the 'Decorators' team. This takes hours of practice!

Above are two examples showing the diversity in kite design. Below is the Neos Omega, the original design by the Hadzicki brothers of San Diego. Bottom: the 'Revolution' proved to be a prophetic name.

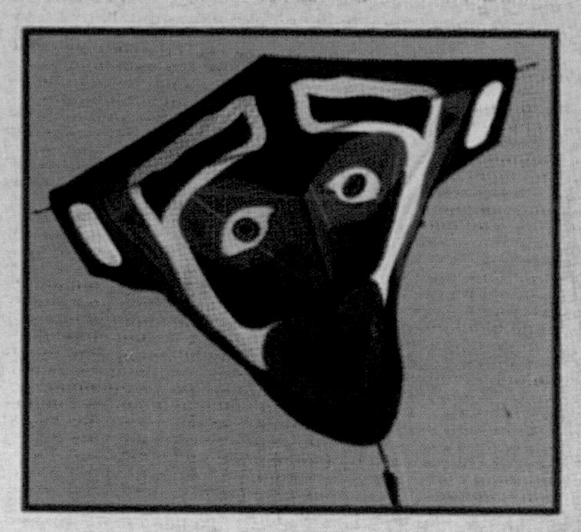

Main picture and above we see two striking examples from Don Mock.

Below: home built quad-liner is a very attractive exercise in appliqué, complete with mini-kites and hot-air ballon! Right:John Robson's Bird kites were well known in Kensington Palace Gardens near the Round Pond, London.

Main picture: Oliviero Olivieri's Batman design. Bottom left is Alan and Carole Peacock with their appliqué 'patchwork' design, quickly identified in the sky!
Right: the Tokyo Museum embraces several rooms with kites collected by Shingo Modegi and Masaaki Modegi.

Bottom right: Jorgen Moller-Hansen's multi-cell kite, made in Denmark, now flown in the USA (Creda Axton).

Left: harnessed and on his way, Ted is hauled aloft - in this instance without a carrier! Bottom left we see BMISS on parade at an Old Warden festival.

Left: multiple drops created by Greg Locke used novel tube containers for the parachutes. A helper is setting up a Ruskybear here so that when the harness falls from the carrier it will also withdraw the parachute from the tube.

Right: the messenger on its way up the kite line, carrying a 'Big Ted' this time. Bottom right we see a bat-winged collapsible messenger being used to lift a Polaroid single-shot camera up to a stop on the line at Scheveningen, Holland.

Main Picture shows Bobo the Gorilla being lofted on Jan Fischer's Parakite at Dieppe.

Above: Betty Boop expresses her feelings on this appliqué Edo prizewinner at Westport Washington State (Creda Axton).

The main picture shows the Rare Air Flash Angel showing off its vented and open areas. Top right we see a pair of Rokkaku kites, each displaying the way the sail forms in flight. Left: it's that man from the telly again! - cartoon faces add to the fun of the Rokkaku challenge.

Tom Van Sant's Concorde

In Chapter Four we credited Tom Van Sant for the introduction of yachting ripstop nylon, and glass fibre rods to kite-making. Up to the time of his exhibition at the ICA in London during the summer of 1976, most kite designs in the Western world were influenced by straight lines, as witness the plethora of delta shapes. Tom changed that thinking with his use of the fabric to induce curvaceous outlines as here in his Concorde.

It is no more than an ogival Delta. It has the same shape of wing as on the Fairey Delta jet which established a world speed record, and was a functional testbed for the eventual Concorde, also an ogival delta. As on simple Delta kites, the leading edges and the spine are not joined, and the spreader or cross-spar is linked to the leading edges by polythene tube, free of the sail and above the spine so that there is a natural dihedral.

At 2440mm, or 8ft span, it's a large kite, and Tom always flew it with a tail streamer. Whether this was for effect or necessity we did not discover. The most interesting aspect was the way in which the hem pockets for the leading edges held the curved outline. An aerofoil camber was also induced by the upper line of the deep keel. This, too, contained a hemmed pocket along the chord centreline, bringing a positive curvature which turned to reflex at the rear. Added symbolic decor, with a droop-snoot 'head' and some frilly fringe material around the 'tail' (to simulate jet haze maybe?), completed a big white bird with a difference.

We are repeating this design from our earlier book not just because of its novel variation on the Delta, but more as an acknowledgement to Tom Van Sant for his reawakening of interest in kite flying after a long period of stale complacency.

Tom Van Sant's Trampoline kite

This inspiring design also appeared in our 1978 book, but it has often been misinterpreted so this time there are extra sketches to explain.

The 'Trampoline' name has arisen since it was introduced as an 'open keel elliptical kite', and the reason will be obvious whenever the kite is seen in flight. The outline and its laced binding to the sail with an air gap between is emphasised by a bouncy performance in the air. The ellipse tends to stretch or distend, each change producing an aerofoil curvature in the sail. Meanwhile the open keel 'tail', being attached to the sail trailing edge, floats up and down as it stabilises the kite into wind and the extended spine bobbles in sympathy with the movement.

This is very much an artist's kite, allowing considerable scope for decoration or modification using for example a trailing rotating tube or other devices on the tip antennae. Tom joins multiple 'tramps' in train with links between the keel tow point and the tips as shown. The originals were 1370mm or 4ft 6in span which is a compact size. Length is not critical, other than to support the open keel 'tail'. What *will* determine construction, however, is the availability of tapered fishing rod blanks in your locality.

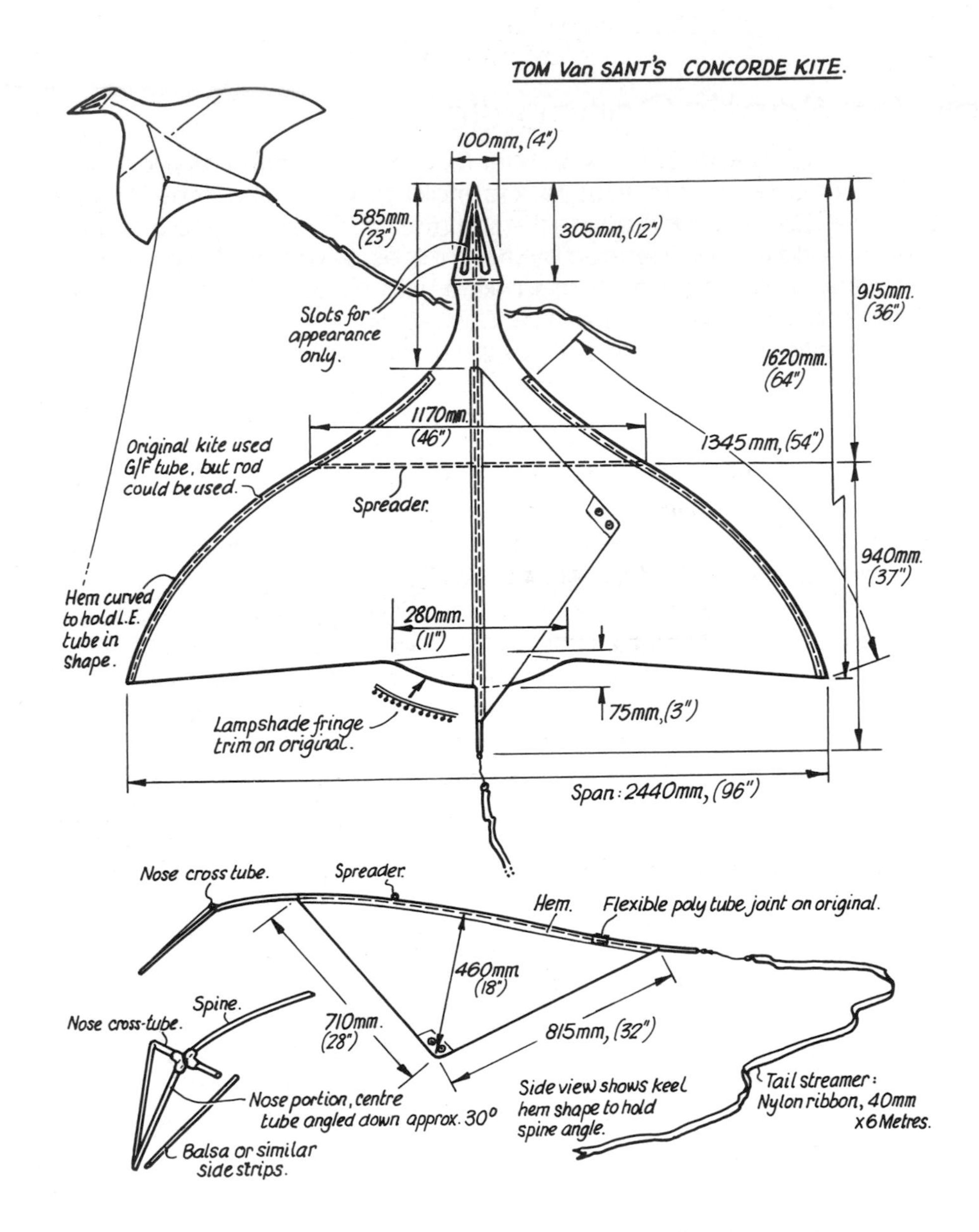

They will have to be in the order of 1070mm (42in) long, depending on the length of antenna you want.

Glass fibre rod and tube can be used instead by having three outside diameters. The central length should be tube, into which another piece of tubing is fitted, on either end. To form the sharply curving tips, and to get oscillation on the antennae, 2mm glass fibre rod inserted in the smaller tubing can complete the required length. It will not be the same as a root-to-tip taper, since the abrupt changes of diameter and consequent stiffness will be noticeable. The drogue is apparently only necessary on a single trampoline, but we would be tempted to use one on each kite in train, at least for the first outing.

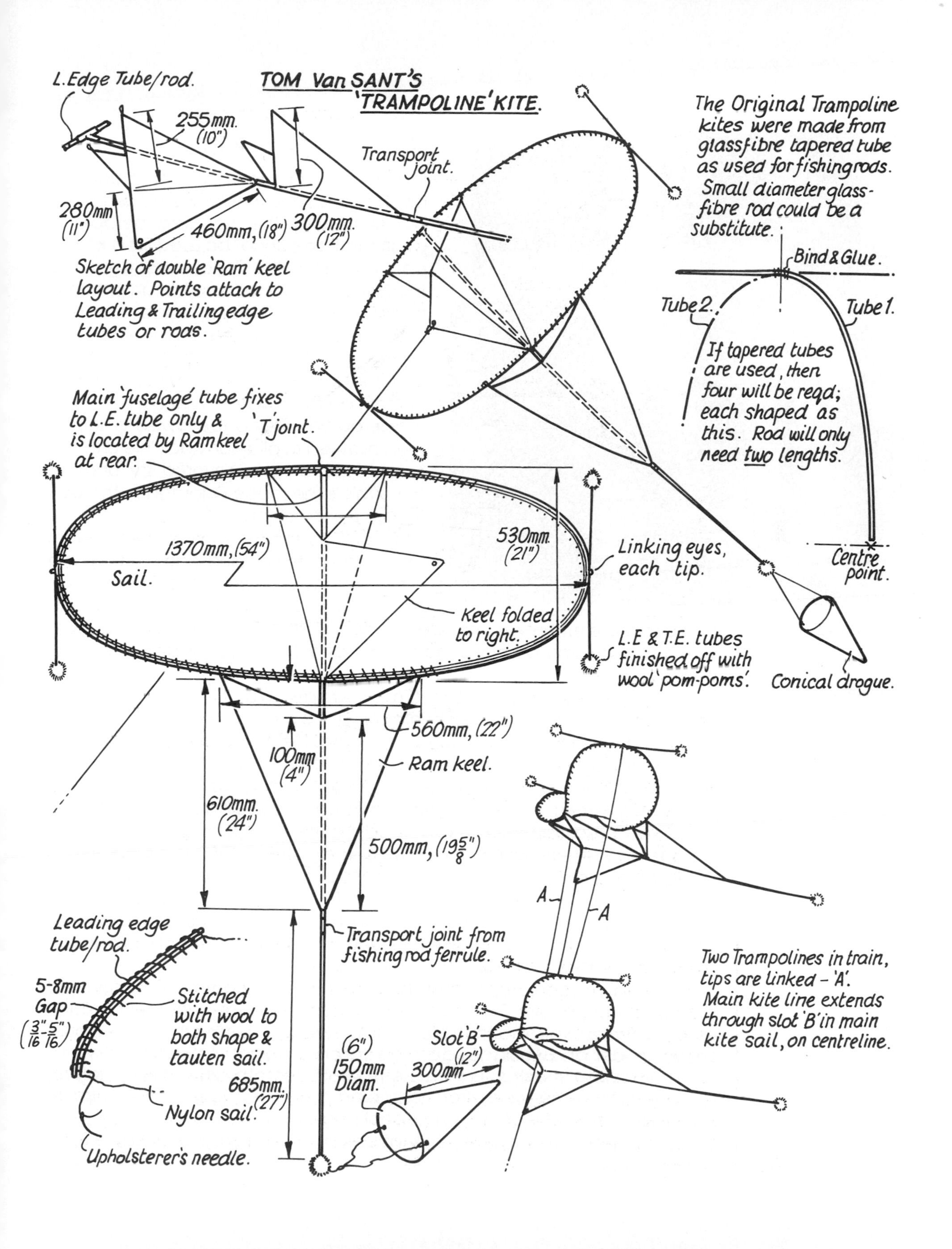

TOM Van SANT'S 'TRAMPOLINE' KITE.
L. Edge Tube/rod.
255mm. (10")
280mm (11")
460mm, (18")
300mm. (12")
Transport joint.
Sketch of double 'Ram' keel layout. Points attach to Leading & Trailing edge tubes or rods.
The Original Trampoline kites were made from glassfibre tapered tube as used for fishing rods. Small diameter glass-fibre rod could be a substitute.
Bind & Glue.
Tube 2.
Tube 1.
If tapered tubes are used, then four will be reqd; each shaped as this. Rod will only need two lengths.
Centre point.
Main 'fuselage' tube fixes to L.E. tube only & is located by Ram keel at rear.
'T' joint.
1370mm, (54")
Sail.
530mm (21")
Linking eyes, each tip.
Keel folded to right.
L.E & T.E. tubes finished off with wool 'pom-poms'.
Conical drogue.
560mm, (22")
100mm (4")
Ram keel.
610mm. (24")
500mm, (19 5/8")
Leading edge tube/rod.
Transport joint from fishing rod ferrule.
A
A
5-8mm Gap (3/16" - 5/16")
Stitched with wool to both shape & tauten sail.
Two Trampolines in train, tips are linked - 'A'. Main kite line extends through slot 'B' in main kite sail, on centreline.
Slot 'B'
(6") 150mm Diam.
300mm (12")
685mm. (27")
Nylon sail.
Upholsterer's needle.

Bill Bigge and one of his many aeroplane kites which have set records.

Aeroplane Kite

Many kite flyers have an aeromodelling background, and they've asked us numerous times why the standard aeroplane configuration cannot be used for a kite. It can, with difficulty. For outdoor flight, the tail surfaces have to be set at an extreme negative angle (leading edge down) in order to obtain some stability when the wings are held at a large angle of attack, almost at the point of stalling. It is possible to fly more easily when the wind is at zero, and the aeroplane kite can be towed to height, then allowed to glide on slack line until the height has to be recovered by another tow. This is the mode adopted in the USA for kite flight *indoors*.

Over the years there have been several exponents of this unusual side to kiting who have set incredible durations. Bill Bigge, an expert model flyer with microfilm models, and also a kite enthusiast, moulded his twin obsessions into an ultra-light kite which would climb on hand movement and could be sustained for long periods.

Carl Brewer evolved a 660mm (26in) span aeroplane kite weighing eight grammes (0.28oz) and, given overnight access to the Seattle Kingdome, created a world record of 9 hours 13 minutes before Washington's leading kite personages as official observers. Flown on a silk line, the flight was only terminated when, after a line break, the kite touched down before the line ends could be rejoined. As it was then 5:56 am, maybe it was appropriate to adjourn for breakfast! We hasten to add that the record holding kite was *not* the one we see here, which appeared at the October 1980 AKA Convention in Seattle for the *outdoor* event limited to kites of 625 sq in (4032 sq cm) where it won second place. With a wingspan of 1830mm (6ft) and structure in small cross section balsa wood, this is not the kind of kite we suggest should appear outdoors unless those arrows on the TV weather forecast predict high pressure and no more than 5mph winds, if any at all.

That said, you could wait for the 10pm weather programme, and then start building for tomorrow! Flat wings and tail can be made extremely quickly, especially with cyano 'instant' glue. The only real problem is that this is a one-piece 'model' and would need a large car or van to get it to the field. It would be a pity to upset the structure while en route! External bracing is essential, not only to restrain the wings from clapping their tips together, but also to maintain a symmetry of the wing panel angles.

Try one when that idyllic calm that aeromodellers pray for and kite flyers hate, prevails for a few summer days.

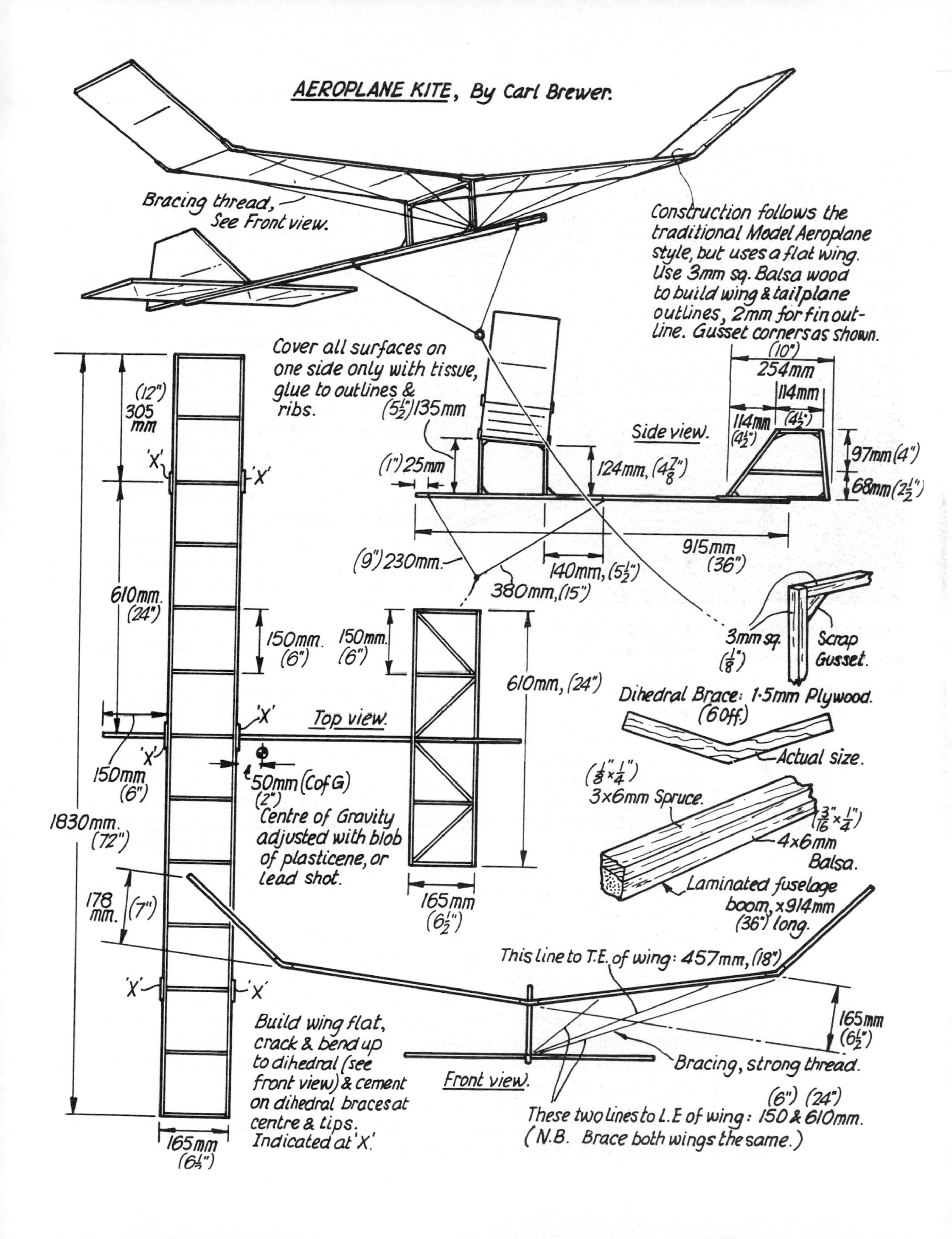

AEROPLANE KITE, By Carl Brewer.
Bracing thread, See Front view.
Construction follows the traditional Model Aeroplane style, but uses a flat wing. Use 3mm sq. Balsa wood to build wing & tailplane outlines, 2mm for fin outline. Gusset corners as shown.
Cover all surfaces on one side only with tissue, glue to outlines & ribs.
(12") 305 mm
'X'
'X'
610mm. (24")
150mm. (6")
150mm. (6")
610mm, (24")
Top view.
150mm (6")
50mm (C of G) (2")
Centre of Gravity adjusted with blob of plasticene, or lead shot.
1830mm. (72")
178 mm. (7")
165mm (6½")
165mm (6½")
Build wing flat, crack & bend up to dihedral (see front view) & cement on dihedral braces at centre & tips. Indicated at 'X'.
(5½") 135mm
(1") 25mm
124mm, (4⅞")
Side view.
(10") 254mm
114mm
114mm (4½")
(4½")
97mm (4")
68mm (2½")
915mm (36")
(9") 230mm.
140mm, (5½")
380mm, (15")
3mm sq. (⅛")
Scrap Gusset.
Dihedral Brace: 1·5mm Plywood. (6 off.)
Actual size.
(⅛"×¼") 3×6mm Spruce.
(3/16"×¼") 4×6mm Balsa.
Laminated fuselage boom, x914mm (36") long.
This line to T.E. of wing: 457mm, (18")
165mm (6½")
Bracing, strong thread.
Front view.
(6") (24")
These two lines to L.E of wing: 150 & 610mm.
(N.B. Brace both wings the same.)

Richard Hewitt's Flexkite

When Richard Hewitt unwrapped this device for the first time at the Spring Kite Festival of BKFA, Old Warden, in 1978, he started something that has gone around the world, full circle.

There were no drawings for it - Richard just created it from five lengths of ripstop nylon, sewn side by side to make a 3.80m or 12ft 6in panel which he cut to a quarter-moon profile. The curved leading edge was hemmed into a continuous pocket so that it could accept the longest length of glass fibre rod which happened to be 5 metres or 16.4ft (as today, 14 years later) and the centre panel became a form of sled kite with two triangular keels at its edges and battens at the junction.

Richard Hewitt's original flexkite with the sail distorted under wind pressure.

It needed a drogue for stability; but from the moment it soared to height, every onlooker realised that Richard had made a new discovery in kite design. Its shape changed with the wind strength. From the basic slice or segment of a circle shape, it became a crescent. When caught by a very strong gust it almost became a full circle! Like some of the larger Deltas, it would adapt its shape, billowing upwards to keep the fabric taut as the wind speed increased.

Pat made his measurements, we christened it the 'Flexkite' and included it in the BKFA *Kitelines* newsletter. In no time at all the design was adopted by other kite enthusiasts in the USA, and, since those earlier days, many have been made with several variations. Hank Szerlag of Detroit was among the first. He scaled it down to 8ft span, using glass fibre fishing rod sections because long lengths of rod were not easy to find at that time. Trailing a Union Jack flag between two drogues, Hank's version promoted the new shape to the extent that others took up the idea of making the leading edge spar tapered.

By having a constant thickness 6mm or 1/4 in centre rod and fishing rod or bicycle flagpole ends joined on by tube ferrules, one obtains the best combination of tapering flexibility and weight distribution. Substitution of glass fibre or even carbon tubes for the chordwise dowels also makes for better durability. Keep those bridle lines long, and preferably linked to a single swivel, to stop them winding into a wrap from liveliness in the kite line.

So there's something both large and simple. It will hoist a banner or national flag to add further colour in your skies and, with its constantly changing shape, will always attract attention.

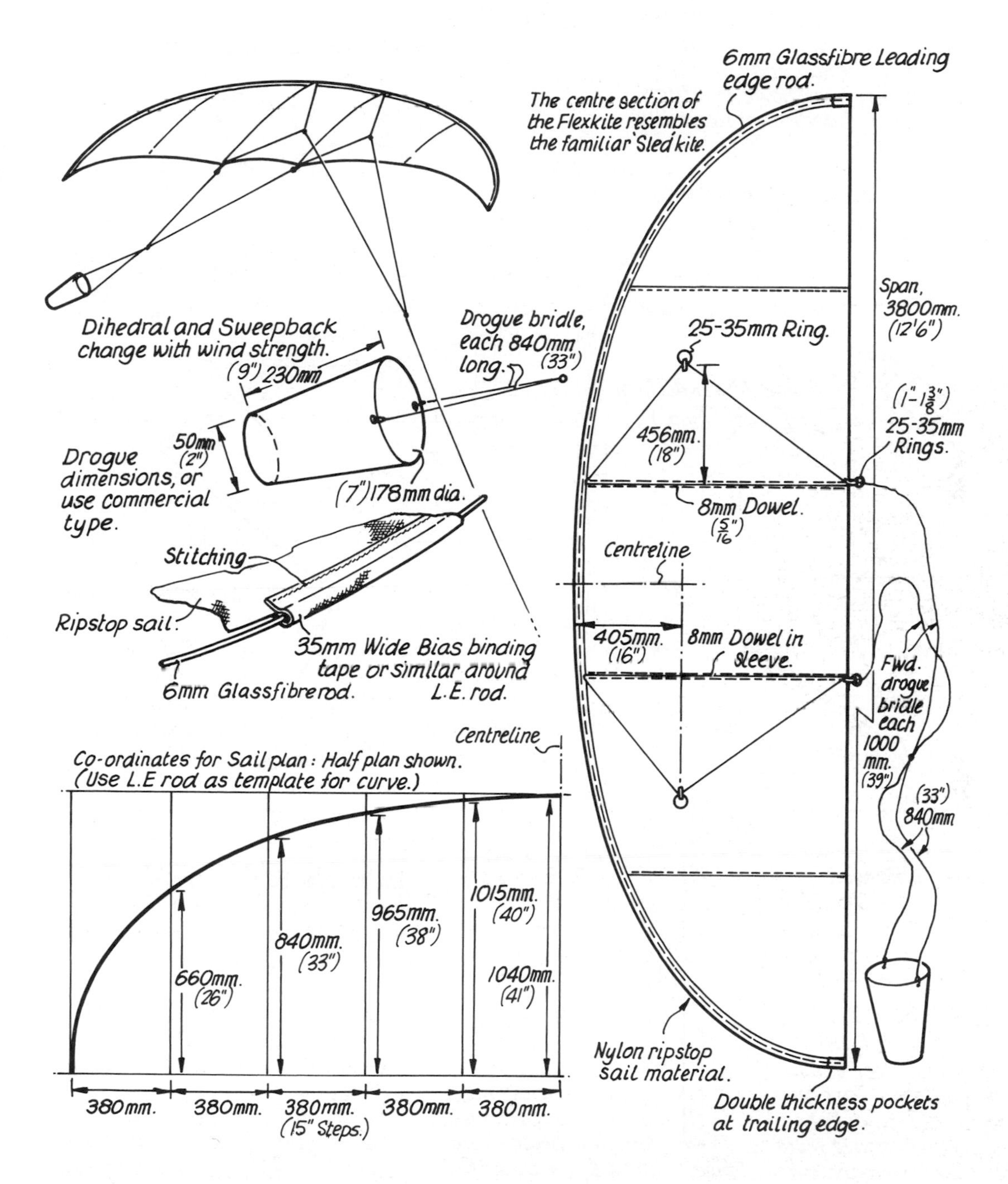

That completes our selection of mixed traditional and modern structures. Dedicated devotees will have to excuse some repetition of designs from our earlier work; but we're sure they will agree that there are none better than old faithful favourites when suggestions are required for novice kite-makers.

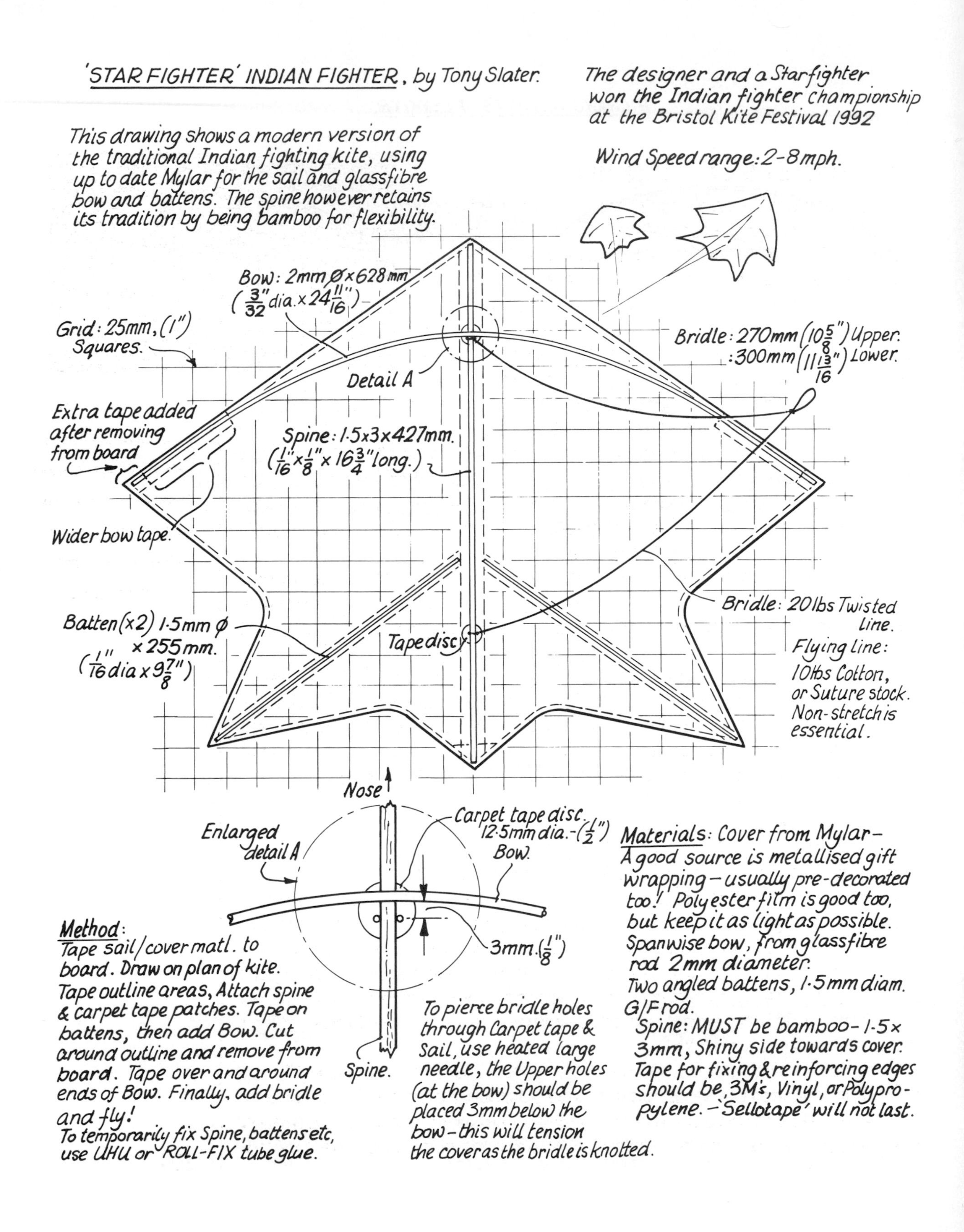
'STAR FIGHTER' INDIAN FIGHTER, by Tony Slater.
The designer and a Starfighter won the Indian fighter championship at the Bristol Kite Festival 1992
This drawing shows a modern version of the traditional Indian fighting kite, using up to date Mylar for the sail and glassfibre bow and battens. The spine however retains its tradition by being bamboo for flexibility.
Wind Speed range: 2-8 mph.
Bow: 2mm Ø x 628 mm (3/32" dia. x 24 11/16")
Grid: 25mm, (1") Squares.
Bridle: 270mm (10 5/8") Upper. : 300mm (11 13/16") Lower.
Detail A
Extra tape added after removing from board
Spine: 1.5x3x427mm. (1/16" x 1/8" x 16 3/4" long.)
Wider bow tape.
Bridle: 20lbs Twisted Line.
Batten (x2) 1.5mm Ø x 255mm. (1/16" dia x 9 7/8")
Tape disc
Flying line: 10lbs Cotton, or Suture stock. Non-stretch is essential.
Nose
Enlarged detail A
Carpet tape disc. 12.5mm dia. - (1/2")
Bow.
3mm. (1/8")
Spine.
Method: Tape sail/cover matl. to board. Draw on plan of kite. Tape outline areas, Attach spine & carpet tape patches. Tape on battens, then add Bow. Cut around outline and remove from board. Tape over and around ends of Bow. Finally, add bridle and fly!
To temporarily fix Spine, battens etc, use UHU or ROLL-FIX tube glue.
To pierce bridle holes through Carpet tape & Sail, use heated large needle, the Upper holes (at the bow) should be placed 3mm below the bow - this will tension the cover as the bridle is knotted.
Materials: Cover from Mylar - A good source is metallised gift wrapping - usually pre-decorated too! Polyester film is good too, but keep it as light as possible.
Spanwise bow, from glassfibre rod 2mm diameter.
Two angled battens, 1.5mm diam. G/F rod.
Spine: MUST be bamboo - 1.5x 3mm, Shiny side towards cover.
Tape for fixing & reinforcing edges should be, 3M's, Vinyl, or Polypropylene. - 'Sellotape' will not last.

Tony Slater's Star Fighter

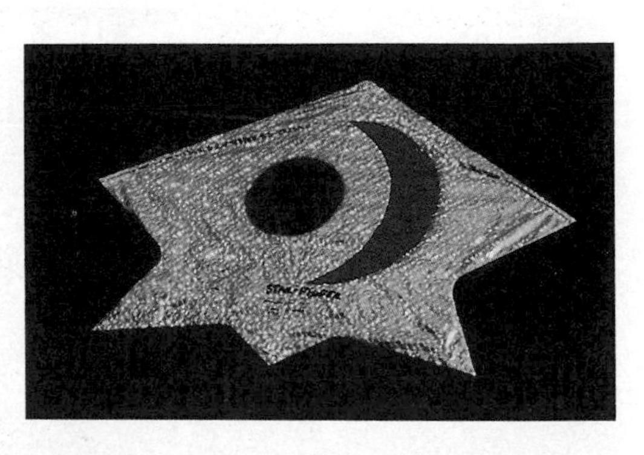

The Star Fighter with a holographic Mylar sail reflects all the colours of the rainbow as it flits in bright sunlight. Use an 'Indian' reel (see page 47) when flying any fighter.

This represents a change as it introduces a European approach to a type of kite that has for centuries been the 'trademark' for kite flying all over India and South East Asia.

Tony Slater, in earlier years, was known as 'The Arm' for his prowess at hurling model gliders made to very high standard from balsa wood. He was British Champion and record holder in this class. His fighter kites have also benefited from the dexterity of his arm, and from many designs he created in the course of development, this represents a fine combination of performance and ease of construction.

Although the Star Fighter departs from Asian tradition with its Mylar sail and glass fibre rod bow and forked battens, Tony emphasises that the central spine must be made from a flat strip of bamboo for its flexibility. There is no substitute, even in the modern world of plastics, which can offer the same qualities essential to the flight performance.

With their two line bridles, fighter kites are restrained when under tension of wind, or a pull on the line from flexing along the centre section of the spine, but all of the rest of the kite must be able to flex both longitudinally and spanwise.

It is this quality, and their light weight which enables them to fly in little or no wind at all. Since the latter so rarely applies in maritime nations like the UK,Tony's Star Fighter is best suited to his recommended wind speed range of 2 to 8 mph. Note the suggestion that metallised gift wrapping can make an ideal sail material. You'll need a piece 60 x 45cm (24 x 18ins.).

Cambodian Kleng-Èk

All day, from first arrivals at 9am to last departures at 6pm for a whole 3-day weekend, visitors to the Dieppe festival in September 1994 could hear a steady humming sound. Most were completely baffled at first and it was not until the wind dropped and the majority of the other kites flopped down that the source became obvious. For very high above all others, at maybe 330m or 1000ft, a single kite held its steady position, trailing an enormously long pair of streamers. No-one seemed to know whose it was, or initially, why it made such a constant hum.

By the second day, the secret (if it ever was that to those who had visited S. East Asia) was revealed. When the kite was brought down by its flyers from Cambodia there were even more surprises for Western eyes. It appeared to be covered in nothing more than brown manila parcel wrapping paper, had the minimal

Above: The Kleng-Èk which was so impressive at Dieppe in 1994 that it was awarded the trophy for outstanding performance.
Below: Dek Sarin from the Cambodian Ministry of Culture with the presentation Kleng-Èk in 1996.

bent bamboo frame reminiscent of the Wau Bulan kites from Indonesia, and the tails which were all of 60m, or 190ft long were bunches of a similar material to the main sail covering.

As each day progressed, the 'Brown Cambodian' became the chief topic of conversation (and admiration) among even the most expert and sophisticated of the international kite flyers at Dieppe.

As for that deep sound of the 'hum, it was created by no more than a strip of rattan, pulled very tight between the tips of the bow by cord strainers at each end. The bow itself was in the region of 3m or 10ft long, longer than the span of the kite itself by a wide margin and made of bamboo with a doubler over the centre portion.

At each end of the rattan hummer were blobs of wax to aid a steady oscillation, and hence the constant sound which had impressed so many visitors and created the initial mystery.

Because the kite was always airborne, and very high, it was not until the prizegiving in the last hour that we had any opportunity to study what was the clear winner of the Dieppe Trophy for the most outstanding performance of 1994

Even then, the problems of language and the fact that the Cambodian team had no literature to offer made it impossible to produce a sketch of our observations.

This situation was resolved when the Cambodians returned to Dieppe in 1996 and they brought with them a specially decorated Kleng-Èk for presentation to the festival organisers.

Thanks to Dek Sarin, Vice-Director in the Ministry of Culture and Fine Arts in Phnom Penh, we were able to measure this kite (as opposite) and note the construction. All of the bamboo framework is tapered – this reduces weight and rigidity at the extremities, for just like the fighter kites, this one has to be flexible and equally so about its centreline.

Because it was made to represent the Kingdom of Cambodia on behalf of the team, it is much reduced from the stunning flyer of 1994. However, by doubling all dimensions there is no reason why a 2280mm or 90in-replica should not sing like a hundred bumble-bees in your skies!

TRADITIONAL CAMBODIAN KLENG-ÈK KITE.

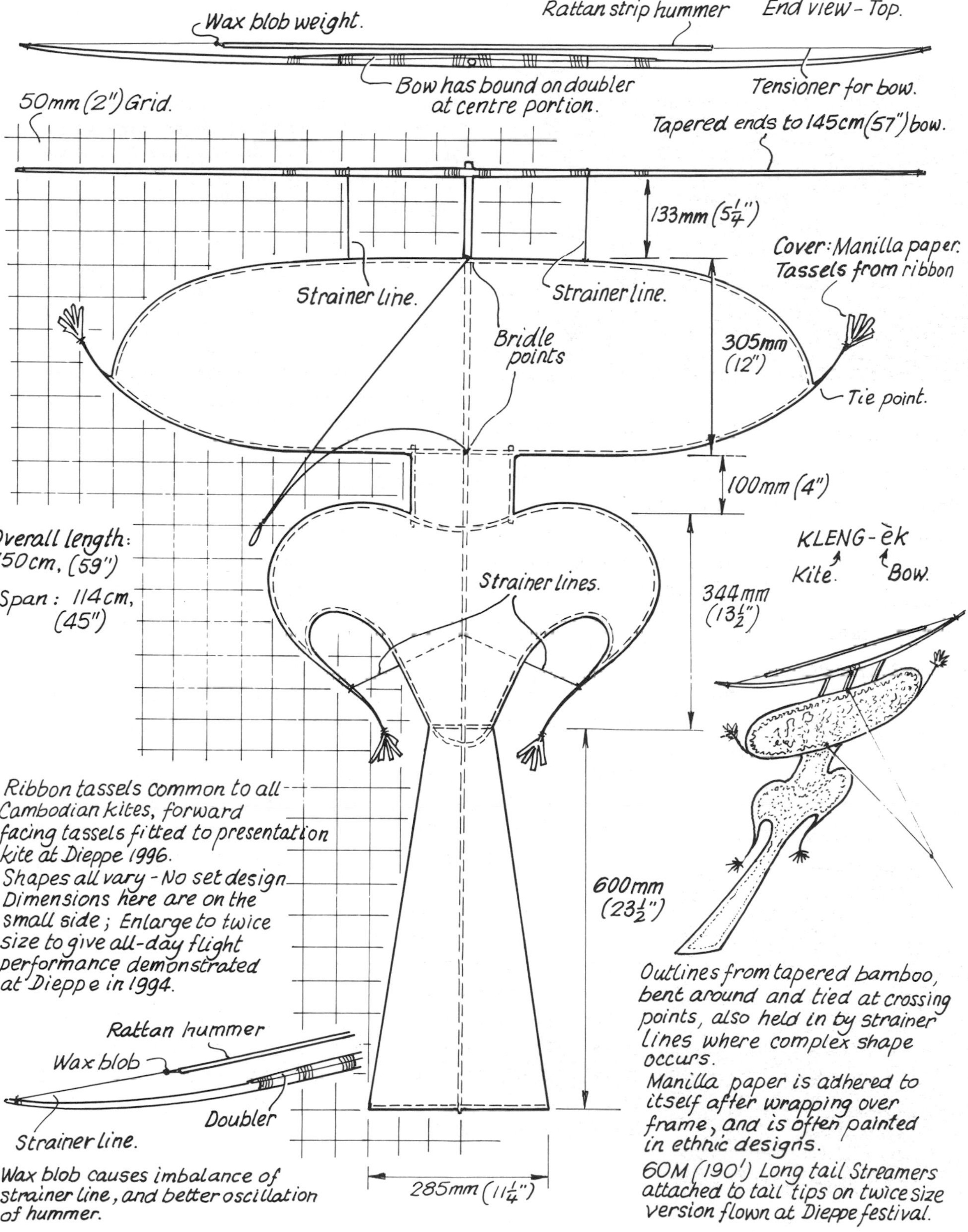

Wax blob weight.

Rattan strip hummer

End view - Top.

Bow has bound on doubler at centre portion.

Tensioner for bow.

50mm (2") Grid.

Tapered ends to 145cm (57") bow.

133mm ($5\frac{1}{4}$")

Cover: Manilla paper. Tassels from ribbon

Strainer line.

Strainer line.

Bridle points

305mm (12")

Tie point.

100mm (4")

Overall length: 150cm, (59")

Span: 114cm, (45")

KLENG - èk
Kite. Bow.

Strainer lines.

344mm ($13\frac{1}{2}$")

600mm ($23\frac{1}{2}$")

Ribbon tassels common to all Cambodian kites, forward facing tassels fitted to presentation kite at Dieppe 1996.
Shapes all vary - No set design. Dimensions here are on the small side; Enlarge to twice size to give all-day flight performance demonstrated at Dieppe in 1994.

Outlines from tapered bamboo, bent around and tied at crossing points, also held in by strainer lines where complex shape occurs.
Manilla paper is adhered to itself after wrapping over frame, and is often painted in ethnic designs.
60M (190') Long tail Streamers attached to tail tips on twice size version flown at Dieppe festival.

Rattan hummer

Wax blob

Doubler

Strainer line.

Wax blob causes imbalance of strainer line, and better oscillation of hummer.

285mm ($11\frac{1}{4}$")

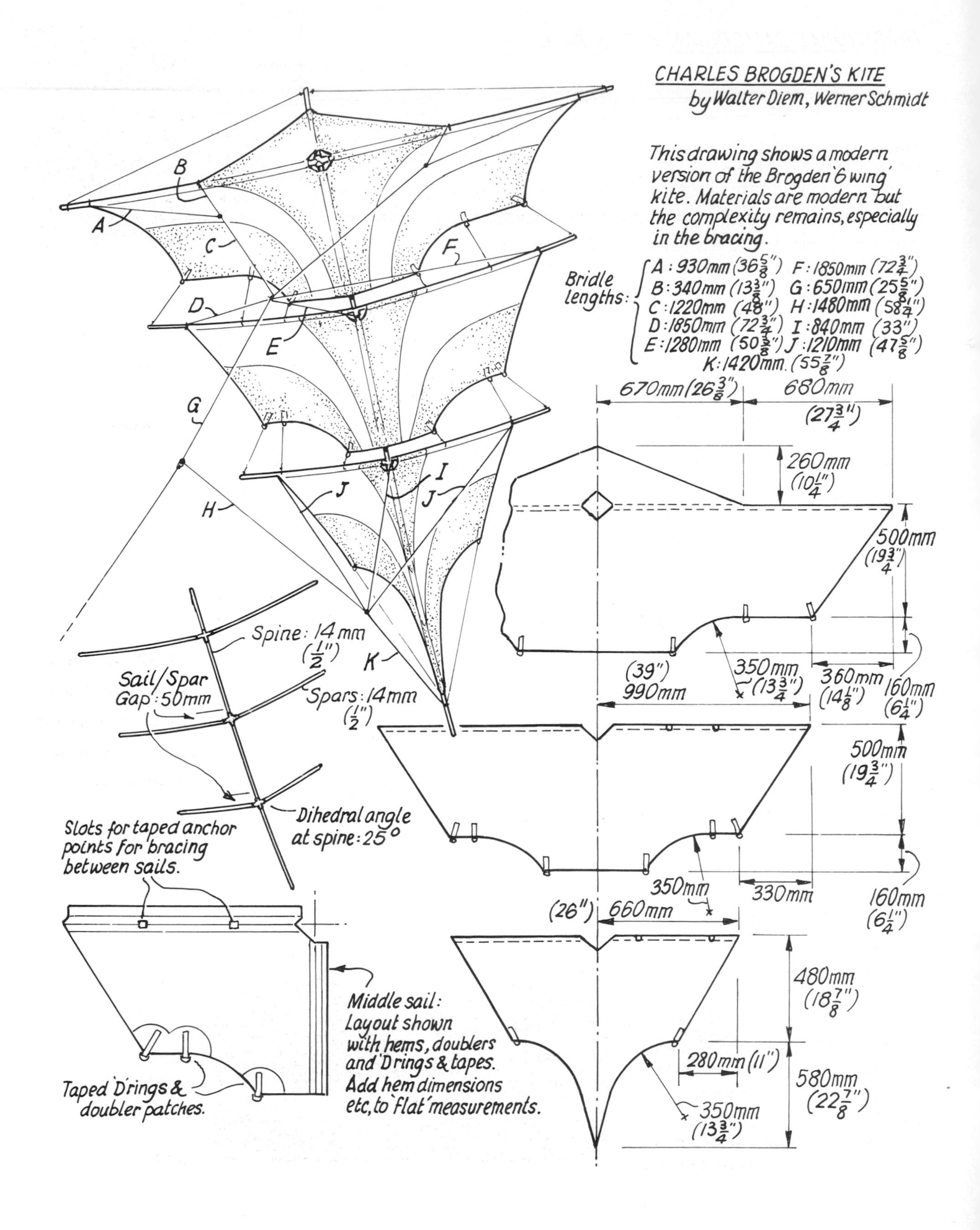
CHARLES BROGDEN'S KITE
by Walter Diem, Werner Schmidt
This drawing shows a modern version of the Brogden '6 wing' kite. Materials are modern but the complexity remains, especially in the bracing.
Bridle lengths:
A: 930mm (36 5/8")
B: 340mm (13 3/8")
C: 1220mm (48")
D: 1850mm (72 3/4")
E: 1280mm (50 3/8")
F: 1850mm (72 3/4")
G: 650mm (25 5/8")
H: 1480mm (58 1/4")
I: 840mm (33")
J: 1210mm (47 5/8")
K: 1420mm. (55 7/8")
A
B
C
D
E
F
G
H
I
J
K
670mm (26 3/8")
680mm (27 3/4")
260mm (10 1/4")
500mm (19 3/4")
350mm (13 3/4")
360mm (14 1/8")
160mm (6 1/4")
(39") 990mm
500mm (19 3/4")
350mm
330mm
160mm (6 1/4")
(26") 660mm
480mm (18 7/8")
280mm (11")
580mm (22 7/8")
350mm (13 3/4")
Spine: 14mm (1/2")
Sail/Spar Gap: 50mm
Spars: 14mm (1/2")
Dihedral angle at spine: 25°
Slots for taped anchor points for bracing between sails.
Middle sail: Layout shown with hems, doublers and 'D' rings & tapes. Add hem dimensions etc, to 'flat' measurements.
Taped 'D' rings & doubler patches.

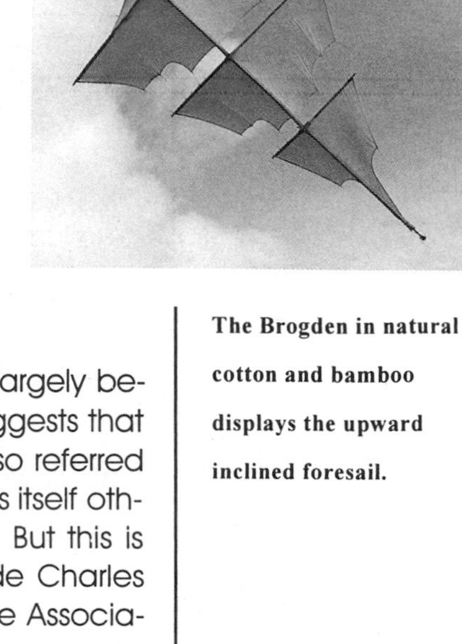

The Brogden in natural cotton and bamboo displays the upward inclined foresail.

Charles Brogden

A prominent figure and regular winner of altitude competitions throughout the first decade of the twentieth century. Charles Brogden developed this multi-winged kite which was unique for its time. He might well have been influenced by the work of William A. Eddy who created the 'Bow' which was to take his name and owed its stability to the upward and outward sweep of its wings, or dihedral.

Brogden's multi-winged kites were to be known as 'dihedral kites' largely because of their steep angle of the wings. The overall shape too, suggests that they were extended variants of the Eddy, and since they were also referred to as 'Burmas', that represents another clue because the Eddy was itself otherwise known as a 'Malay', from where the original shape arose. But this is speculation, and not meant to detract from a design that made Charles Brogden a champion of early days in the Kite & Model Aeroplane Association at Wimbledon Common, S. London.

Unlike most other kites, Brogden's would also glide – well enough to cover three times its own length! It was initially conceived as part of his experimental flying machine in 1890 and he was in esteemed company throughout his work. When he won the Royal Aeronautical Society competition on 25 June 1903 at Findon on the Sussex Downs he was flying against Samuel Franklin Cody, his son Leon, and S. H. R. Salmon – all accomplished flyers and also seeking a way to designing the man-carrying flying machine.

Left: Charles Brogden and his original kite at the Findon competition (Royal Aeronautical Society photo)

With the best average of measured heights at 1,555ft or 474m Brogden's performance was 15% better than that of S. F. Cody, his nearest rival. In a report on a K&MAA National competition held at Wimbledon in September 1910 Charles Brogden was a judge, along with Major B. Baden-Powell and Walter Brooke, the founder of 'Brookite' (still going strong). After the event Brogden demonstrated his 'Burma' (as it was then known) to a height that took out line 2¼ miles or 3.62km! The kite was out of sight for 45 minutes.

So much then for its background. Despite its successes, few drawings or descriptions have ever been published although the six-winged form has been adopted by many present-day enthusiasts. Our drawing is of a 'half-size' version by Walter Diem and Werner Schmidt which they described in their fine book *Drachen mit Geschichte* (*Kites with History*). Instead of the original 17ft length, this is 9ft or 2736mm and modern materials – glass fibre spine and spars and ripstop nylon sails are used. What is not obvious is the fact that it is really three units which plug together plus a foresail which is inclined at 5 degrees upwards. The joiners may be a problem, Werner's were a work of art in metal turning.

The Original Box Kite

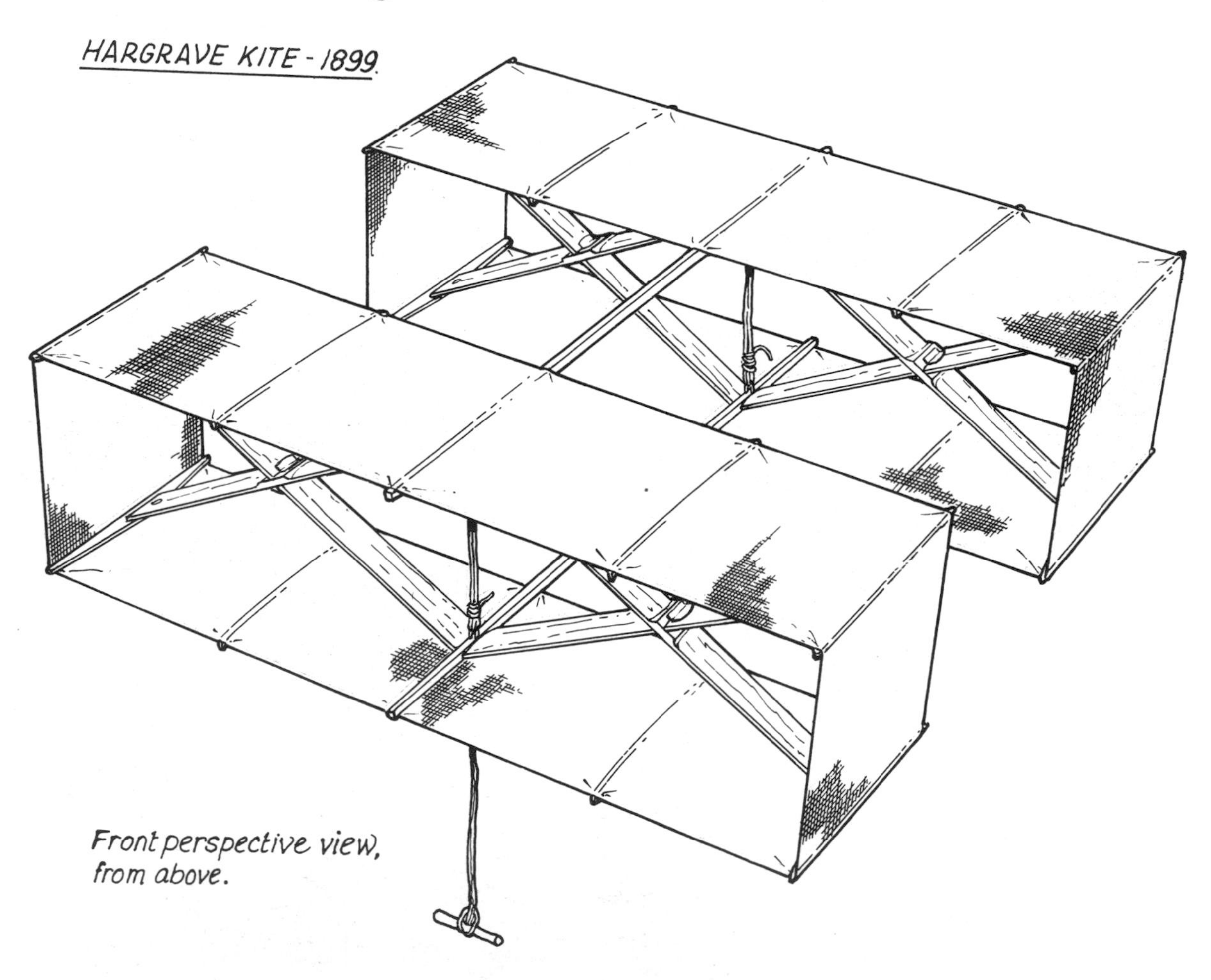

The following illustrations and dimensions are based on the example displayed in London's Science Museum - Aeronautical Gallery.

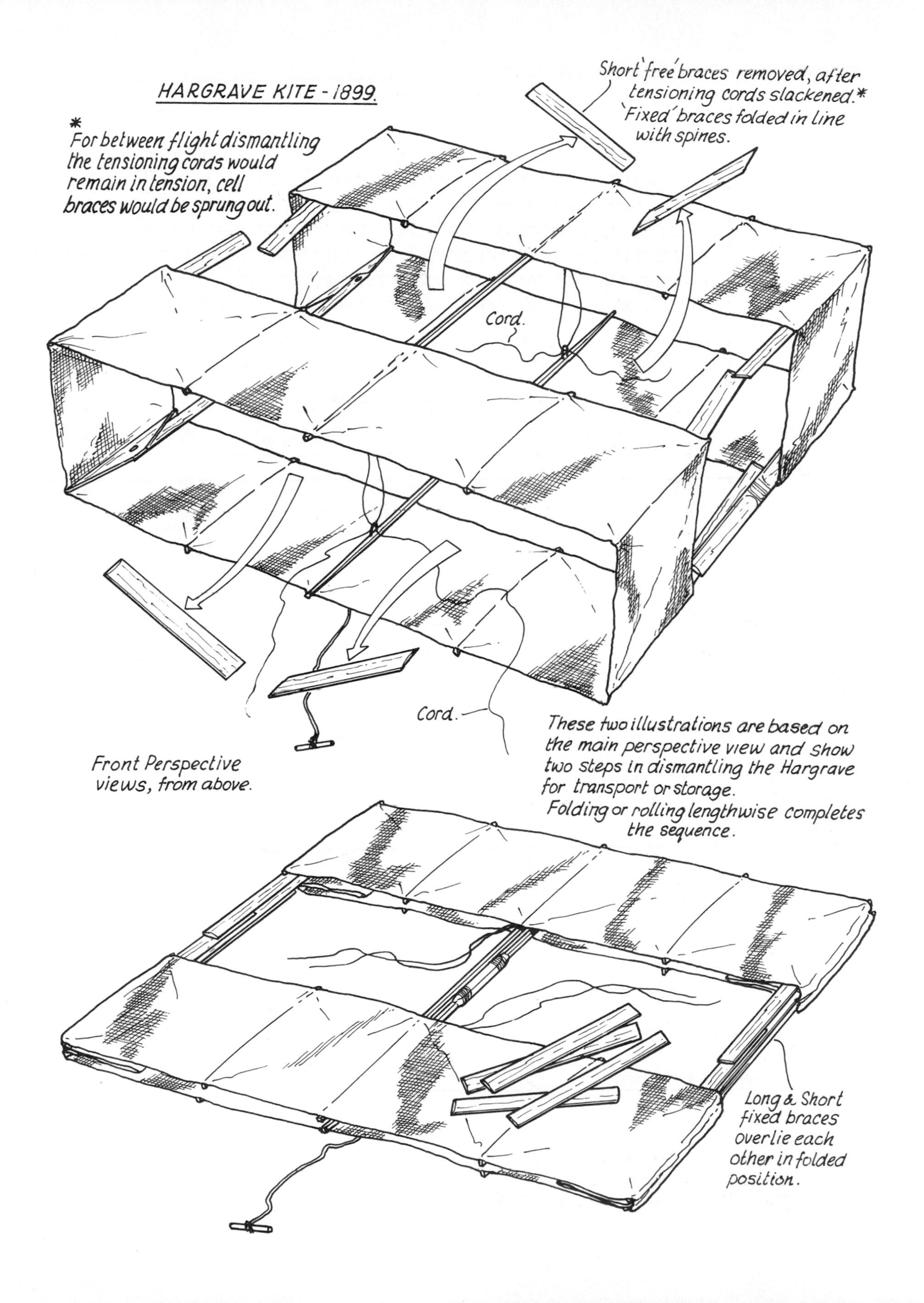
HARGRAVE KITE - 1899.
Short 'free' braces removed, after tensioning cords slackened.*
'Fixed' braces folded in line with spines.
* For between flight dismantling the tensioning cords would remain in tension, cell braces would be sprung out.
Cord.
Cord.
Front Perspective views, from above.
These two illustrations are based on the main perspective view and show two steps in dismantling the Hargrave for transport or storage.
Folding or rolling lengthwise completes the sequence.
Long & Short fixed braces overlie each other in folded position.

Lawrence Hargrave – Pioneer

The relatively simple structure of a frame supporting two sets of wings, locked into rigidity by the tension of cord lanyards and known as the Hargrave 'box kite' became one of the foundation stones of manned flight.

This design was only one of many inventions produced in New South Wales, Australia by Lawrence Hargrave in the search for the way to achieve flight. His kites were not created for pleasure, they were part of Hargrave's serious studies, among which was a vision of a vertical train of lifting surfaces, under

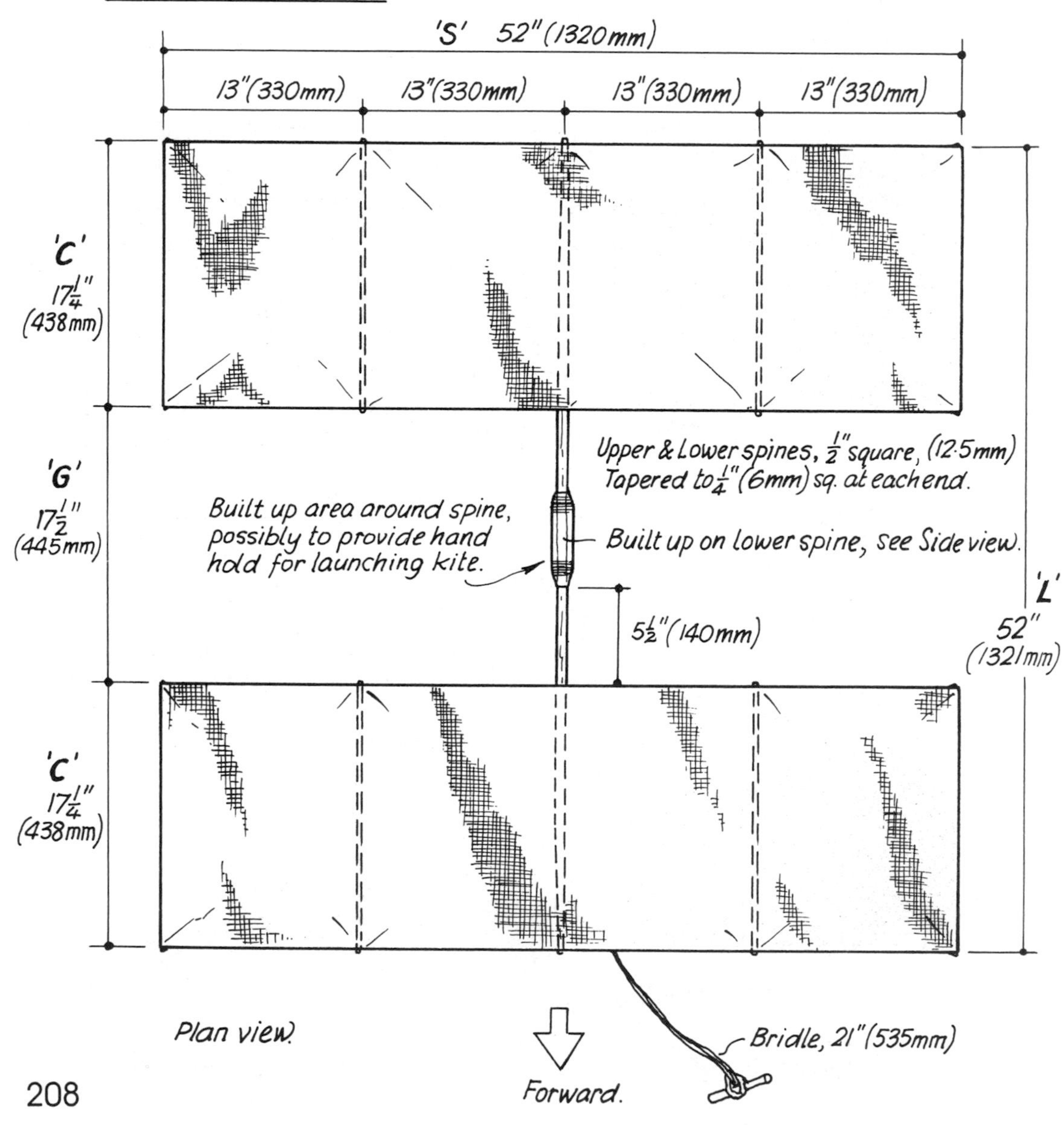

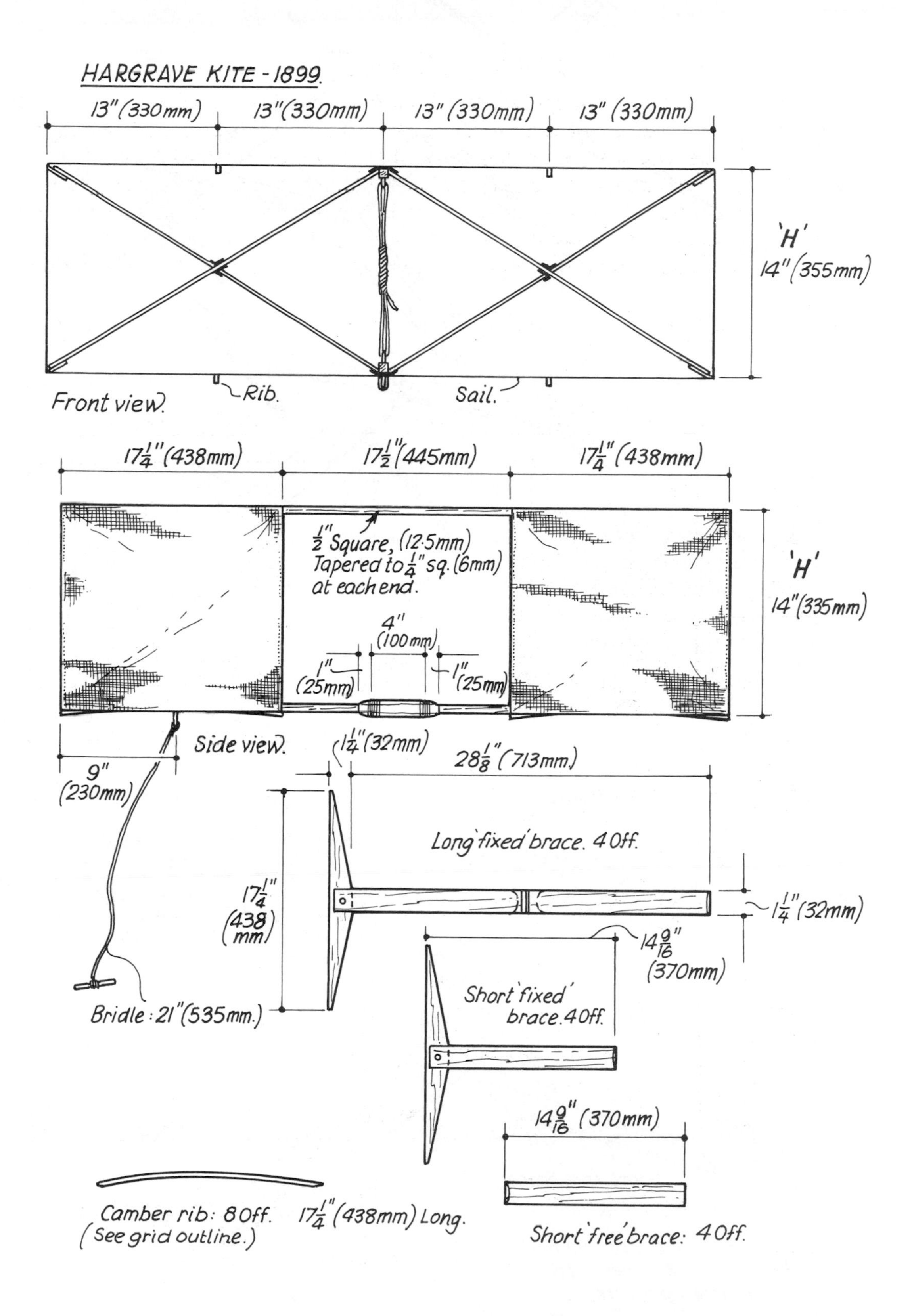
HARGRAVE KITE - 1899.
13"(330mm)
13"(330mm)
13"(330mm)
13"(330mm)
'H'
14"(355mm)
Front view.
Rib.
Sail.
17¼"(438mm)
17½"(445mm)
17¼"(438mm)
½" Square, (12·5mm)
Tapered to ¼" sq. (6mm)
at each end.
4"
(100mm)
1"
(25mm)
1"
(25mm)
'H'
14"(335mm)
Side view.
9"
(230mm)
1¼"(32mm)
28⅛"(713mm)
Long 'fixed' brace. 4 Off.
17¼"
(438 mm)
1¼"(32mm)
14 9/16"
(370mm)
Short 'fixed' brace. 4 Off.
Bridle: 21"(535mm.)
14 9/16"(370mm)
Camber rib: 8 Off.
(See grid outline.)
17¼"(438mm) Long.
Short 'free' brace: 4 Off.

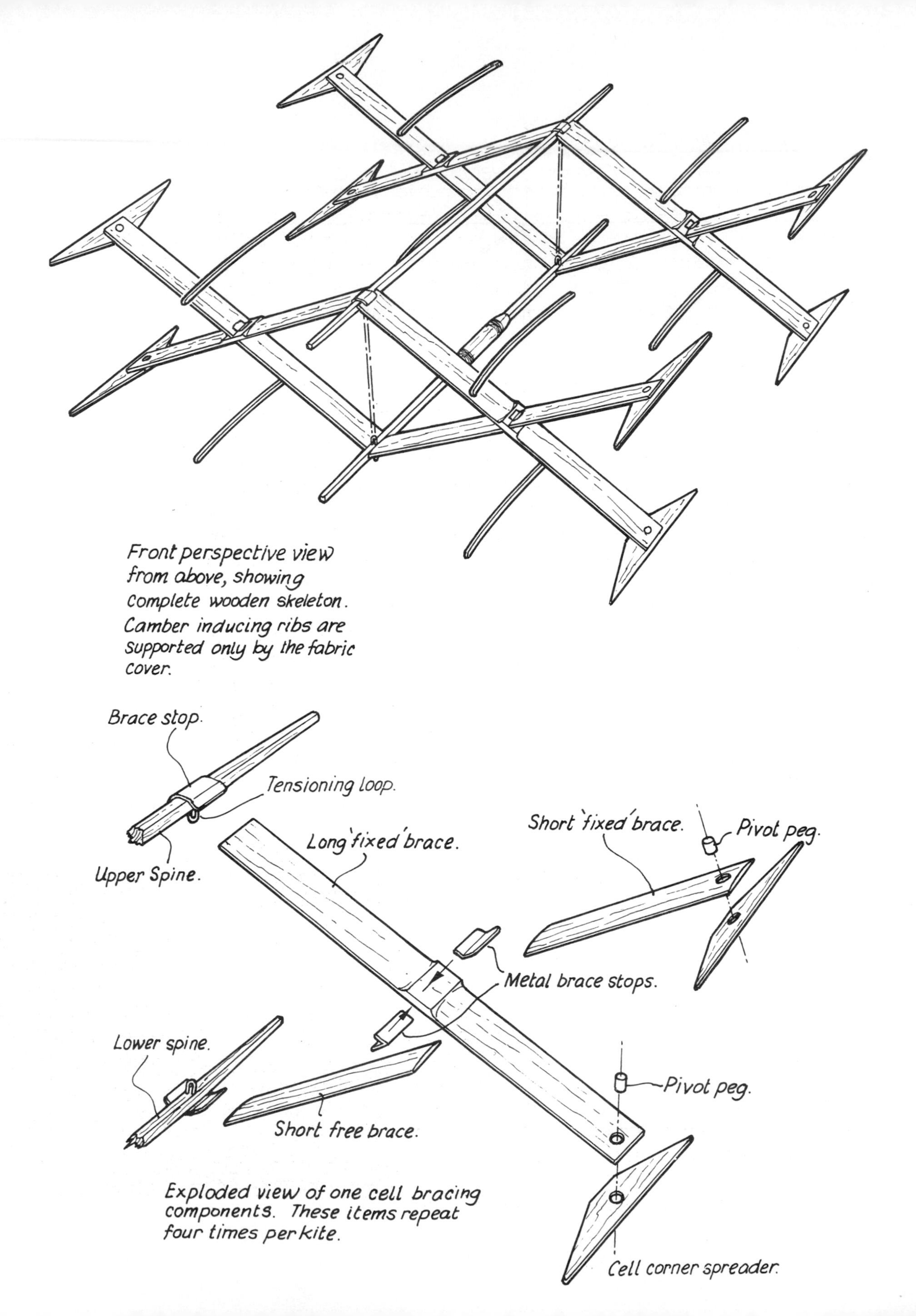

Front perspective view from above, showing complete wooden skeleton. Camber inducing ribs are supported only by the fabric cover.

Exploded view of one cell bracing components. These items repeat four times per kite.

which the flyer would be suspended with a power driven propeller.

Born in 1850 at Greenwich,the son of a barrister, Hargrave sailed in a clipper ship to Sydney at 16 to join his father who had left 10 years earlier and had risen to become a senior judge. The three month journey had a great influence. Far from using his good mathematics and following his father into Law practice, young Hargrave had watched the albatrosses, studied the sails and was soon persuaded by a fellow passenger to join an expedition which circumnavigated the Australian Continent.

This journey, lasting from March to October, was only a start to his 49 years in Australia and a career worthy of an adventure film script. He was shipwrecked, became an astronomer,invented the rotary engine, a jet engine, an ornithopter and made many experimental flying models in his search for success in flight.

As one of the coterie of pioneers then seeking the evolution of the flying machine, Hargrave's papers were freely distributed and among his correspondents were Octave Chanute in Boston, Percy Pilcher and Sir Hiram Maxim in London, William Eddy in New York and S.P. Langley at the Smithsonian in Washington. These great personalities were first to recognise the potential of an isolated, but gifted, fellow visionary.

Meanwhile his father had become wealthy and bought 1325 acres at Coalcliff,30 miles south of Sydney, subsequently named Stanwell after the town in Middlesex and built a house there. Later he gave part to the older son Ralph, and 420 acres to Lawrence so that he could develop a colliery.

Ralph also had a house at Stanwell, and in 1887 built another very substantial 72ft or 22m square property with surrounding veranda and six basement rooms beneath large living quarters. It was called *Hill Crest* and as soon as it had been completed, Ralph left for Japan and England but died in mysterious circumstances in Hong Kong.

We've related this background to Hargrave's location and widespread interests in abbreviated form (there are several excellent biographies, notably that by W. Hudson Shaw & Olaf Ruhen 1988 UQP ISBN 0 7022 2157 0) to set the picture of where, from August 1893 to April 1899 Lawrence Hargrave made the box kites for which he has become internationally famous.

Lawrence inherited *Hill Crest* with its facilities for kite construction and large beach for tests. He enlisted the help of James Swaine who lived on the estate as a caretaker, and little more than a year after moving in from Sydney made the man-carrying kite train ascent that was to establish his reputation more than any other of his achievements.

On 12th November 1894, Hargrave and Swaine took

Replica of Hargrave's 1894 man-lifting train in the Sydney Powerhouse Museum illustrates the simple seat.

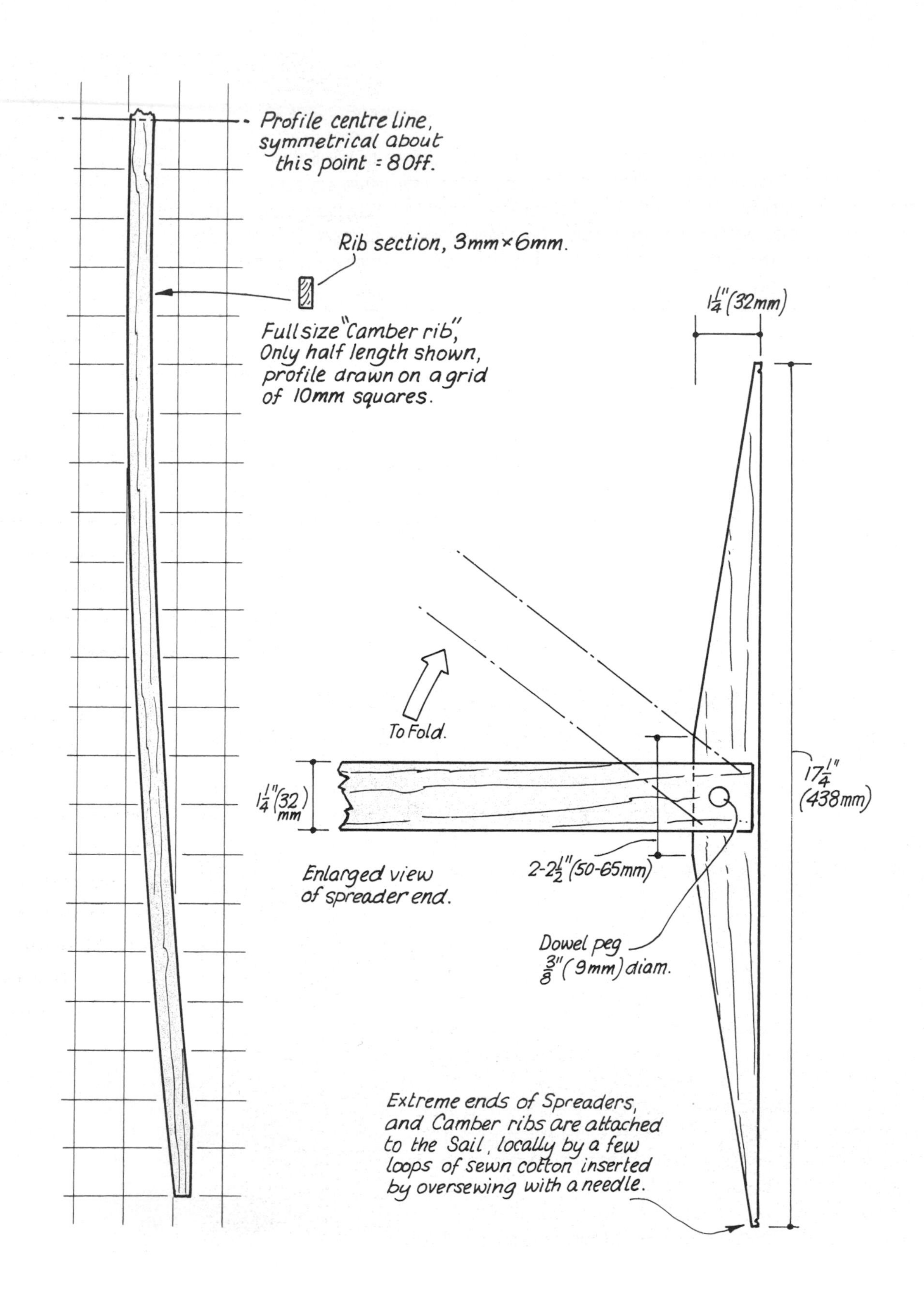
Profile centre line, symmetrical about this point = 8 Off.
Rib section, 3mm × 6mm.
Full size "Camber rib", Only half length shown, profile drawn on a grid of 10mm squares.
1¼" (32mm)
17¼" (438mm)
To Fold.
1¼" (32 mm)
2-2½" (50-65mm)
Enlarged view of spreader end.
Dowel peg ⅜" (9mm) diam.
Extreme ends of Spreaders, and Camber ribs are attached to the Sail, locally by a few loops of sewn cotton inserted by oversewing with a needle.

five kites of different sizes (see table at end) to the beach and waited for the wind to freshen. 'Hargrave had fashioned a sling to support himself with a forked base as a seat and this was to be 6ft or 2m below the lowest, and largest of the kites.

We can do no better here than to quote Hargrave's own hand-written caption to slide 38 in his lecture notes from papers on 'The Cellular Kite' he was later to give to the Royal Society of New South Wales on August 5th 1896.

> "The 4 kites I managed to lift myself with James Swaine who helped me. The reel and line. Bags (filled with sand from the beach). The spring balance to read the pull when Swaine had let out all the line. The anenometer to read the wind velocity. The fishing line to plumb the vertical and determine the lift and drift.The slung seat (better than any basket). The barrow to convey the gear down to the beach at Stanwell Park"

These items can be seen in the posed photograph which was taken by Charles Bayliss after the event. The collapsed kite 5 is at left foreground.

Replica Hargrave Cellular kite on exhibition at the Science Museum in Canberra. Framework of the original kites was fashioned in American Redwood. Note that the Canberra replica excludes the basic features of the Hargrave design!

Left: Famous posed photo (see Hargrave's own caption in text) taken to verify the sucessful man-lift with Lawrence Hargrave, and assistant James Swaine.

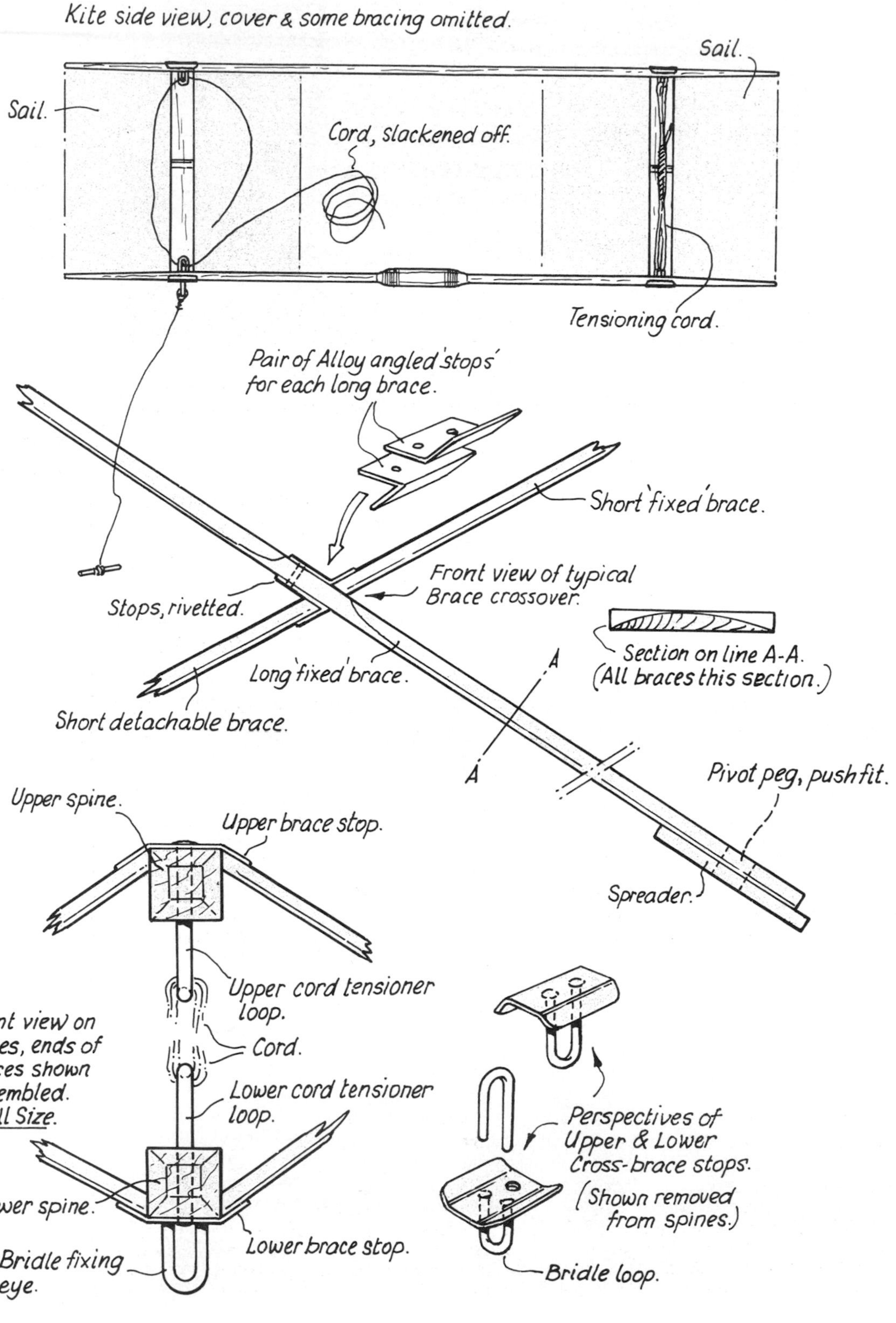

Kite side view, cover & some bracing omitted.

Sail.

Sail.

Cord, slackened off.

Tensioning cord.

Pair of Alloy angled 'stops' for each long brace.

Short 'fixed' brace.

Front view of typical Brace crossover.

Stops, rivetted.

Section on line A-A. (All braces this section.)

A

A

Long 'fixed' brace.

Short detachable brace.

Pivot peg, push fit.

Spreader.

Upper spine.

Upper brace stop.

Upper cord tensioner loop.

Front view on spines, ends of braces shown assembled. <u>Full Size</u>.

Cord.

Lower cord tensioner loop.

Lower spine.

Lower brace stop.

Bridle fixing eye.

Perspectives of Upper & Lower Cross-brace stops.

(Shown removed from spines.)

Bridle loop.

Stanwell Park, viewed from Bald Hill. *Hill Crest* is in the largest group of houses to the right of centre. The beach is where the man-lift was done in 1894. A centenary attempt by Simon Freidin was frustrated here but successful later in Queensland.

Comparison with the dimensions table below will show that the kite which Pat has drawn here, is about half the size of the 'lifter' and at 25 sq ft only 28% of the area. This is a kite sent to G. H. R. Salmon in exchange for cotton cloth in 1901. It was later given to the Royal Aeronautical Society and then, in 1928, presented for safe keeping to the Science Museum and it was there that Pat was able to measure and sketch for these pages. There is some evidence that Hargrave Cellular kites of similar size went on sale. Certainly there was a need for improved income during the family visit to London from April to May 1899. Obliged to live in rooms far removed from spacious *Hill Crest*,

Hargrave's kite train, November 1894

Kite	A (pilot)	B	C	D	E (Lifter)
Overall length (L)	71" 180.3cm	74" 188cm	107" 271.7cm	102" 259cm	108" 274.3cm
Overall span (S)	60" 152cm	60" 152cm	92.5" 235cm	78" 198cm	108" 274.3cm
Chord of each cell (C)	23" 58.4cm	23" 58.4cm	27" 68.6cm	30" 76.2cm	30" 76.2cm
Height of each cell (H)	22.5" 57.1cm	22.5" 57.1cm	22.5" 57.1cm	27.5" 69.8cm	30" 76.2cm
Gap between cells (G)	25" 63.5cm	28" 71.1cm	53" 134.6cm	42" 106.6cm	48" 121.9cm
Tow point to nose	19" 48.2cm	19" 48.2cm	32" 81.2cm	27" 61cm	34" 86.3cm
Weight of kite	87oz 2.27kg	94oz 2.66kg	152oz 4.3kg	144oz 4.08kg	232oz 6.57kg
Surface area	38.5ft^2 3.58m^2	38.5ft^2 3.58m^2	69ft^2 6.41m^2	65ft^2 6.04m^2	90ft^2 8.36m^2

On 12 December 1894, a train of kites, A, B, D and C (C withdrawn due to collapse of St'b'd cells in forward section) lifted Lawrence Hargrave 16 feet (4.87m), a total weight of 208lb 5oz (94.6kg) in a wind velovity of 21 mph, 33.8 kph (9.4 m/sec). Total lifting surface area 232ft^2 (21.56m^2). Kites flown on manila yarn, anchor to lifter 42ft (12.8m); 46ft (14m) between E, D & D, B; 52ft (15.8m) B to A.

Hill Crest, where Hargrave made his kites. The missing veranda has been replaced since our visit in 1994.

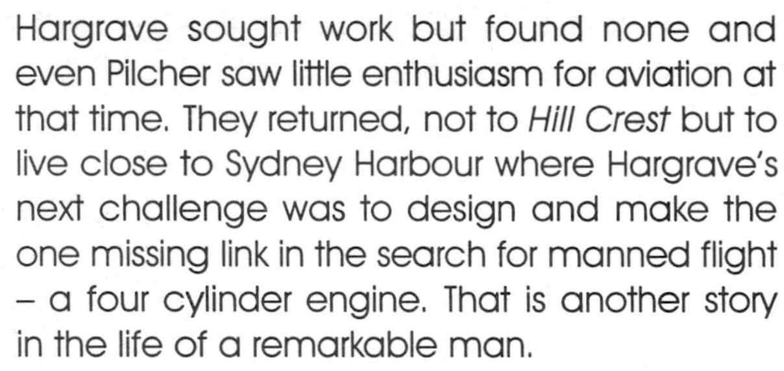

Hargrave sought work but found none and even Pilcher saw little enthusiasm for aviation at that time. They returned, not to *Hill Crest* but to live close to Sydney Harbour where Hargrave's next challenge was to design and make the one missing link in the search for manned flight – a four cylinder engine. That is another story in the life of a remarkable man.

Right: Worthy memorial to Australia's hero is sited above Stanwell Park on Bald Hill from where hang-gliders regularly soar. Hargrave would have been pleased.

S. F. Cody Kites

The legendary 'Colonel' Cody who produced innumerable kites in his progress towards making the first British aeroplane to fly was possibly influenced by the Hargrave kite he bought for his son Vivian. The central frame of the Cody kite suggests that, but the use of four spines or longerons and the dihedralled wings that distinguish the Cody series from all others are characteristic. Cody has many adherents, and among these is Paul Chapman who has contributed the two designs here.

While it is generally known that Cody's sons Vivian and Leon carried on their father's interests, working at Farnborough with kite experiments, it was not until comparatively recently that Air Ministry drawings for these well tested kites were revealed.

Their purpose, either in erecting a cable barrage or lofting aerial bombs which might be snared by intruding aircraft was seriously explored at Scapa Flow, Exeter and Mullion and is the subject of two fascinating booklets by David Hughes.

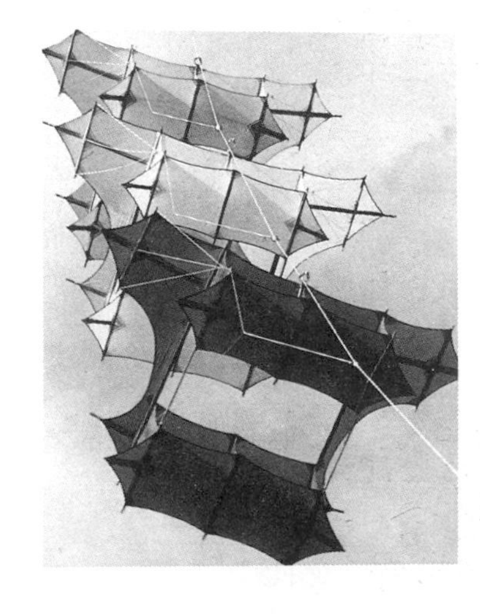

A train of 3 Cody's slip up to the pilot, each engaging a stop on the way.

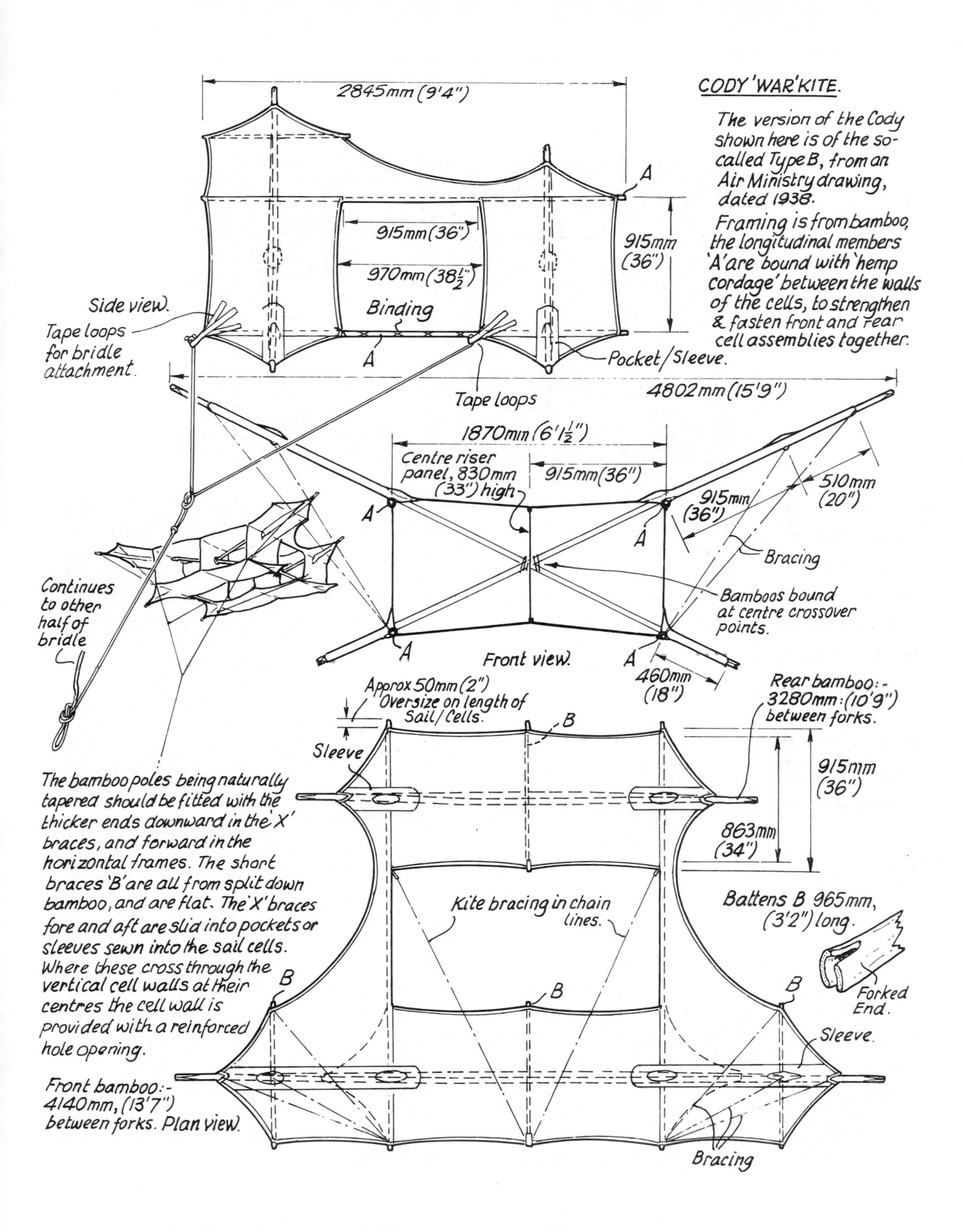

CODY 'WAR' KITE.
The version of the Cody shown here is of the so-called Type B, from an Air Ministry drawing, dated 1938.
Framing is from bamboo, the longitudinal members 'A' are bound with 'hemp cordage' between the walls of the cells, to strengthen & fasten front and rear cell assemblies together.
2845mm (9'4")
A
915mm (36")
970mm (38½")
915mm (36")
Side view.
Tape loops for bridle attachment.
Binding
A
Pocket/Sleeve.
Tape loops
4802mm (15'9")
1870mm (6'1½")
Centre riser panel, 830mm (33") high
915mm (36")
510mm (20")
915mm (36")
A
A
Bracing
Continues to other half of bridle
Bamboos bound at centre crossover points.
A
Front view.
A
460mm (18")
Approx 50mm (2") Oversize on length of Sail/Cells.
B
Rear bamboo:- 3280mm: (10'9") between forks.
Sleeve
915mm (36")
863mm (34")
The bamboo poles being naturally tapered should be fitted with the thicker ends downward in the 'X' braces, and forward in the horizontal frames. The short braces 'B' are all from split down bamboo, and are flat. The 'X' braces fore and aft are slid into pockets or sleeves sewn into the sail cells. Where these cross through the vertical cell walls at their centres the cell wall is provided with a reinforced hole opening.
Kite bracing in chain lines.
Battens B 965mm, (3'2") long.
B
B
B
Forked End.
Sleeve.
Front bamboo:- 4140mm, (13'7") between forks. Plan view.
Bracing

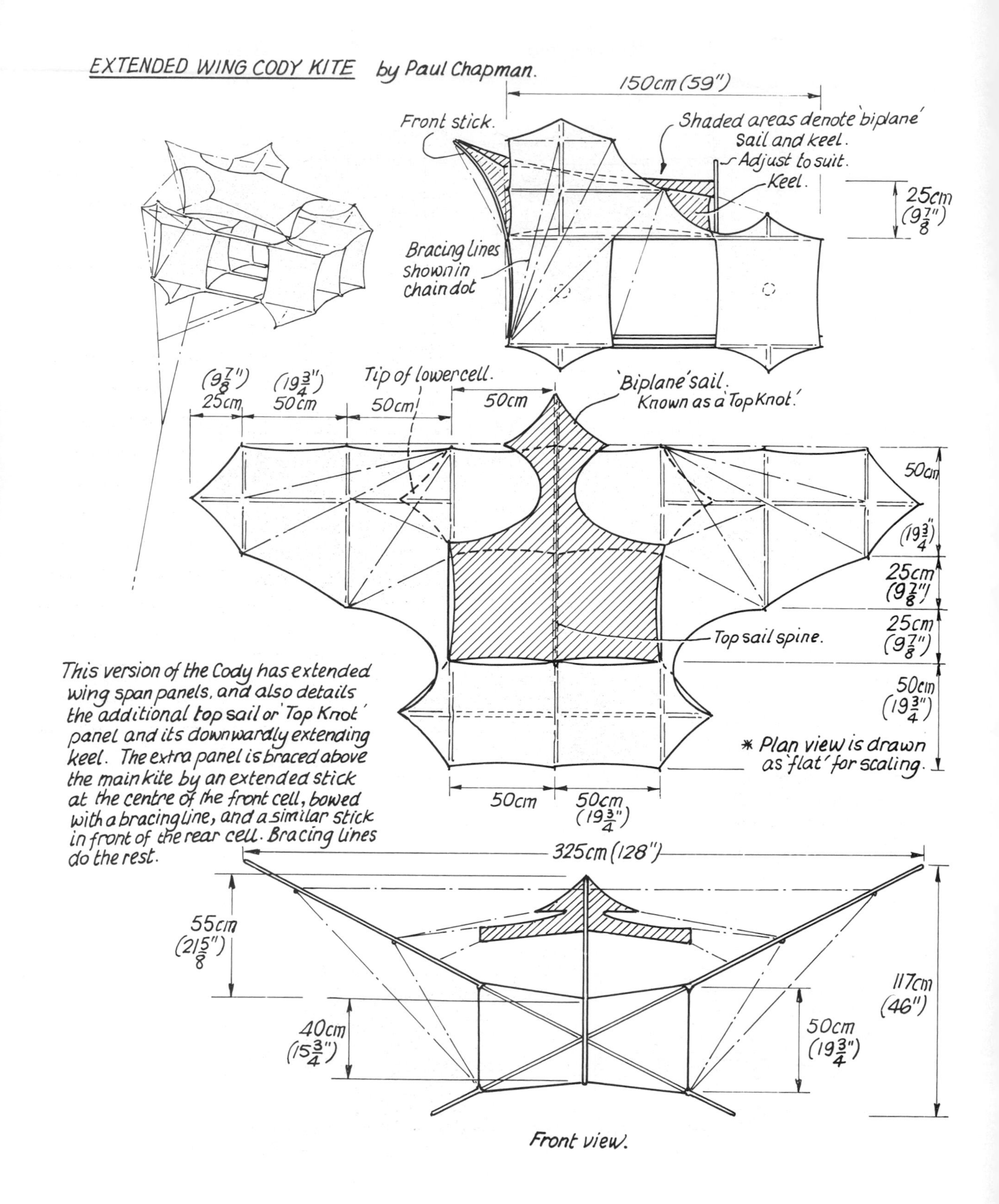
EXTENDED WING CODY KITE by Paul Chapman.
150cm (59")
Front stick.
Shaded areas denote 'biplane' sail and keel.
Adjust to suit.
Keel.
25cm (9 7/8")
Bracing lines shown in chain dot
(9 7/8")
(19 3/4")
25cm
50cm
Tip of lower cell.
50cm
50cm
'Biplane' sail. Known as a 'Top Knot.'
50cm
(19 3/4")
25cm (9 7/8")
25cm (9 7/8")
Top sail spine.
50cm (19 3/4")
This version of the Cody has extended wing span panels, and also details the additional top sail or 'Top Knot' panel and its downwardly extending keel. The extra panel is braced above the main kite by an extended stick at the centre of the front cell, bowed with a bracing line, and a similar stick in front of the rear cell. Bracing lines do the rest.
* Plan view is drawn as 'flat' for scaling.
50cm
50cm (19 3/4")
325cm (128")
55cm (21 5/8")
117cm (46")
40cm (15 3/4")
50cm (19 3/4")
Front view.

Few realise that the Air Ministry "B" Type, price £15 in 1939 was in the front line of the defence of the UK through the first two years of the 1939–1945 war. Some of these kites are now in museums. Perhaps their most useful asset was to make the enemy devise and fit cable cutters! The next stage was, of course, to develop cut resistant cable. Most of these kites were known as the 3ft Cody and to avoid confusion we should explain that this was the standard dimension of the side panels. The extended wing version drawn is considerably scaled down in size, in fact two-thirds the span and little over half the length so it is the more practical subject.

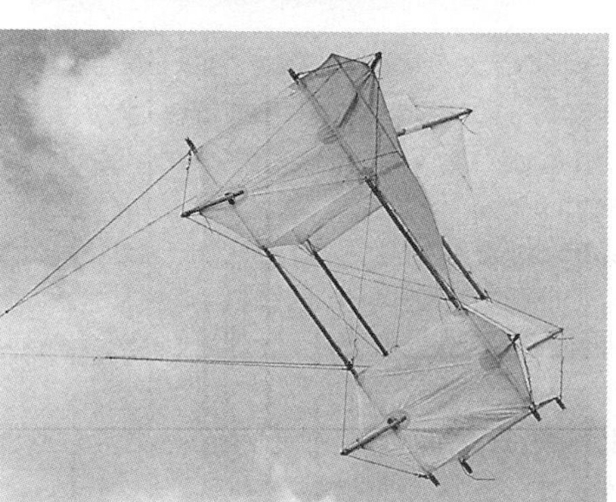

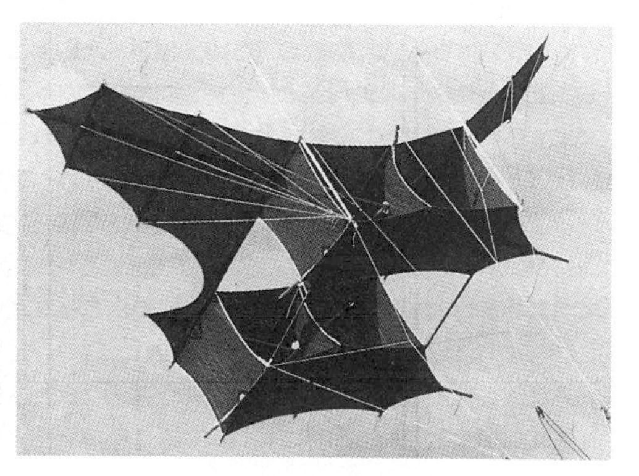

Top: Trio of Cody fans, Hugh Andrew, Nico Van den Berg and Paul Chapman.
Below left: Steep initial angle of Paul Chapman's Cody at take-off.
Bottom left: Extended wing Cody shows the extra rigging required. This is actually a lifter with masts at front for the bridle links to the train, and pulleys for control from the basket.
Bottom right: One of the "Cotton Club" Cody's shows the steep initial angle. Originals used cotton or silk while many modern Cody's have Ripstop nylon sails.

Two Viking Longships in full sail, in company with a pair of Dutch Rings. Together, they made an outstanding impression at Dieppe 1996. The simplicity of an unbraced sheet on two booms across a central mast and a stabilising 'bow' beneath was an inspiration.

Swedish SVEA Viking Longship

The sight of these full rigged Viking ships soaring in a seaside breeze impresses not only for their symbolism but also as a study of how they are made, and their extreme simplicity. They originated from a challenge between Stefan Andersson and Andreas Agren who were making folded paper kites inspired by the Chinese rigid wings, but with a 30cm or 12in limitation.

The two belong to the Sala Kite & Tango Party and they were planning their trips across the North Sea to festivals in Monmouth and Washington. The vision of Viking ships in the sky led to a small prototype with a 3-dimensional hull which worked well, and in a short time Per Bystrom made his up to 2.5m span. It flew well enough to get the Silver Prize at the Stockholm Festival. This was the first SVEA. Per then made Gota and Vendel, each named after ancient Swedish territories.

Andreas was then inspired by that mysterious force that acts after an accident. His longship fell from a shelf onto his head! Looking at the offending kite he realised just what was needed to give the impression of an invading Viking ship.

In no time at all, a triple sail 'Vasa' was made, then after the Portsmouth Festival, the 'Mary Rose' became another of Andreas's "cut-wedge-and-tape" paper kite projects as he described in *KiteLines* 12/1. The scene was set, then on to Dieppe 96 and the invasion of Normandy.

The secret of the design is in the hourglass shape of the sail. As the sail fills, so its profile gives a dihedral effect while the hull, which is braced away from the central mast, acts as a stabiliser and, being formed into a vee shape, holds the kite on station.

Add a decorative figurehead on the bowsprit and the impression of a ship in full sail is complete.

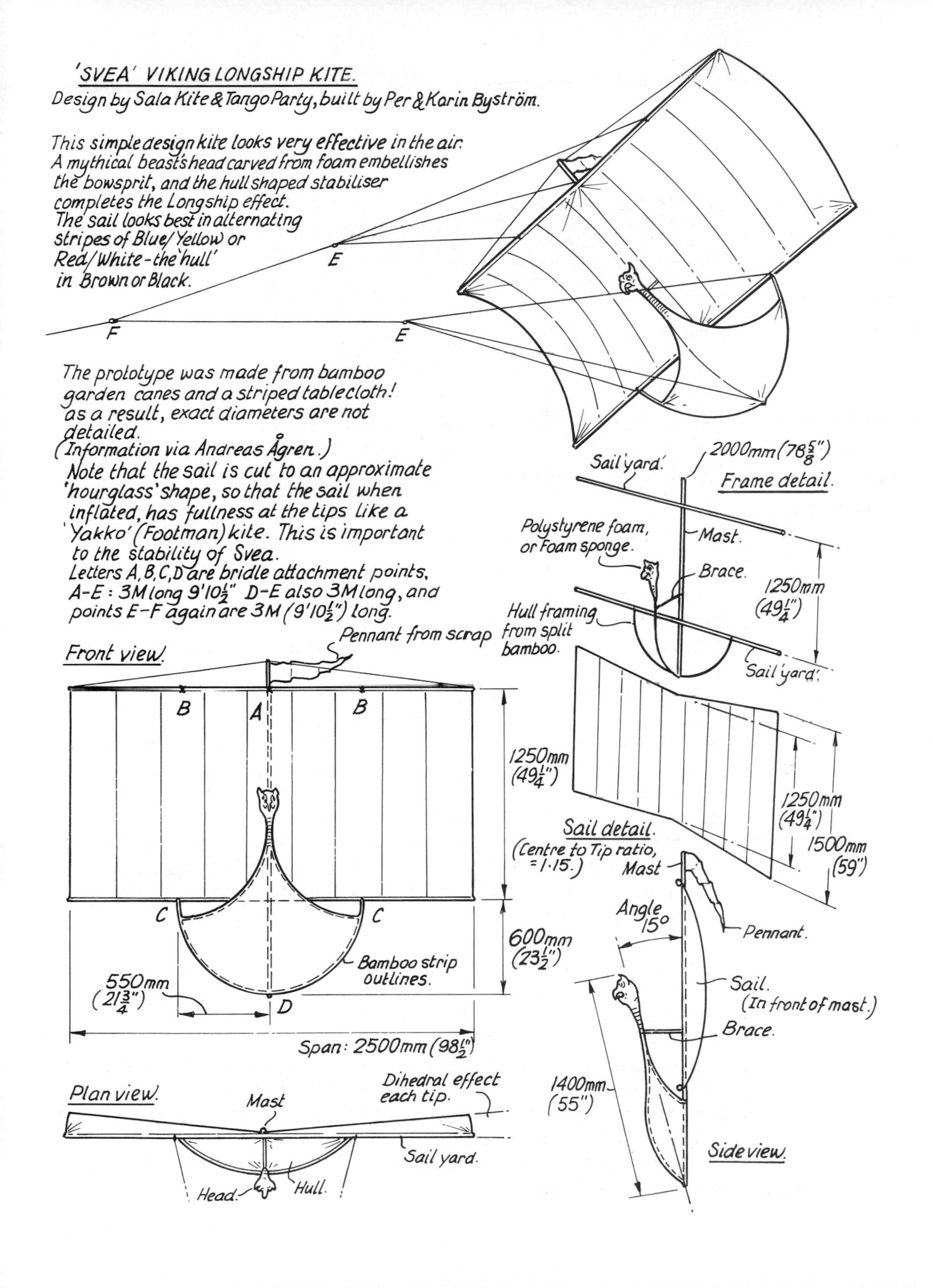

'SVEA' VIKING LONGSHIP KITE.
Design by Sala Kite & Tango Party, built by Per & Karin Byström.
This simple design kite looks very effective in the air. A mythical beast's head carved from foam embellishes the bowsprit, and the hull shaped stabiliser completes the Longship effect. The sail looks best in alternating stripes of Blue/Yellow or Red/White - the 'hull' in Brown or Black.
E
F
E
The prototype was made from bamboo garden canes and a striped tablecloth! as a result, exact diameters are not detailed.
(Information via Andreas Ågren.)
Note that the sail is cut to an approximate 'hourglass' shape, so that the sail when inflated, has fullness at the tips like a 'Yakko' (Footman) kite. This is important to the stability of Svea.
Letters A, B, C, D are bridle attachment points, A-E: 3M long 9'10½" D-E also 3M long, and points E-F again are 3M (9'10½") long.
Sail 'yard.'
2000mm (78⅝")
Frame detail.
Polystyrene foam, or Foam sponge.
Mast.
Brace.
1250mm (49¼")
Hull framing from split bamboo.
Sail 'yard.'
Pennant from scrap
Front view.
B
A
B
1250mm (49¼")
C
C
600mm (23½")
Bamboo strip outlines.
550mm (21¾")
D
Span: 2500mm (98½")
1250mm (49¼")
1500mm (59")
Sail detail.
(Centre to Tip ratio, = 1.15.)
Mast
Angle 15°
Pennant.
Sail. (In front of mast.)
Brace.
1400mm (55")
Side view.
Plan view.
Mast
Dihedral effect each tip.
Sail yard.
Head.
Hull.

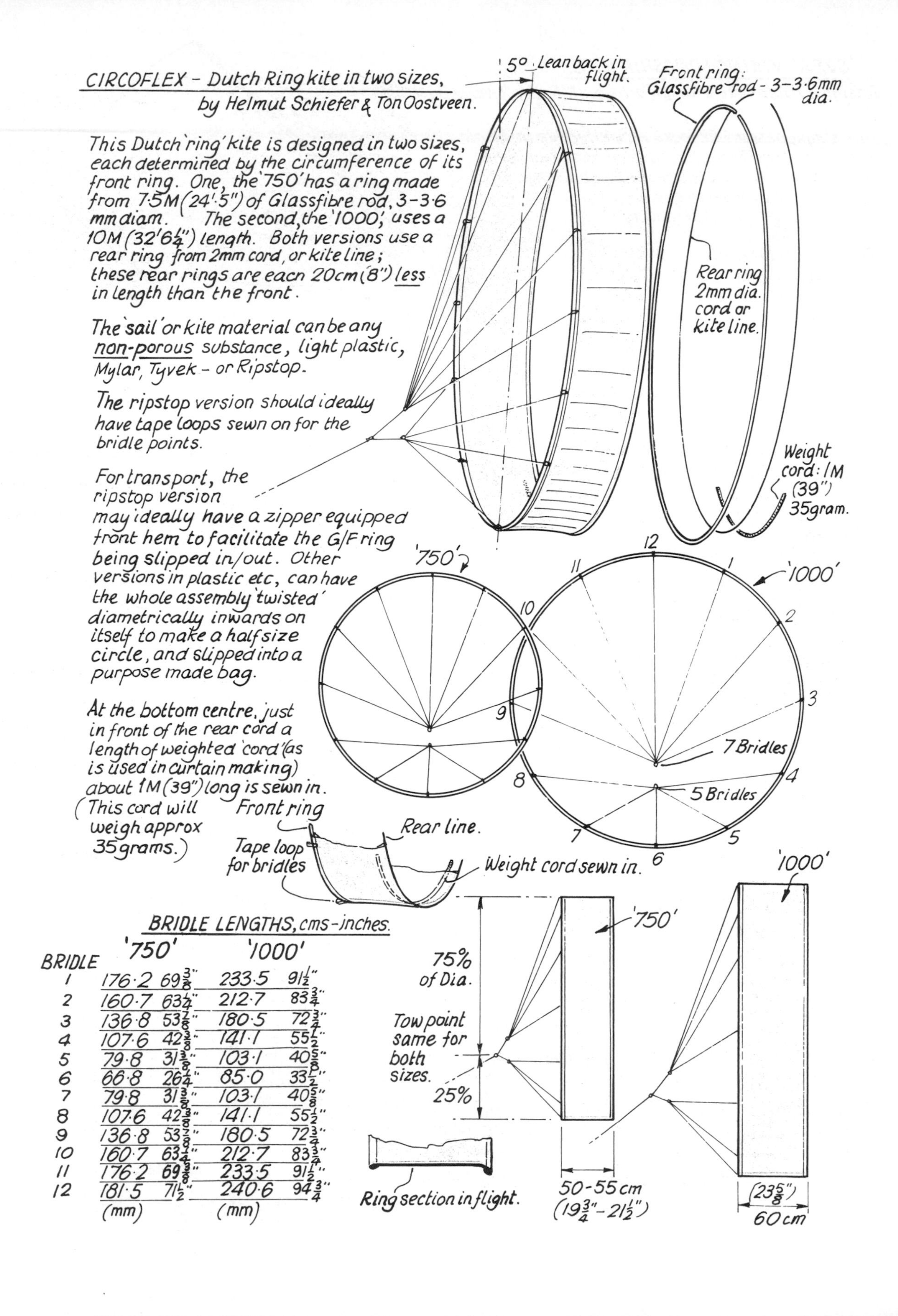

BRIDLE	'750'		'1000'	
1	176·2	69⅜"	233·5	91½"
2	160·7	63¼"	212·7	83¾"
3	136·8	53⅞"	180·5	72¾"
4	107·6	42⅜"	141·1	55½"
5	79·8	31⅜"	103·1	40⅝"
6	66·8	26¼"	85·0	33½"
7	79·8	31⅜"	103·1	40⅝"
8	107·6	42⅜"	141·1	55½"
9	136·8	53⅞"	180·5	72¾"
10	160·7	63¼"	212·7	83¾"
11	176·2	69⅜"	233·5	91½"
12	181·5	71½"	240·6	94¾"
	(mm)		(mm)	

Circoflex

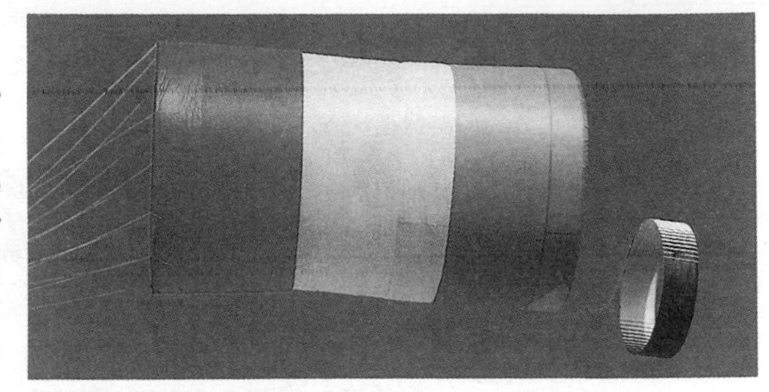

To say that the 'Dutch Rings' by Helmut Schlefer and Ton Oostveen were the most innovative kites of 1996 would be an understatement. As everyone would agree when they were awarded the Trophy for best new design at Dieppe, the Circoflexes were not only a joy to see in the sky, their constant performance earned the admiration of kite flyers from 30 nations.

The concept originated at Easter 1996 when flying the tube kites. These in themselves were a triumph in bridling. Most tubes depend on being tethered to a line from a carrier kite but Helmut and Ton had worked out a way to divide the bridle into two units of equally spaced lines and linking them so that the tension held the mouth of the tube near to vertical.

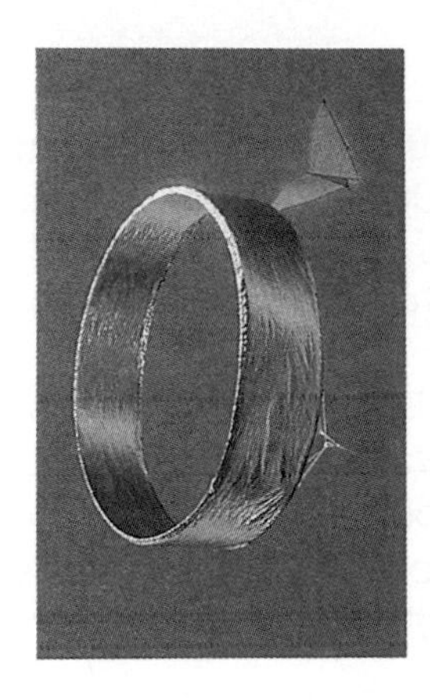

Why not, they queried, shorten the chord of the tube and increase the diameter? The idea worked, and so the rings were made. We're certain it could not have been that easy. The simplest things are hardest to devise, and it took a number of faxes that Pat exchanged with Helmut before all the construction became clear, even though we had measured and made notes at Dieppe.

There are two sizes. The popular one is the smaller for the reason that it it calls for 'only' 7.5m or 24ft 5in of glass fibre rod. Even if you are able to get to a factory where such rod is made it is still unlikely that you can purchase this in one length of 3mm or 3.6mm! For the large ring, the length increases to 10m or 36ft 6¼in to make it even more improbable so the answer is to join the rod with ferrules.

Apart from that, cost and availability of materials make the rings very attractive – they cost very little!

Master and offspring. The original Dutch tube kite in national colours with a Circoflex for comparison, seen also at left in company with a delta. Note the inward deflection of the trailing edge – an important feature.

Six Circoflexes, all at a steady 5 degree backward leaning angle, all silently confounding the aerodynamicists with their remarkable performance.

Covering is in any light plastic, the more reflective or colourful the better. Naturally the strip has to be the same or greater than that long GFK rod, so it will inevitably have to be joined from shorter lengths. If a plastic is used it can be turned back over the rod and sealed. The bridle lines can be tied permanently. Ideally, follow the note for a ripstop version, using a hem and zipper, with tape loops for the lines. Now the real trick! Inside the rear hem there's a string 20cm or 8in shorter than the rod when it is tied so that it pulls the trailing edge inwards. Add the curtain weight and you're almost ready. How about a pair, in gold, to fly at that wedding reception?

CHAPTER TEN

The law and kite flying

One of the great attractions of kite flying is the freedom one feels when the line tautens and the kite soars with the wind. It is a recreation which takes one far away from the problems of the day, removing any stressful cares and replacing them with pleasure and contentment. Why then, should we be concerned with the law?

Air Navigation Order (UK)

You might have wondered why the figure of 60-metre line length should appear so often in literature concerning kites in the United Kingdom. This length of line is a standard for all ready-made kites as sold complete with handle and line. It is also the length of line ready-wound onto handles sold as accessories.

The reason for this length is explained by what is known as the Air Navigation Order, in which Article 76(1) makes clear that kites may not be flown higher than 60 metres or within an aerodrome traffic zone without written permission from the Civil Aviation Authority. In addition, the Order requires under Article 56 that the flyer shall not recklessly or negligently cause danger to any person or property.

Exceptions may be granted for flight above 60 metres. Article 55 requires that the flyer shall not endanger an aircraft or its passengers. Rule 14 of the Rules of the Air 1996 stipulates that markings in the form of 2-metre tubular streamers at 100 metre intervals be attached to the line.

Sounds complicated? It shouldn't really, as we have actually simplified the wording as generally understood by the kite flying fraternity in the UK. The complete extracts are to be found at the end of this chapter. What it means to the greater majority of British kite flyers, possibly 99.9% of the whole international movement, is that the rules of air safety require you to be outside the traffic pattern of any aerodrome by a good margin, and that you have to be satisfied with a standard line length for any local flying.

Exceptions

Written permission, as mentioned in the official wording of the Air Navigation Order 1996, Article 76(1), is not such an ugly obstacle as might at first be required.

The gentlemen at the Civil Aviation Authority are well acquainted with the situation, and over the years have assisted the kite movement in obtaining exclusions from the restraints. Altitude record attempts in West Wales, man-

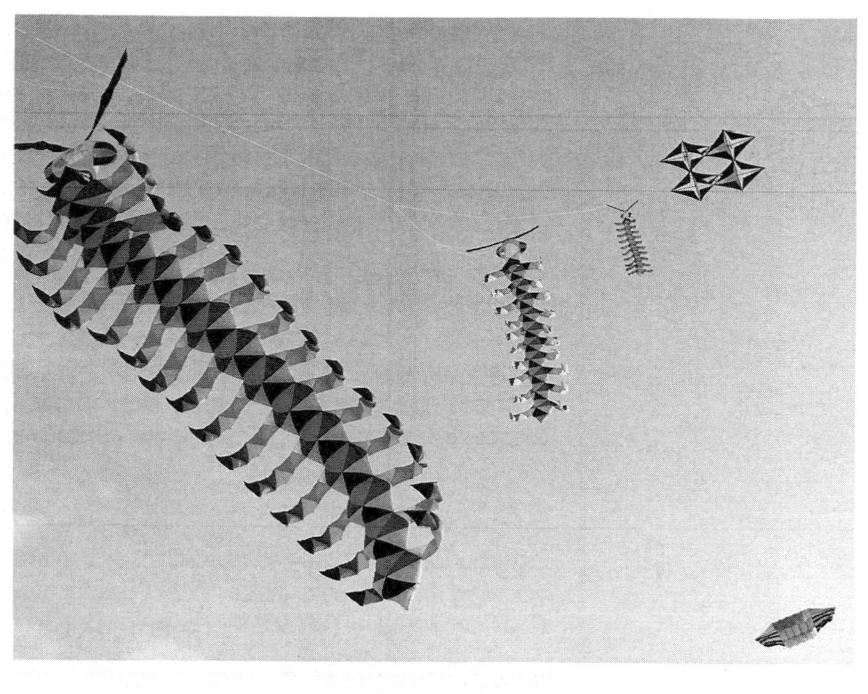

The decorative inflated kite, either insect shaped as here, or in the form of a giant windsock, requires a good lifter. This train of centipedes is by Peter Lynn, topped by one of his reliable Tri-D triangular box kites. Such a train needs to exceed 60m height, so exceptions are required.

carrying Cody kite train operations and aerial photography of industrial sites are but three examples of special circumstances.

Kite rallies and festivals take care of the regulations by arranging for a blanket exclusion to an agreed altitude often as high as 305 metres (1000ft). In some instances, the rallies have been well within the traffic zone of an aerodrome if not actually on the aerodrome! Collaboration with Air Traffic Controllers has enabled the two activities to operate safely.

Application to the CAA, Directorate of Airspace Policy, Hillingdon House, Uxbridge, Middx. UB10 0RL will provide any enquirer with an explanatory letter and application form, as reproduced at the end of this chapter.

Common sense

The restraints as described have not seriously affected kite flying. They reflect a national concern for safety in a tightly-packed country where there is comparatively dense aerial activity.

The CAA needs to know if kite flyers want to fly high, or to be near an aerodrome, and they have only barred operations on a few rare occasions, and for good reasons. Other countries do not have exactly the same stringencies. Airspace is less cluttered. There are, nevertheless, many other rules which apply although perhaps not with the same power of legislation. One can embrace these under the general term of 'common sense'.

Self protection ought to be a priority, yet it is so often ignored. The obvious risks are conveniently obscured by avid enthusiasm. So there ought to be a self-regulatory 'law' to save us from ourselves.

Third party risks

Although we extol the kite as a beautiful object, there are occasions when it can be a visual, audible or physical risk to nature and, in turn, to mankind.

Let's look at what *has* happened and establish more rules to safeguard the future. Horse riders have been thrown as a hacking group came into view of a stunt kite. Hikers have complained that the noise of multiple stunt kites disturbs the peace of the countryside. A motorcyclist suffered a cut throat when riding into a discarded line stretched across the highway. A prize-winning classic car was expensively damaged by a falling kite in a kite festival car park. Broken lines have damaged grass cutting equipment. A young child in a pushchair was hit by a rigid framed kite.

This chapter of accidents confirms the need for caution. While such incidents are rare, the risk is always there and we return to that Article 56 of the Air Navigation Order, namely 'A person shall not recklessly or negligently cause or permit any aircraft (kite) to endanger any person or property'.

It is a rule that can be enforced in the UK courts of law. So let's establish some behavioural regulations for ourselves.

Rule 1: WEAR GLOVES

Use stout leather gloves as sold for gardening or rubbish clearance to protect yourself from burns or cuts, particularly when nylon, polyester or Aramid (Kevlar) line is in use off the reel for a single line.

Rule 2: STAY CLEAR OF HIGH TENSION WIRES

Those electricity cables are routed over open spaces, and at a height on pylons for a purpose. Even so, they are insulated from ground contact. Kitelines will conduct the high current if they are damp and you will be the earth contact if the kite is flown near enough to attract a spark across the gap. Actual contact with HT lines could be suicidal. Stay away if you don't want to be a spark plug!

Rule 3: BEWARE OF ELECTRICAL STORMS

Thunder indicates a discharge somewhere nearby. Come down! The atmosphere must be dampening in such conditions and, as Benjamin Franklin discovered in 1752 when he conducted electricity down a hemp line from a spike aerial on his kite, there's an explosive charge waiting to happen!

Rule 4: WATCH YOUR FEET

It does not happen often but, when it does, it can more than ruin the day when you twist an ankle, maybe do much worse when running backwards. Beware the burrows!

Rule 5: SEEK SPACE TO FLY

Ensure that there is adequate room around you for safe flying, especially for stunt kites. Do not overlap car parks, bridle paths or footpaths.

Rule 6: RESPECT NATURE

Avoid areas where the visual appearance of a moving kite can disturb animals, especially sheep and horses which can panic, or brooding birds which may abandon their eggs.

Rule 7: DO NOT OVERFLY HIGHWAYS
If there is a road downwind, then limit line length or move upwind. If a line breaks, recover the free end *quickly* if it has crossed any roadway.

Rule 8: RESPECT THE PEACE
Some kites create a lot of noise, enough to disturb the pleasures of the countryside or seaside. Avoid quiet zones such as naturist beaches or bird sanctuaries.

Rule 9: BEWARE PEOPLE!
Watch for intruders. Unwitting walkers, other kite flyers skygazing as they cross the flight line, especially family groups with baby carriages. Remember - lines can burn or cut and composite spars can bruise.

Rule 10: BE TIDY
Carry away all your rubbish, broken spars, line, tails or picnic remnants. They're not only unsightly, they can also kill animals likely to eat plastic parts.

Man lifting kites must fly high, the'pilot' kite will go to 300m (1000ft) which is another case for exemption.

Local authority regulations

You may have noticed when entering the local park to fly your kite that there is a notice board close by the entrance. Somewhere in the long lines of many words, there *may* be a reference to kite flying, along with ball-games, fishing or other pleasurable pursuits. It would be unlikely, as the great majority of local authorities regard kites as ecologically acceptable. There are some exceptions but, in each case, where a local authority proposes to incorporate a restriction or to prohibit totally any hobby or sport, it is bound by the Home Office to publish a public notice and to accept appeals.

If the hobby, say kite flying, has been habitually practised in that area over previous years then there would be an enquiry resulting from the appeal. Ultimately, should a ban be established, the local authority would be bound, under the terms of the Home Office, to provide another site.

Common land

It so happens that the most suitable areas do not fall within urban zones and are in the countryside, generally referred to as 'commons'. They are large tracts of heathland, mostly untended and unsuitable for agriculture. Governed by 'commoners', these attractive kite sites can vary from mown grass adjacent to racecourses, as at Epsom, York, Bath, Newbury and Cheltenham, to undulating ground like the slopes of the Malvern Hills or Dunstable Downs.

Much as they appear to be free to use, they are still the responsibility of some kind of Board of Governors and therefore administered so that the area is preserved from development or misuse. So, no matter how much the kite flyer might be disturbed at having to give space to an occasional golfer engaged in practice driving, or a dog lover exercising a pack of mongrels, he or she has to remember that it's a case of equal rights - share and share alike.

Although we've concentrated on the situation as it applies to the UK with specific place names, similar situations arise throughout the world, with changes according to national procedures.

Relevant Sections of the Air Navigation Order (UK)

Rule 14 The Rules of the Air

(1) A kite while flying at night at a height exceeding 60 metres above the surface shall display lights as follows:

(a) a group of two steady lights consisting of a white light placed 4 metres above a red light, both being of at least five candela and showing in all directions, the white light being placed not less than 5 metres or more mg than 10 metres below the lowest part of the kite;

(b) on the mooring cable, at intervals of not more than 300 metres measured from the group of lights referred to in sub-paragraph (a), groups of two lights of the colour and power and in relative positions specified in that sub-paragraph, and, if the lowest group of lights is obscured by cloud, an additional group below the cloud base; and

(c) on the surface, a group of three flashing lights arranged in a horizontal plane at the apexes of a triangle, approximately equilateral, each side of which measures at least 25 metres; one side of the triangle shall be delimited by two red lights; the third light shall be a green light so placed that the triangle encloses the object on the surface to which the kite is moored.

(3) A kite while flying by day at a height exceeding 60 metres above the surface shall have attached to its mooring cable either:

(a) at intervals of not more than 200 metres measured from the lowest part of the kite, tubular streamers not less than 40 centimetres in diameter and 2 metres in length, and marked with alternate bands of red ana white 50 centimetres wide;or

(b) at intervals of not more than 100 metres measured from the lowest part of the kite, streamers not less than 80 centimetres long and 30 centimetres wide at their widest point and marked with alternate bands of red and white 10 centimetres wide.

CIVIL AVIATION AUTHORITY **Air Navigation (Amendment) Order 1996**

Kite/s - Application for Permission to Fly
Above 60 metres/ Above 30 metres within an ATZ

Note: 28 days notice MUST be given of the event. To avoid delay, include all information requested below.

1. Event____

2. Kite Operator____

3. Location of flight **IMPORTANT:** Attach a copy extract of 1:50,000 Landranger OS Map showing site marked CLEARLY.

OS Grid Ref: Map No:_______ Grid Letters_________ Easting (3fig)___________ Northing (3fig)__________

Full Postal Address of Site____

____

Contact telephone number at site (if available)____

4. Nearest Aerodrome (if applicable)____

5. Date/s of flight/s_________________ Daily Period (in local time)From__________ To__________

6. Height above ground level of kite/s____

7. Has the Event Organiser's permission been obtained? Yes/ No *

8. Has the Landowner's permission been obtained? Yes/ No *

Landowner's Name_________________________ Date permission obtained__________________

9. Have the Police any objections to the flight? Yes/ No *

Full address and telephone number of Police Station concerned____

____

Name and Rank of Officer consulted____

....

DECLARATION

I declare that I have checked the above information and that to the best of my knowledge it is correct: and that I am aware of my obligations under the Air Navigation (Amendment) Order 1996 as the Operator of a Kite/s.

Signature of Applicant_____________________________ Date____________________

Name (BLOCK CAPITALS) and status____

Address____

____

____

Telephone Number (state if business or private)_____________________ Fax No_________________

CHAPTER ELEVEN

The fun in kites

In an age when 'fun' is an added description to so many sporting accessories and we have the Fun-Bike,the Fun-Car and aeromodellers have their 'Fun-Fly' national events for unconventional and highly acrobatic radio controlled models it's surprising that we have yet to learn of a '"Fun-Kite'.

Maybe that is because most of us consider that all kites generate fun, albeit in differing ways. The Sport kite can be called a fun kite, until that is, it becomes one used for very serious competition. The 'Trick' Sport kite is very much a funster with its ability to go inverted and be recovered. The fun comes in learning how to do this and other tricks.

Our own thoughts go back 60 years (ahem!!) to the Atalanta hexagonal kites created by the remarkable Sophocles Xenophon Pantcheff. Those tissue covered kites came with literature that listed accessories which would run up the line then drop away. Message carriers, parachutists, or a really effective 'buzzer' would occupy our young minds for hours.

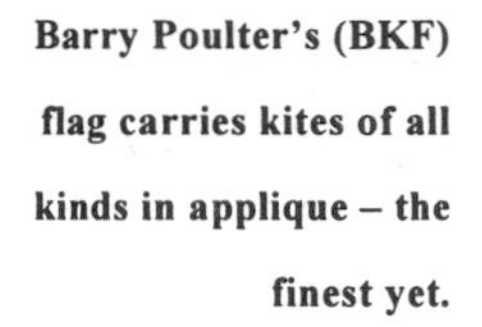

Barry Poulter's (BKF) flag carries kites of all kinds in applique – the finest yet.

Possibly such distance in time has coloured our memory but that early experience with kites that did something other than just fly remains with us today, even though we have made giant strides in the variety of kite designs.

Each time we return to base from a kite festival there's a flush of inspiration. So much to see sometimes that scarcely more than one or two of our kites are brought out for a flight. We do know of at least one flyer who takes hours each time, selecting and packing which kites to take to a festival, then spends the whole day socialising and not one of his kites has been aired!

The moral here is that there are times when our kites seem 'ordinary'. They have plenty of air time at the local site and that old adage 'familiarity breeds contempt' comes true when we're faced with a galaxy of new shapes, colour blends and performances to be seen at the festivals that draw flyers from all parts around.

Noveltíes have always been with us, and they emerge periodically as though in cycles. As we write, it is the turn of the Autogyro kite

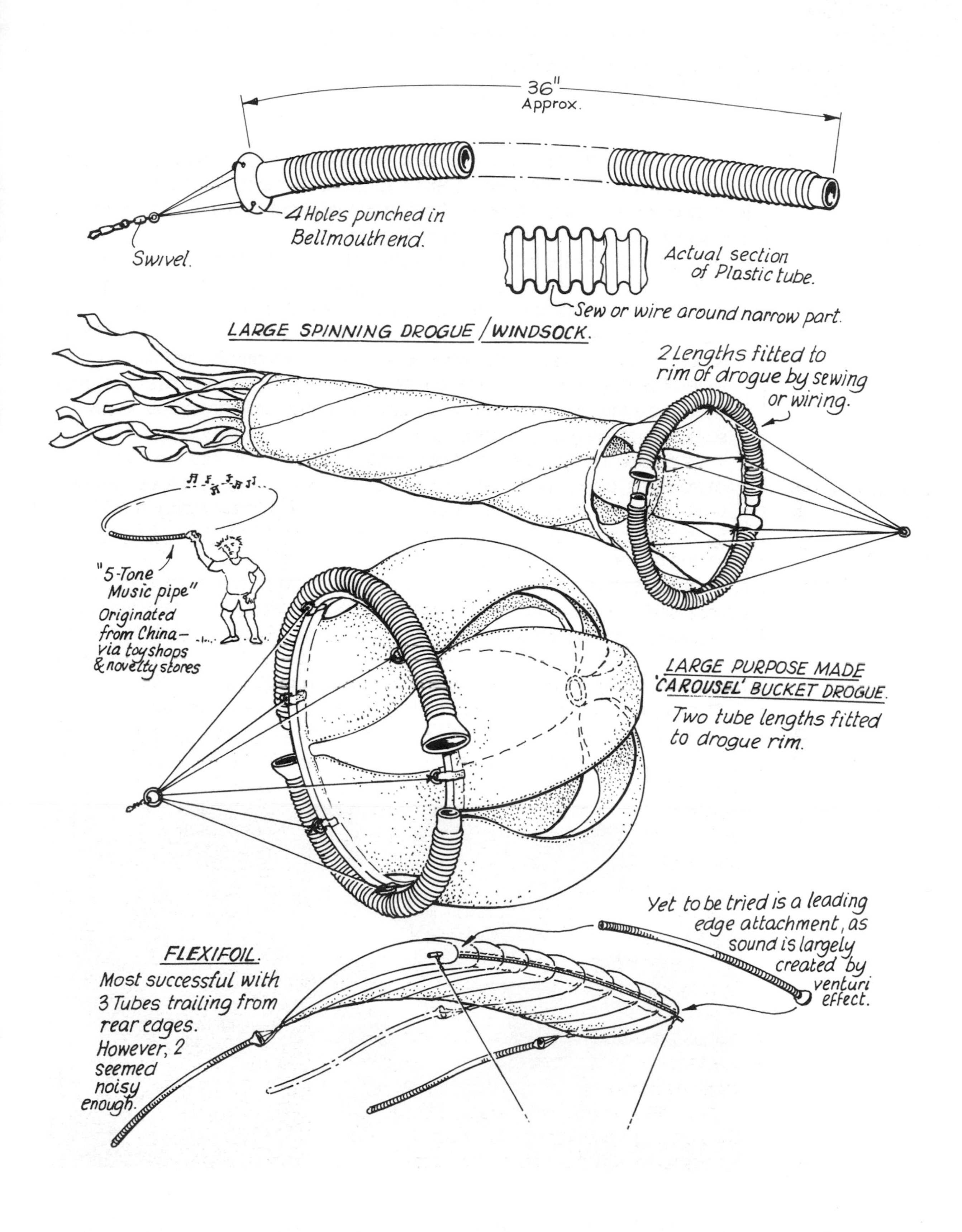
36" Approx.
4 Holes punched in Bellmouth end.
Swivel.
Actual section of Plastic tube.
Sew or wire around narrow part.
LARGE SPINNING DROGUE / WINDSOCK.
2 Lengths fitted to rim of drogue by sewing or wiring.
"5-Tone Music pipe"
Originated from China – via toyshops & novelty stores
LARGE PURPOSE MADE 'CAROUSEL' BUCKET DROGUE.
Two tube lengths fitted to drogue rim.
Yet to be tried is a leading edge attachment, as sound is largely created by venturi effect.
FLEXIFOIL.
Most successful with 3 Tubes trailing from rear edges. However, 2 seemed noisy enough.

with revolving rotor(s) replacing sails. The rotary kite has had a full share of publicity when one covered in radar sensitive material was flown (without official clearance) into an airfield traffic zone and confused the controllers. Not exactly the best way to make friends!

Inflatables like the famous legs from Martin Lester and the enormous turtle by Peter Lynn are clearly fun kites although it is the grounded blowfish which amuses the youngsters most as they can participate in trying to get it to levitate on its tether. Much smaller air-filled mammals – even pigs and cows, or fishes – are flown as shaped windsocks. They always attract attention and add to the bewilderment of the uninformed onlookers who never seem to understand how these figures can fly so high. The rather ordinary carrier kite is often so high that it remains unnoticed.

Of course the greatest fun of all comes in Bear-land where the lovable miniature 'Teds' congregate before performing their dramatic drops. Motorised winch lifts have appeared to speed up the process of elevating the skydivers. Such is the demand at charity events where the dropniks operate a fund-raising scheme for donors who pay-per-drop, and like it.

The ever inventive Roger Pike of Essex Kite Group 'has a novel suggestion to add to the fun of kite flying by using a mock cockpit with a joystick to control a 2-line Sport kite. Pat's sketch illustrates how commonly available materials can be used to construct the device. Throw in your own ideas and who knows, we may have found a way for tired old authors to fly!

There's another way of adding to the fun quotient at ground level. Club, or group flags have been with us for years. They first displayed the initials of

All of Eric de Monternal's kites are aeroplane shapes and few of them follow conventional kite structures. They fly well, and typically, Eric dons leather blouson and helmet.

REMOTE CONTROL FOR STUNT PILOTS: by Roger Pike.

This illustration is to scale, but is not intended to be slavishly copied - you may have similar items already to hand in your loft, garage or workshop. Interpret these basics how you wish.

All Roger's framing & control yoke is from 15mm diam. copper plumbing tube (from D-I-Y stores). Right angles etc are from the "Tees" & "Elbows" for the same purpose - use the type which already contain a solder ring. "Yorkshire" is the trade name. Just slip on and heat.

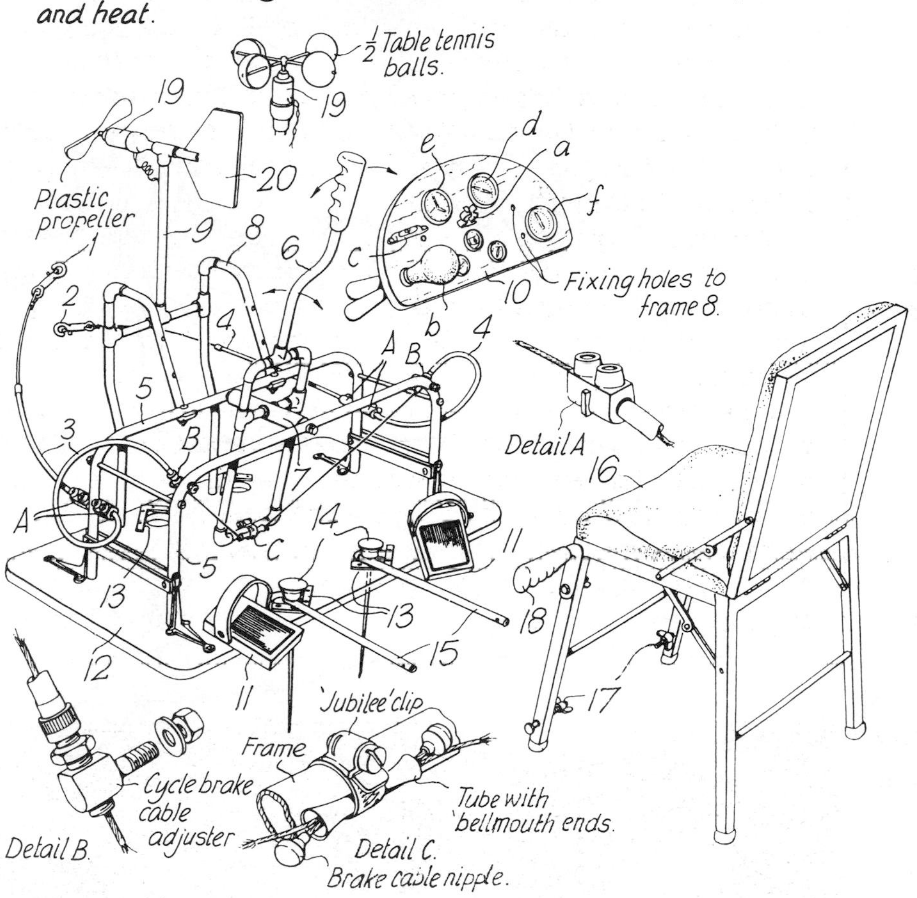

Legend.—

The Instrument panel contains :- a: A parking clip for the joystick. b: A bulb horn for the unwary, c: Spirit level, d: Airspeed indicator, (driven by generator/motor 19) e: Clock, f: Compass.

Main references indicate :- 1&2 Doglead clips for kite lines. 3&4: Cable outers to smooth out cable runs. 5: Tubular frame. 6: Control column. 7: Gimbal unit. 8: Secondary frame unit. 9: Upright to carry Airspeed generator. 10: Ply panel. 11: Footrests. 12: Baseboard. 13: Ground anchor holes. 14: Anchor rods. 15: Spacer tubes for chair legs. 16: Chair (folding). 17: Wing nuts for 15. 18: Steady handle. 19: Small electric motor (battery). 20: Wind vane.

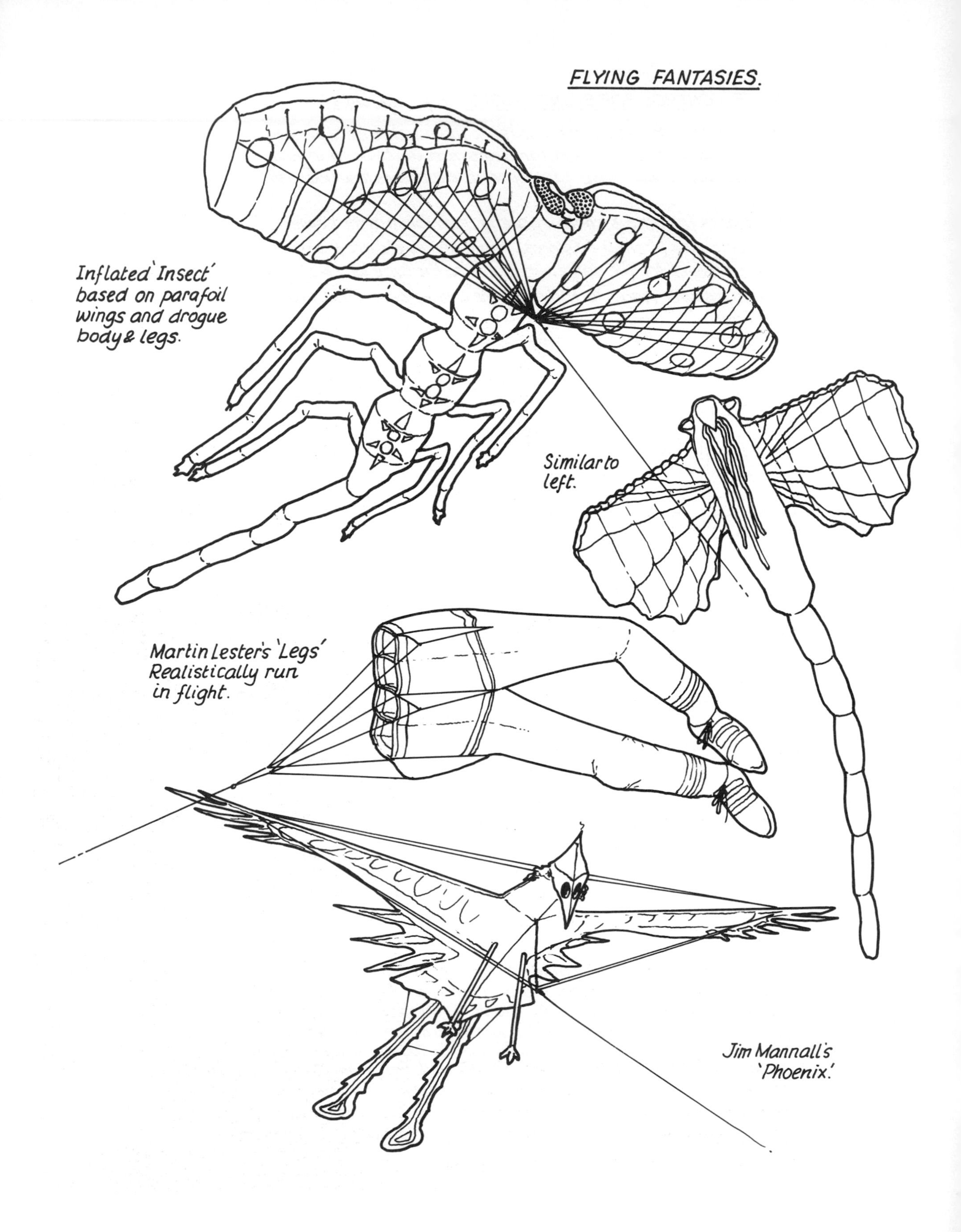
FLYING FANTASIES.
Inflated 'Insect' based on parafoil wings and drogue body & legs.
Similar to left.
Martin Lester's 'Legs' Realistically run in flight.
Jim Mannall's 'Phoenix.'

The ephemeral kite is a French speciality-This one (left) has just one central mast. None of the colours of Normandy touch the mast. They are held in place by the bridle, and fly well. Below: Another French classic. Three separate kites flown in formation by Michel Gressier's team, each symbolising artist's paints.

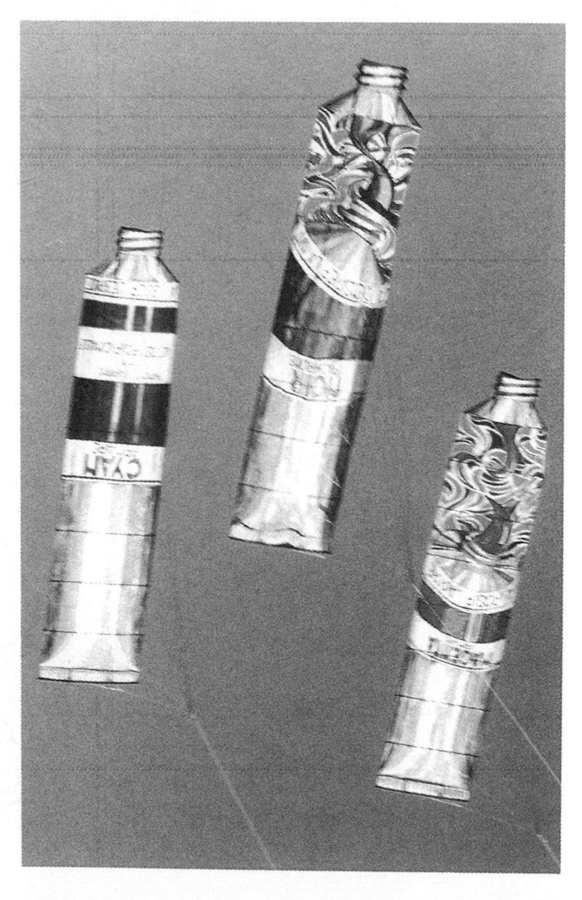

the group as an identity for location on the field. Vertical sails, held to a roach pole are tapered and bend with the wind. As their popularity grew and the skills of the applique specialists improved, the flags became more colourful and pictorial. A veritable forest of flags was erected at the 1996 festivals, and with them, a trend towards making wind farms arrived. These included rotating vanes, symbolic weather vanes and trailing streamers to add to the fun of the festivals.

While there is no mention of competition between the groups for the most spectacular example,there is no dispute over the finest to date. Barry Poulter of the Brighton Kite Flyers produced the ultimate challenge in applique with every type of kite displayed on his tall 'square rigged' pictorial. Many personalities and their kites are recognisable in his thought-provoking flag, see page 230.

Fun? There's no question that all kite flyers become totally immersed in this wonderful hobby and benefit from all of its charms. Our world is up there on the line. The kite can bring us serenity, or excitement. Planning the design, making the kite and then testing it for that

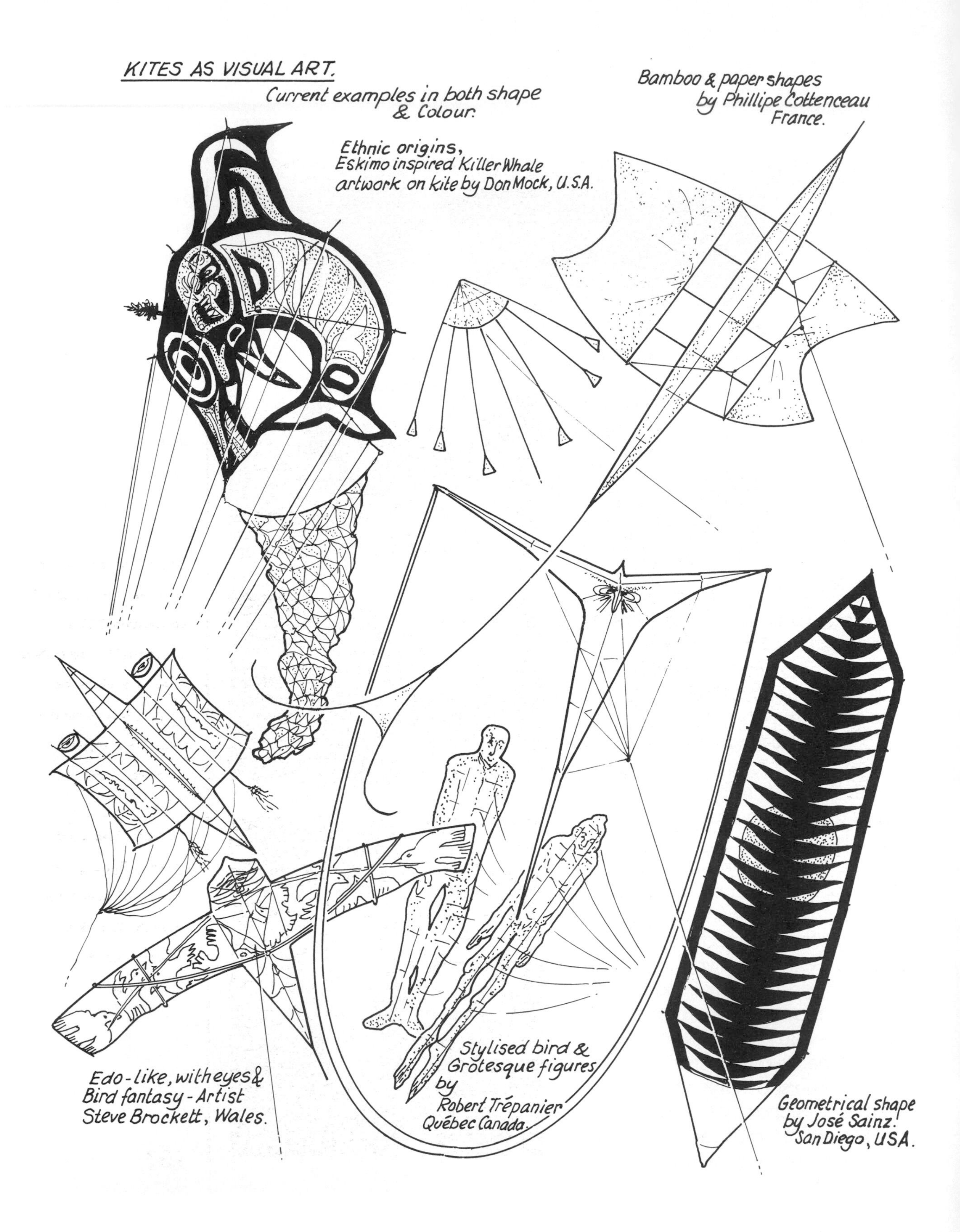
KITES AS VISUAL ART.
Current examples in both shape & Colour.
Ethnic origins, Eskimo inspired Killer Whale artwork on kite by Don Mock, U.S.A.
Bamboo & paper shapes by Phillipe Cottenceau France.
Edo-like, with eyes & Bird fantasy - Artist Steve Brockett, Wales.
Stylised bird & Grotesque figures by Robert Trépanier Québec Canada.
Geometrical shape by José Sainz. San Diego, USA.

Many kites and not one of them could be called conventional-The tubes,airship and train of fishes are windsocks trailing from a carrier line. The octopus supports itself.

first flight is an exhilarating experience we should all value highly. The fun element is a spice that distinguishes the colourful, the original or the spectacular performer from the ordinary.

We hope that inspiration will be found in the pages of this book.

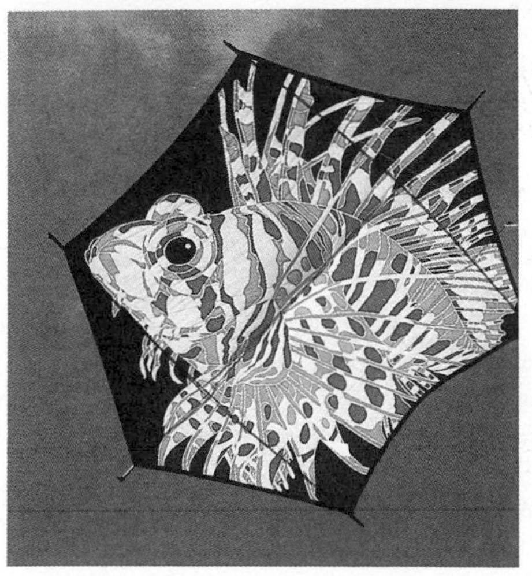

Of many decorated Rokaku fighting kites, this by Michael Alvares from Perth with its Scorpion fish is spectacular.

APPENDIX 1

International Kite Flying Organisations

AUSTRALIA

Australian Kite Association
H Bushell
10 Clayton Court
Springvale
Victoria 3171

Adelaide Kiteflyers
4 Wilpena Ass Close
Eden Hills
S. Australia 5050

Australian Stunt Kite Association
51 Glenhuntley Road
Elwood
Victoria 3184

Brisbane Kite Group
6 Colbourne Street
Accacia Ridge
Queensland 4110

Australian Kite Flyer's Society Inc
Godfrey Gamble
197 Bobbin Head Road
Turramurra
New South Wales 2074

Cairns Kiteflyers
PO Box 252
Trinity Beach
Queensland 4879

Canberra Kite Club
No. 1 Townhends Street
Phillip
ACT 2606

AUSTRIA

Vienna Kite Festival
H Georgi
Argentinierstrasse 16
A-1040 Vienna

Phoenix
Drachenbau und Fliegerverein
Reiner Kendl
Eisenhandstr 38
4020 Linz

BRAZIL

Zeca Das Pipas
Rua: Sao Raimundo SO
CEP 21760-270-R.J.

BELGIUM

Cerf Volant Club de Belgique ASBL
Draken Club van Belgie VZW
rue Defacqz 33
1050 Bruxelles

Le Nouveau Cervoliste Belge
Jacques Durieu
45 rue de la Houssiere
B-1435 Hevillers

KAPWA (Aerial Photography)
M Dusariez
14 Ave Capt Piret
1150 Bruxelles

CAMBODIA

Cultural Ministry
11 Street 57
Sangkat Boeung Keng Kang
Khan Chamkar Marn, Phnom Penh

CANADA

The Toronto Kitefliers
280 Wellesley Street
701, Toronto
Ontario M4X 1G7

British Columbia Kitefliers Ass.
Rosemary Meyer
3991 Puget Drive
Vancouver BC V6L 3V3

CHINA

Chinese Kite Association
9 Tiyuguan Rd
Beijing
102300

Nantong Kite Fliers
No 5 Quinnian Road
East Nanton
Jiangsu

International Kite Federation
Secretariat and Weifang Kite Association
92 Shengli Street
Weifang
Shandang Province 261041

CIS

Galine Zverik
5 Home 9 Sadovara St
PO Box 110, 325000 Kherson

CURACAO

c/o Carol Jansen
Sloep 37
9732 CB Holland

COLOMBIA

Club de Cometeros Maliky
Cra 27a 53-06 (of 406)
Bogota

FRANCE

Federation Francais de Cerf Volant (FFCV)
4 Rue de Suisse
06000 Nice

Cerf Volant Club de France
BP 186-75623
Paris, Cedex 13

French Kan Kite
2 Rue Hermann
La Chapelle
75018 Paris

Cerf-Volant Club Celtique
La Croix Fleurie
35240, Retiers

Le Ciel Pour Simaise
2 ter rue Rabelais
3700 Tours

Ephemeres Millenaires
58 rue Jean Bodin
4900 Angers

Icare
Kerclequinet-Brillac
56370 Sarzeau

GERMANY

Drachenclub Deutschland ev.
Postfach 350127
D-40443 Dusseldorf

DADL Hannoverscher Drachen Club
Weetzener Landstrasse 117
D-3005
Hemmingen 1

Fesseldrachenclub
Otto Lilienthal
Stargarderstr 62
Berlin 1058

VTD (Vlieger Team Dortmund)
Wodenstr. 2
D-44359
Dortmund 15

Drachen & Drachensachen
Eissenacher Str 81
D-1000
Berlin 62

HOLLAND

Nederlands Vliegergezelschap
Meendaal 39
6228 GE
Maastricht

HONG KONG

Hong Kong Kite Association
1A, 7th Street
Tai Wai Village
Shatin, N.T.

HUNGARY

Hungarian Kite Club
Istvan Bodoczky
Kiss Ljos u.30
Budapest
H-2092

Alfodi Sarkanyereszto Club
Magyar Gabor
Debrecen, Vasavari Pu.12
H 4028V

APPENDIX 1

INDIA

Ahmedabad Kite Museum
Bhanu Sha
28 Manager, Opp Nehrunagar
SM Road
Ahmadabad 380015

Golden Kite Club
Raghavi Nivas
10 Dadi Seth 1st Lane
Balbunath, Bombay 400-007

India Kitefliers
3126 Lal Darwaza Bazaar
Sita Ram
Delhi 6

Indian Fighter Kite Fliers Assoc.
Flat No. 313/D Sagar Appts
Sonapurlane
LBS Marg, Kurla (West)
Bombay 400 070

Kangi Kite Club
470/2 Laketown
Block B, Calcutta 700089

Pink City KFC
402b 'Narita'
Yari Rd, Versova
Bombay 400-061

INDONESIA

Indonesian Kitefliers
Perstuan Penggemar
Layang 2 Jakarta Barat
Jalan Pinangsia Gang Asem No 18
Jakarta Barat

Bali Kite Association
Seni Rupa Udayana University
J1 Jendra Sudirman Denpasar

ISRAEL

Israel Eli Shula
Y 17 Barkochva St
Jerusalem 97875

ITALY

Associazione Italiana Aquilonisti
Via Dandolo 19
00153 Rome

Le Aquile D'Urbino
Palazzo del Collegio
61209 Rafaelo
Urbino

Cervia Volante
Via Pinarella 26
48015 Cervia

Riminivola
Via Roma 70
47087 Forli

JAMAICA

Kite City
25 Auburn Avenue
Patrick City, Kingston

JAPAN

Japan Kite Association
c/o Taimeiken
No 12-10
1 Chome Nihonbashi
Chuo-Ku Tokyo 103

KOREA

Korean Kite Fliers Association
23 Hyochang-Dong
Yongsan-Ku
Seoul

MALAYSIA

Malaysia Kitefliers
Kampong Pasir Gajah
Jalan Ayer Putih
Kemanam
Trengganu

MALTA

Malta Kitefliers Association
A Damenia Gay
3/107a Rudolphe St
Sliema

NEPAL

NirmalMan Tuladha
CNAS-Itu Kirtipur
Pono 3757
Kathmandu

NEW ZEALAND

New Zealand Kite Association
c/o Janet Malcolm
35 Wairiki Road
Mt Eden 4, Auckland

New Zealand Summertime Kite Day
D Pitfield
61 Main North Road
Woodend, Canterbury

NORWAY

Det Norske Dragelskap
Norwegian Kite Society
Josefinesgt 38
Oslo, 2

POLAND

Polish Kite Club
82-100 Wagrowiec
ul Zeromskiego 20

SINGAPORE

Singapore Kite Association
Shakib Gunn
601 LKN Building
135 Cecil Street
Singapore 0106

SOUTH AFRICA

South African Kite Angling Assoc
11 Fairbridge Street
Boksburg 1460

Kite Club of SA
Box 201293
Durban N4016

SOUTH AMERICA

Cometeros de la Villa
AA 11059
Medellin
Colombia

Associacion Ecologica Cometeros
Del Volador
Apartedo Aereo 051201
Medellin
Colombia

SPAIN

Club D'Estels
C/Senyor Rafel 52.2
07420 SA Pobla
Mallorca

Majorcan Kite Club
Falko Haase
C/Caltrava 18
17001 Palma de Mallorca

SWITZERLAND

Drachenclub Basel
Postfach 19
CH 4123
Allschwil 2

TAIWAN

Kite Association of Taiwan ROC
4 Fl No 10 Lane 205
Hou Chu Wei Street
San Chung City
Taipei, Taiwan, ROC

TASMANIA

TKFA
K Stevenson
8 Summerleas Road
Fern Tree
Tasmania 7101

THAILAND

Thailand Kite Fliers
c/o Ron Spaulding
16/147 Soi Suayai-Utis
Radachadaphisek Road
Bangkok 10900

TURKEY

Egup Kardes
Ucurtmalari
Selansiz cd 69/9
Uskudar/Istanbul

UNITED KINGDOM

The Kite Society of Great Britain
PO Box 2274 Gt. Horkesely
Colchester, Essex CO6 4AY

UNITED STATES OF AMERICA

American Kitefliers Association
352 Hungerford Drive
Rockville, MD 20850-4117

Drachen Foundation
1907 Queen Anne Ave north
Seattle WA 98109

VIETNAM

Hue Kite Club
15a Rue Mac Dinh, Chi Hue.

APPENDIX 2

Kite Museums

Musems dedicated solely to kites

Above: a mass of shapes and colour greets the visitor to the Taimaken Kite Museum in Tokyo.
Below: A beautiful Chinese kite, one of many in the World Kite Museum, Long Beach, Washington State.

CHINA

Weifang Kite Museum, Shanding Province.
(Also Secretariat for International Kite Federation.)

GERMANY

Deutschen Museum, Munich 26, D-8000.
(Early German kites, plus Chinese and wartime sea-rescue box kite and transmitter in 'Fallschirme und Drachen' section.)

INDIA

Kite Museum, Kochrab, Ahmendabad 380-006.
(Centre of Indian Fighting Kites.)

JAPAN

Taimaken Kite Museum, 12-10, 1-Chome, Nihonbashi, Chou-ku, Tokyo 130.

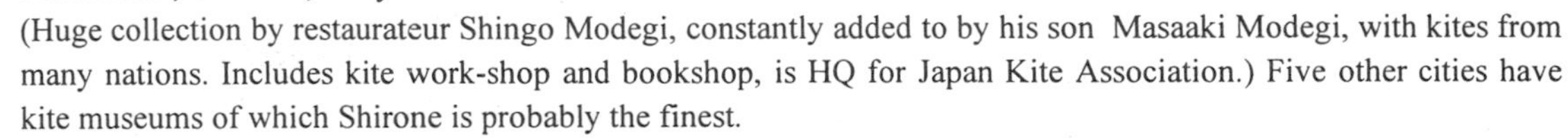

(Huge collection by restaurateur Shingo Modegi, constantly added to by his son Masaaki Modegi, with kites from many nations. Includes kite work-shop and bookshop, is HQ for Japan Kite Association.) Five other cities have kite museums of which Shirone is probably the finest.

UNITED STATES OF AMERICA

World Kite Museum, PO Box 964, 3rd Street NW, Long Beach, Washington 98631.
(Includes a Hall of Fame, honouring kiting pioneers and incorporates the late David Checkley's collection from 14 nations and 31 US states. It has the 'finest and most representative collection of Japanese kites' according to one source. Over 230 Chinese kites and video displays of kite fighting in Thailand, Korea, Japan and India were on view.)
The Drachen Foundation 1907 Queen Anne Avenue North, Seattle WA 98109.

(An educational foundation to promote knowledge of kites, holding archives of historic material including Cody and Jalbert collections.)
NASM, Smithsonian Institution, Paul Garber Facility, Silver Hill, Washington, DC 20560.
(Early kites from America including original Woglam, Chinese and the Paul Garber Target Kite, some in NASM, others at Silver Hill.)

UNITED KINGDOM

RAF Museum, Hendon, London NW9 5LL
Fleet Air Arm Museum, Yeovilton, Somerset BA22 8HT
Museum of Army Flying, Middle Wallop, Hants SO20 8DY
(Each has a representative Cody kite, and the last has Cody memorabilia.)

International Kite Festival Calendar

Some of the following are programmed annually for particular days of the month, others are related to festivals. Specific dates are therefore not quoted. The month is a general guide for travellers on the international circuit. Refer to Appendix 1 for addresses of the national organisers who usually have details and registration forms ready for circulation 8 weeks before the event.

January

INDIA	Asian Kite Flying Championship, Bombay
	International Kite Festival, Ahmedabad

February

SINGAPORE	Annual Kite Festival, Marina South, Singapore
AUSTRALIA	Stunt Kite Championships and Kite Festival, Melbourne
GREECE	PanHellenic Kite Carnival, Patras
NEW ZEALAND	Summertime Kite Day, Christchurch

March

CHINA	Nantong Kite Festival, Nantong, Jiangsu
	Beijing International Kite Fair
HAWAII	Sport Kite Championships, Honolulu Hawaii
NEW ZEALAND	Naitonal Kite Festival, Eltham
THAILAND	Royal Thai Kite Competition, Bangkok
UK	Blackheath Kite Festival, London
USA	Great Delaware Kite Festival, Lewes
	Smithsonian Kite Festival, Washington
MALAYSIA	Pasir Gudang Kite Festival Johor

April

AUSTRIA	Vienna Kite Festival
CHINA	Weifang Conference
CURACAO	Antilles International
FRANCE	Le Touquet Kite and Aerial Photography Festival
	Rencontres Internationales de Cerf-Volants, Berck sur Mer
JAMAICA	Kite Festival, Negril
UK	Kite Society Spring Kite Festival, Old Warden, Biggleswade, Beds
USA	Maryland Kite Festival, Falls State Park, Baltimore
	Reno International Kite Festival, Rancho San Raphael Park
	April Fools Kite Fly, Mercer County Park, Trenton NJ
	Spokane Kite Festival

May

BELGIUM	Belgian and European Stunt Kite Championships, Oostduinkerke
CANADA	Touch the Sky Festival, Lebreton Flats
	National Stunt Kite Championships, Kortright Center, Toronto
	Pacific Rim Kite Festival, Vanier Park, Vancouver
DENMARK	Copenhagen Kite Festival, Hjortekjaergate
GERMANY	Fruhlinsdrachenfest, Bundesgartenscahu, Berlin
	Ascension Day Kite Festival, Hannover
ITALY	Castiglione del Lago
JAPAN	Hammatsu Kite Festival Nakatajima Beach
	Zama Giant Kite Festival, Sagami River
	Echu Daimon Kite Festival, Daimon
MALAYSIA	Kelentan International Kite Festival
UK	Kite Society Annual Convention
	Brighton Kite Festival
	Weymouth Kite Festival, Dorset
USA	Bucks County Kite Fly, Core Creek Park, Langhorne PA
	EAA Air Adventure Museum Kite Fly, Oshkosh WI
	Great Lakes Stunt Kite Championships, Grand Haven State Park, MI
	East Coast Stunt Kite Champs., Wildwood NJ
	Junction Kite Retreat, Texas University, Lubbock

APPENDIX 3

June

CANADA	Verdun Montreal
DENMARK	DiAero International Kite Fliers Meeting, Fano
FRANCE	Fete du Vent, Dunkerque
ITALY	Art in the Sky, Cesane, Urbino
	Cervia Volante International, Senigallia
	Sky over the Sea International, Volumedia
JAPAN	Giant Rokkaku Kite Fight Meeting, Sanjo
	Shirone Giant Kite Festival, Shirone
NETHERLANDS	International Kite Festival, Scheveningen, The Hague
POLAND	International Kite Festival, Wagrowiec
SPAIN	Mallorca Festival
UK	National Stunt Kite Championships, Blackheath, London
	Midland Kite Flyers International, Cofton Park, Longbridge
USA	Rogallo Kite Festival, Jockey's Ridge Park, Nagas Head NC
	Mackinaw City Kite Festival, Airport, Mackinaw MI
	Chicago World Rokkaku Championships, Ned Brown Woods, Schaumberg IL
	Westport International, Westport Beach, WA

July

BELGIUM	Ostend Kite Festival, Beach, Ostend
CANADA	Canada Day Festival, Waterloo, Ontario
JAKARTA	International Kite Festival, Jakarta/Yogya Karta and Bali
NORWAY	Dragefest, Lighthouse Rock Island, Stangholen, Risor
UK	International Festival of the Air, Northern Area Fields, Washington, Sunderland
	Blackheath Summer Festival, Blackheath, London
USA	Fourth of July Celebration, Ocean City NJ
	Berkeley Kite Festival & California Stunt Championships, N. Waterfront Park
	North Coast Stunt Kite Games, Toledo OH

August

GERMANY	Kite Festival, Stolln, Rhinow
FRANCE	Montpellier Kite Festival
SWITZERLAND	Kite Party, Lenzerheide, Valbella
UK	Kite Society Summer Kite Festival, Middle Wallop, Hampshire
	Chelmsford Spectacular, Chelmsford, Essex
	Bournemouth Kite Festival, Hengistbury Head
USA	Washington State International, Long Beach WA

September

AUSTRALIA	Festival of the Winds, Bondi Beach, Sydney
DENMARK	Paul Henningsen Memorial Day, Holbaek Marina
FRANCE	Dieppe Ciels du Monde International (even number years)
	European Wind Meeting, Prado Beach Park, Marseilles
GERMANY	Lenkdrachen-Meisterschraft, Freizeit-Park, Marienfelde, Berlin
INDIA	Laketown Maiden Calcutta
SWITZERLAND	Basler Drachenfest, Grun 80, Basle
UK	Bristol International Kite Festival & World Cup Sport Kite Championships, Ashton Court, Bristol
USA	Lincoln City International Kite Festival, D River Wayside, Lincoln City OR
	Pymouth Kite Day, Plymouth Beach, MA
	Cleveland Kite Festival, Edgewater Park OH
	Sunfest Kite Festival, Ocean City Beach, Maryland MD
	Family Day Kite Festival, Shoreline Park, Santa Barbara CA

October

EVERYWHERE	'One Sky, One World' Kite flyers celebrate on 2nd Sunday

AUSTRALIA	Stunt Kite Championships, Cairns Victorian Kite Championships, Wyndholm Reserve, Wendouree VIC
BELGIUM	Wanne International, Chateau de Wanne, Trois-points
SWITZERLAND	International Stunt Kite Festival, Silvaplana Zurich Kite Festival, Allmend, Brunau
UK	Kite Society Autumn Kite Festival, Old Warden, Biggleswade, Beds.
USA	AKA Convention (location changes) Top Gun Sports Kite Champs., Ford Island, Pearl Harbour HI Carolina Kite Festival, Atlantic Beach NC Cranberry Festival Kite Fly, Long Beach WA

November

GUATAMALA	All-Saints Day Kite Festival, Santiago de Secatepequez
SPAIN	International Kite Festival, Fuerteventura, Canary Islands
USA	Niagara Festival of Lights, Maple Leaf Tower, Niagara Falls International Open Peanut Butter Cookie Kite Fly, Seaside NJ

December

THAILAND	Buriram Kite Festival, Central Stadium, Buriram

NB: This listing is intended as a general guide to established kite festivals where international participation is known to have been welcomed in past years, prior to publication. There are many other events in national calendars. These can be found in association newsletters and specialist magazines.

Typical of any major kite festival is this combination of Deltas, Parafoils, Tubes, Flow-forms and a Rokkaku in the fore-ground, as caught by Creda Axton at the June 1992 Westport, Washington International.

APPENDIX 4

Bibliography

In fact, this is not a full bibliography. We have over 90 different kite titles on our shelves and, by the time this book appears, there is little doubt that this number will have expanded further. The majority of our collected books concern kites as they are flown around the world today. None of them are over 25 years old and yet, regrettably, only a few of them are currently in print.

There are specialised sources, for example the Kite Lines magazine bookstore in the USA (PO Box 466 Randallstown MA 21133-0466 USA), through which one can obtain abroad selection of kite books, and many kite suppliers carry the more popular titles. Library sources are very helpful, too, but that is no substitute for having any of these useful references immediately to hand when most wanted.

Taking the view that it would be better to expand on a few rather than list bare data on all, we have produced an abridged list of titles. Their selection is based on two virtues: firstly, they have an affinity with the nature of this book and, secondly, their text and illustration is both original and inspiring.

Listed, we hasten to add, in no particular sequence of merit, here is our recommended further reading. Please note that prices are only a guide, and were correct at time of purchase.

EXPERTISE SANS FRILLS

Cottrell, Mark **Swept Wing Stunt Kites** Self-produced, DTP, short run (very) perfect bound A4 sheets, self-distributed by the Kite Store, 48 Neal Street, London WC2H 9PA. 1990. Limited edition $11.95 or £3.95. Not a book, but produced well enough to last as long and certainly, with 44 typed pages, containing more information than many another that has gone through the printing presses. The cover (protected by a stout, clear plastic) is a harbinger of the informative gems inside, showing the family tree from 90-degree Delta to 110-degree Stealth Bomber shapes. As a Kite Store partner, Mark is always ahead of the game. He was a stunt champion in his earliest days, now he tells how to design and understand why things are the way they are. A new edition in 1997 will be completely updated and is expected to be printed professionally.

INDIAN FIGHTERS

Bahadur, Dinesh **Come Fight a Kite** ISBN 0-8178-5927-6

Harvey House Publishers, New York (UK Distributors A&C Black) 1978. Paperback and bound editions. Price $5.95 or £2.95.

This landscape, 56-page B&W instruction book on taming the Indian type of fighting kite is by the founder of America's famous 'Come Fly a Kite' shops. Large type and illustrations give an impression that it is aimed at juveniles, but we did not mind that, as it explains in full all aspects of those fighters that swarm over Ahmedabad and Bombay each Utran (winter solstice). Brief mention of other kites.

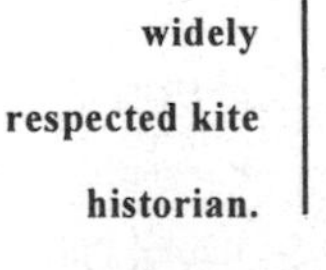

Clive Hart, widely respected kite historian.

DUTCH TREAT

Botermans, Jack and Weve, Alice **Kite Flight** ISBN 0-03-008518-7

Henry Holt & Co New York. (Originally Vliegers Maken Plenary Publications, The Netherlands) 1986. Paperback. An Owl book. Price $9.95.

A large format, square book created by Jack with kites, etc. from Alice, but real credit must go to David van Dijk, the photographer, who has filled 105 out of the 119 pages with brilliant graphics. It's Jack's page design, the first and best halftone and line presentation of about two dozen designs that we can ever expect to see. The cover claims 40, the other 16 must be in the suggested multiple box combinations. Ranging from Malay to Cody, all drawings are neat and *without* dimensions. Maybe this is because they have all been converted from the original metric to Imperial and, for convenience, they are in the text which is by Rob van den Dobbelsteen. Some conversions are rather off – 60 metres has become 24 inches and the Cody is 7ft tall and 10ft 20" span. Never mind. Metrication will win in the end (we hope) and the photos make up for the slips.

COMPLETE HISTORY

Hart, Clive **Kites, An Historical Survey** ISBN 0-911858-38-5.
Published by Paul P Appel, Mount Vernon, NY. 1982. Paperback $13.95 or $29.95 with a hard cover if you are lucky enough to get one.
First published by Faber & Faber in 1967 as a hardback edition at only £3.50, then revised with additional information. This is THE book to have for research on the history of kites. The bibliography alone covers 14 of the 210 pages. Well illustrated, though without kite plans, it encapsulates almost all one could find in dusty archives around the world.

COLOURIFIC

Ha, Kuiming & Ha, Yiqi **Chinese Artistic Kites** ISBN 962-07-5034-7.
Published and printed in Hong Kong. Paperback and cased editions. Price: $16.95 paperback or £24.00 hardback in the UK.
Our copy of this amazing example of colour printing came via a Chinese bookshop in London. An English language version is listed by the Kite Lines Bookshop and quotes Tal Streeter as saying that this is the finest book available on Chinese kites. We agree absolutely. 160 pages in square format gives all one requires on over 80 decorative, but flying, kites. Each is reproduced in perfect colour with fine line structural drawings. A true collector's item.

Similarly beautiful, and with a broader scope is **Kites** ISBN 7-80551-254
Published by Shandong Friendship Book Press, 39 Shengli Street, Jinan, China 250011. 1990. Hardback, with 112 pages and extensive English text, giving the best yet account of kite origins and present-day Chinese kite making. No designs but some beautiful photographs. Likely to be expensive!

DESIGN TOME

Eden, Maxwell **Kiteworks** ISBN 0-8069-6712-9
Sterling Publishing Co. New York (UK distributors Cassell) 1989. Paperback and cased editions. Price $19.95 or £14.95
There are 57 chapters in this American A4, 288-page volume and 31 of them provide most detailed blow-by-blow instructions for around three dozen kite designs. Add data on materials, how to construct, to fly and stunt, and Maxwell Eden's work becomes worthy of the subtitle **Explorations in Kite Building and Flying**.
No metrications except in the conversion table, and some disappointing artistry in airbrushed colour views; but maybe that's the way the general public see kites best. Alan Radom's line drawings make the book a 'must have'. It's 2cm (0.78in) thick and weighs 950g (2lb 1.5oz) as a paperback!

Peter Powell and David Pelham, two boosters for kites through the 80s.

EVERYONE'S FAVOURITE

Pelham, David **Kites, The Penguin Book Of** ISBN 0-14-004117-6.
Penguin Books, London 1976. Paperback. Price $14.95 or £11.00.
It is a measure of the value of this book that it is *still* in print, unaltered since its release in May 1976 as Penguin's lead title. Many reprints have been made and, thankfully, the imperfect binding of the first edition has been cured. This makes a copy with loose pages among the 228 more valuable as a collector's item! It is good because it has illustrations on every page, many in colour. The first section summarises history then, for 110 pages with vivid background colour, we get knots, structures, reels etc and 100 kite designs. No dimensions, but a metric scale comes with each copy. We've often wondered how kitemakers have managed to work it all out; but so many flyers have credited David's book when asked where they found plans for otherwise obscure designs.

APPENDIX 4

KNOW HOW

Gomberg, David **Stunt Kites** (No ISBN) self published via Cascade Kites, Salem, Oregon 1988. Paperback only. Price: $11.95 or £8.95

David was first off the mark with a book that is exclusively concerned with stunters. DTP and computerised illustration puts the reader behind the handles with guidance from no less than 18 of America's leading exponents. A book to take you from a plastic sail basic trainer through to complex manoeuvres with the latest carbon framed investment. No photos because it doesn't need them.

JAPANESE CRAFT

Hiroi, Tsutomu **Kites: Sculpting the Sky** ISBN 0-241-1001-1.
Elm Tree Books, Random House. 1978. Paperback. Price: $4.95.

Originally published in Japan, we felt lucky to get a signed copy in 1974 although the text was meaningless. Photo sequences and drawings take one through the techniques and provide good plans for many Japanese and a few Western kites. The translated edition now completes the picture. 144 pages with some colour.

TRAVELLERS TALES

Yolen, Will **The Complete Book of Kites & Kite Flying** ISBN 0-671-22191-4.
Simon & Schuster, New York 1976. Hardback.

As a journalist Will Yolen had quite a reputation. Add that to being an explorer, friend of the famous and world-wide traveller, as well as an ardent kiteflyer, and Will's memoirs become fascinating. Francis Rogallo introduced him to kites and, in turn, he recruited Ambassadors, Maharajahs, Prime Ministers and even a Miss Universe into his International Kite Association. As well as the travelogue, there are 25 kite designs drawn by Caleb Crowell on convenient grids. Yolen's 256 American A4 pages may well overlap what can be found elsewhere; but the style is inimitable. What else could one expect from the man who fought authority and won the right to fly in New York's Central Park for everyone else to enjoy from December 8th 1965. Price was $9.95, but you could probably treble that price if one could be found!

FRENCH VIEW

Clément, Gérard **Cerfs – Volants** ISBN 2-86519-119-2 Published by ACLA, Paris 1993.Hardback-Price FF 245, US $54.95.

This is a French language large format coffee table book with superb photography and graphics.As a picture book to admire it carries representative illustrations to impress the general reading public that there is a historical background.That China, Japan, India and Malaysia have long standing traditions in kitemaking. That the Western World has produced a wide diversity of kites of design and purpose. As a high quality production the book is a collector's item though its all upper case text is shall we say, – 'difficult'. Backed by the Musée de l'Air where we bought our copy, it ought to have carried much more from their archives(as yet unseen) on the early days in France.

Will Yolen and his illustrator Caleb Cromwell, a combination of great talents.

PIONEERING DESIGNS

Diem, Walter and Schmidt Werner **Drachen mit Geschichte** ISBN 3-88034-656-9. Hugendubel, Munich, Germany 1993. Hardback. German language US price $22.95.

The title means "Kites with a History" and a sub-title explains that each of the subjects are detailed for self-construction. It is an excellent book, well produced in every aspect except that like the famous Pelham work, it tends to fall apart with frequent study! Thirteen kites are described in the 160 pages, each a historical classic and all with photo illustrated stage-by-stage instructions for a replica with similar structure to the original. The close-up detail is first class. Not only is the reader assured that the kite has been built and tested, he or she is offered the opportunity of purchasing some of the specially machined fittings. The kites range from the Hargrave of 1895 to the Sauls of 1941, the majority from Lecornu, Lamson, Brogden and Lindberger coming around the turn of the century-If you want to join the 'Cotton Club' this is the book.

KITES ILLUSTRATED

Morgan, Helene and Paul. **The Book of Kites.** ISBN 0-86318-785-4 Dorling Kindersley, London 1992 £10.99 Hardback.

Large format, impeccably produced, superb studio work, a lexicon of kite identity photographs with lots of white space around every picture in the true D-K syle. As an introductory book it does the job well, and when you've passed through the catalogue section, the how-to-fly dept is dealt with in the same style, all from studio shots. Five designs complete a book to send to any "just interested but not yet active" friend.

DUTCH TREAT – 2

Horst, Servaas van der/ Velthuizen, Nop **Stunt Kites to Make and Fly** ISBN 90-6868-052~8 Thoth, Amsterdam, Netherlands £12, US $21.95 Paperback.

First published in Dutch, and as a result of demand outside The Netherlands, this has become the number one recommendation for anyone who wants to know all the ins and outs of Sport Kite flying. Now the clever pair have gone another stage further and written "Stunt Kites II" with designs for 8 kites and a Buggy. Four years on from launch of the first book, there's no reason to think that anything has changed in techniques and that's a sign of a good book.

GERMAN TALENT

Schimmelpfennig, Wolfgang.**Phantastische Drachenwelt.** ISBN 3-8068-4513-1 Falken, Niedernhausen, Germany 1991.US $31.95

Not quite coffee table size but with all the quality of high standard photography and printing.Prolific Wolfgang intros the kite types, selects the dozen finest and most creative kiteflyers from around the world, describes the festivals at Berlin, Fano, Scheveningen, Singapore, Bali, Malaysia, Thailand and Hammamatsu then gives us plans on a huge fold out sheet for 4 designs. What more could one ever want?

Schimmelpfennig, Wolfgang. **Making and Flying Kites**. ISBN 0-600-55895-9. Hamlyn Creative Crafts 1988. Paperback. Price $12.95 or £3.95.

The emergence of the craze for kites in Germany has been fostered by numerous books. This one proved its worth to become translated into English (and French). Clear illustrations, excellent photography, good construction detail and a range of popular designs in its 80 pages make it good value.

INSIDE JAPAN

Streeter, Tal. **The Art of the Japanese Kite**. ISBN 0-8348-0088-8. John Weatherhill, New York & Tokyo 1974. Hardbound. Price $23.95 or £13.50.

In two visits we found the Tokyo kite shops but could only understand a little of the national culture. Tal lived in Japan for two years. This enabled him to provide us with a personal view of the folklore and traditions that have influenced the Japanese kite scene for centuries. To read the well-expressed text is to experience the joy of meeting leading kitemakers, to witness festivals and to absorb the atmosphere of an ageless society. It also provides a directory of 50 different designs within the fine selection of photographs in its 181 pages.

IDEAL FOR THE BEGINNER

Baker, Rhoda/Denyer, Miles. **Making Kites** ISBN 1-85076-366-6 The Apple Press London 1993. Hardback £8.95
We first met Rhoda and Miles at a craft fair. They said they were writing a book but as the types of kite they were making and selling were rather basic, and down in price, we thought their chances might be slim. Just shows how wrong one can be! This is a super book for the novice. It bears all the marks of a Dorling Kindersley Studio with the high quality photography and graphic layout, and all of it in colour too! Twelve projects take the reader through the stages from a sled to a keeled diamond, with a serpent, a single line and a sport delta in between. This book has not had the promotion it deserves. As an instruction manual set out in easy to read, photo illustrated form it has no peer.
Just one point jars, the how to fly sport kites section is a straight lift from Helene and Paul Morgan's book but in a different art technique.

INCOMPARABLE HAMAMATSU

Koike, Hiroshi **Hamamatsu Festival** No ISBN, Mr. Pond Company, I.K.E. Japan 1981. Softback.
This is one of our most treasured publications. It is a picture book of 100 pages, all filled with action photographs taken during the annual festival at Hamamatsu by five professional photographers. There are very few pages with any text, but that is a tribute to the picture story which conveys the thrill of the parades that take place from May 3rd to 5th and the rivalry of the 60 or more teams which do battle with their huge kites. The noise and tumult of the exchanges almost shouts its way out of the pictures. When all is over, to the victors the spoils and on each of the three nights the streets are crowded by the revellers. As the English caption states, 'Every face is rosy and wet with too much Sake and a lot of sweat'.
Although not generally available, there is a price of 1850 Yen on our copy. Maybe someone out there does have a stock so that collectors of books which come under the "original and inspiring" category we mentioned at the beginning might with luck locate a copy. We were fortunate to be given one when the Kite Team from Hamamatsu made a visit to Birmingham on 26th July 1987. After flying with Midlands Kite Flyers in Sutton Park, the Lord Mayor invited all to a Civic Reception in the opulence of the banqueting suite of the Council House and as fellow kiteflyers will readily appreciate, even without the Sake, that was quite an occasion.

Lifeblood of the kite movement is within the periodicals and newsletters that disseminate topical information, plus, of course, the telephone and Internet.

Kite Periodicals

GERMANY

Drachen Magazin Postfach 201863, D-20208 Hamburg. Tel: 040 418624. Fax: 040-440561. Bimonthly, 88 pages, A4, Colour, English summaries.

Sport & Design Drachen Verlag fur Technik und Handwerk, Postfach 2274, D-76492, Baden-Baden. Tel: 07221 5087-91. Fax: 07221 5087-52. Bimonthly, 86 pages, A4, Colour. Single issue DM9,00. German language.

UNITED KINGDOM

Kites 11, Hill Drive, Hove, East Sussex BN3 6QN. Tel/Fax: 01273 308 787. A bold 60 page colour throughout venture by Barry Pitman. Bimonthly, but after the fourth issue in 1997 it is likely to go twice yearly in a larger format. The original price for a single issue is £2.50. English language.

Kite Passion Custom Publishing France, 18, rue Horace Vernet, 92136 Issy les Moulineaux, France. UK subscriptions, PO Box 152, Woking, GU21 1FS. Bimonthly, 68 pages, A4, full colour throughout.
The English variant of the parent French magazine. Single issue £2.95. Six issues initially £14.75.

The Kiteflier The Kite Society, PO Box 2274, Great Horkesley, Colchester, Essex CO6 4AY. Tel: 01206 271489. Quarterly, 48 pages, A4, mono only. Single issue £1.75. Also available by membership subscription. Includes STACK, Roman Candle, Brighton and Midlands KF News. English language.

UNITED STATES OF AMERICA

American Kite American Kite Co., 480 Clementina Street, San Francisco, CA 94103-9931. Tel: (415) 896 0830. Fax: (415) 896 0485. Quarterly, 72 pages, American, A4, Colour. Single issue $3.50, four issues $10, foreign $18. English language.

Kite Lines Aeolus Press Inc. PO Box 466, Randallstown, MD 21133-9987. Tel: (410) 922 1212. Fax: (410) 922 4262. Also **Kitelines Bookstore** at the same address. Quarterly, 72 pages, American, A4, Colour. Established Spring 1977 succeeding **Kite Tales**, which was established in 1965. Maintains international calendar, booklist, covers all types. Single issue $4.50, four issues $16, foreign $22. English language. Each volume (4 issues) provides a bookload of topical information.

ITALY

Tempo di Aquiloni-Kitetime Via IV Novembre 15, 47037 Rimini, Italy. Quarterly, 64 pages, A4. International coverage, including art and gastronomy. Single issue $10, four issues $35.

FRANCE

Planette Cerf-Volant 104, rue Edouard Vaillant, 93100 Montreuil, France. 100 pages, A4. Includes hot air balloons, boomerangs and buggies. One year 225FF, two years 420FF.

Cerf-Volant Passion 18 rue Horace Vernet, 92130 Issy les Moulineaux, France. Tel: 4648 6666.
Bimonthly parent of UK variant. Sport kite priority. Single issue 28FF.

ASSOCIATION NEWSLETTERS

The majority of kite groups have their own newsletter. We are unable to include a full listing and, in any case, their content is mostly concerning local events, personalities and domestic notices. The following National Associations provide a substantial newsletter – some of the magazine standard.

Le Lucane Cerf-Volant Club de France, 30 pages, A4. Numerous original and vintage designs. French language.

Hoch Hinaus Drachen Club Deutschland. 68 pages, A5. Reports, results, listings. German language.

Cervi Volanti Associazione Italiana Aquilonisti. 52 pages, A4. Designs, history, experiments and new structures. Sketches, technicalities and original concepts. Italian language.

Kiting American Kitefliers Association. 28 pages, American, A4. Reports, listing, convention data and colour reports of the same. News from regions and calendar of domestic events. English language.

Estel Barcelona Estels Club. 28 pages, A4, magazine style. Spanish language.

Flying High Journal of the Australian Kiteflyers Society Inc. 38 pages, A4. Reports, news from groups. Domestic calendar. English language.

INDEX

ACKNOWLEDGEMENTS

THANKS!

What began in 1989 as a retirement project to produce a compact handbook for kite flyers grew month by month into something more involved as the assembly of notes, photographs and sketches expanded. Now, as we produce a 2nd edition in 1997, the kite movement is twice as large and constantly demanding information.

This reflects in many ways the friendship that inspires kite flyers to exchange their experiences and design developments. Help and encouragement came from all quarters and our grateful thanks are long overdue to those who provided inspiration in the earlier days. Regrettably some of those pioneers are no longer with us.

So first we owe credit to that grand doyen Paul E. Garber of Washington D.C. and those London pioneers Alex Pearson and John Robson, each of whose creations anticipated the so-called 'modern' kite. Then to Robert Ingraham, who founded the first periodical on kites and to Valerie Govig, who took it over and by creating *Kitelines*, accumulated a wonderful information source. Newsletters have similarly contributed, often unwittingly from the Northern, M.K.F., White Horse and Essex Kite Groups in the UK, Cervi Volanti in Italy and le Lucane in France. The Kite Society's *Kiteflier,* AKA's *Kiting* and Chicagoland *Sky Lines* have been a constant help.

Dedicated researchers and innovative designers, all of them remarkable characters whose friendship we value highly, have motivated our purpose. Peter Powell, Don Dunford and Andrew Jones for the first steerable kites. Tom Van Sant for the first use of ripstop nylon and glass fibre. Action kites and the Top-of-the-Line team that pulled it all together. Mark Cottrell, Tim Benson, Rob Brassington, Al Hargus, Lee Sedgwick and Joe Hadzicki, are among those whose original thinking took us further.

We thank the single line flyers who have transformed shapes so dramatically over the past decade. Martin Lester, George Peters, Joel Scholz, Jorgen Møller-Hansen, Oliviero Olivieri, André Cassagnes, Peter Lynn, Yvonne de Mille, Don Mock, Steve Brockett, Helmut Schiefer, Ton Oostveen, Andreas Agren, Tony Slater, Walter Diem, Werner Schmidt, and Richard Hewitt among many.

To those prime movers of finding ways to *use* kites, for photography and Bear-drops, Tom Pratt, Simon Kidd, Raoul Fosset, Garry Woodcock, Maurice Sawyer, Jan Fischer, John Barker and Hugh Andrew, a special thanks for all their information.

Kite suppliers too, Andy King of the Kite Store, London, Pat and Ron Dell of Kiteability, Nick Harrison of Brookite, Carole and Martin Thomas of The Leading Edge, Dvid Clarke of Windy Kites, David and Sarah Green of Greens Kites, Nop Velthuizen of Vlieger Op, John Cochrane of Cochranes of Oxford have all helped enormously.

Suppliers of raw materials, Bainbridge Aquabatten, Carrington Novare, Pultrex Ltd., James Pearsall & Co., RBJ Reinforced Plastics, Quicks and Wiggins Teape willingly provided technical assistance.

Special thanks to those individuals who rallied to requests and who will at last be able to see their contribution in print. John Clarke, Carole and Alan Peacock, Ted Fleming, Paul Chapman, Martin Powell, Tony Cartwright, Bob Peiron and George Maurer.

Most important thanks to ex-colleagues at Nexus Special Interests, Bev Laughlin, Lyn Corson and Evelyn Barrett, each of whom has played more than a little part in seeing this work through its very long process to the final printing.

To Kite Groups and Individual kite flyers everywhere . . .
Tight lines and pleasant breezes!

Ron Moulton and Pat Lloyd